Neural Control of Movement

Neural Control of Movement

Edited by

William R. Ferrell

University of Glasgow
Glasgow, Scotland, United Kingdom

and

Uwe Proske

Monash University
Clayton, Victoria, Australia

Springer Science+Business Media, LLC

Library of Congress Cataloging-in-Publication Data

On file

Proceedings of the Thirty-Second Congress of the International Union of Physiological Sciences (IUPS), held August 1–6, 1992, in Glasgow, Scotland, United Kingdom

ISBN 978-1-4613-5818-3   ISBN 978-1-4615-1985-0 (eBook)
DOI 10.1007/978-1-4615-1985-0

© 1995 Springer Science+Business Media New York
Originally published by Plenum Press, New York in 1995
Softcover reprint of the hardcover 1st edition 1995

10 9 8 7 6 5 4 3 2 1

To my daughter Sophie Ferrell

FOREWORD

Presented with a choice of evils, most would prefer to be blinded rather than to be unable to move, immobilized in the late stages of Parkinson's disease. Yet in everyday life, as in Neuroscience, vision holds the centre of the stage. The conscious psyche watches a private TV show all day long, while the motor system is left to get on with it "out of sight and out of mind." Motor skills are worshipped at all levels of society, whether in golf, tennis, soccer, athletics or in musical performance; meanwhile the subconscious machinery is ignored. But scientifically there is steady advance on a wide front, as we are reminded here, from the reversal of the reflexes of the stick insects to the site of motor learning in the human cerebral cortex. As in the rest of Physiology, evolution has preserved that which has already worked well; thus general principles can often be best discerned in lower animals.

No one scientist can be personally involved at all levels of analysis, but especially for the motor system a narrow view is doomed from the outset. Interaction is all; the spinal cord has surrendered its autonomy to the brain, but the brain can only control the limbs by talking to the spinal cord in a language that it can understand, determined by its pre-existing circuitry; and both receive a continuous stream of feedback from the periphery. Progress is slow in terms of the standards set by Molecular Biology, partly because many of the individual "facts" of neurophysiology only gain useful meaning when they can be put in a wider setting; in contrast, the mapping of one more gene and the determination of its base sequence is a thing in itself, available for immediate practical use. Quite apart from simple intellectual interest, each of us must periodically forsake the compelling immediacy of our own experimentation to take the wider look; failing this, what we all do risks sinking into oblivion, temporarily cited by a few other specialists who share the same jargon but never integrated into a meaningful context. Thus students of the motor system, whatever their particular interests, should both listen to each other and write for each other, and not just relate to their immediate colleagues. And, in fact, our record is good; we are relatively few, we personally do our own experiments, we remain in our chosen area long enough to get to know each other, and our type of work rarely leads to the cut-throat competition found in too many other fields.

Such thoughts may well have been in the mind of the organizers of the 32nd Congress of the Union of Physiological Sciences, held in Glasgow in August, 1993. At any rate, they set aside five full days for the five consecutive symposia on the Motor System, ranging from Comparative Physiology through Proprioception and the Spinal Cord to Cortical Control and Motor Learning. Each symposium was orchestrated by experts in the area, who chose both the speakers and their topics, and then usually led the discussion. The second day honoured the memory of Ian Boyd who had given his life to Glasgow and to the muscle spindle, and who had been instrumental in bringing the Congress to Glasgow. If he had not been suddenly cut down at the age of 60, he would surely have told us something new; he was as active as ever, and with his unique expertise the loss of his continuing contribution is irreplaceable.

As one who sat through the meetings with enjoyment, I can report that the intellectual banquet was overwhelming in both quantity and quality. One problem was that there was practically no time to view and discuss the accompanying posters, though much could be done by avoiding eating. Attendance was good throughout, aided by a wet, windy, and cold Scottish summer. Another problem was the impossibility of taking adequate notes and remembering all that one would wish to remember (an advantage of continuing to teach as one ages is the repeated demonstration that the young suffer similarly). This book with its series of articles on a variety of topics by a variety of speakers should help refresh the memory of those who attended; more importantly, it gives those who were not present an outline of what went on. Thus it should help everybody achieve an overview. But any such book provides only a partial summary and cannot reproduce the experience of attending. For one, the illustrations are inevitably only a carefully selected sub-set of the slides presented at the time. In the present book, there is also selection of the topics included among those presented. All were asked to contribute, but only some responded. Many of those who refused will have had the best of reasons; repeated publication of the same material helps nobody. Those who have written include the energetic and organized, and those with a message that they wish us to hear. They have provided a wealth of material which we can all be grateful for having brought together under a single cover.

Peter B. C. Matthews

PREFACE

This book, entitled "Neural Control of Movement", came about as a result of a series of five symposia on motor control which were held as part of the 32nd Congress of the International Union of Physiological Sciences (IUPS), held in Glasgow, Scotland, August 1-5, 1993. Bringing the Congress to Glasgow was the brainchild of Ian Boyd, Buchanan Professor of Physiology at the University of Glasgow, who sadly died in 1987 and was not able to see the results of all his efforts. A key organizational feature was to bring all the satellite symposia into the Congress under particular themes. Under the theme of Motor Control, five symposia were held on consecutive days, covering a broad range of topics and preparations from *Crustacea* to man. We were privileged to be asked by the IUPS Programme Committee to organize a symposium in honour of Ian Boyd entitled "The Ian Boyd Symposium." Early on during the planning of this symposium we had agreed that we would choose our speakers under two broad headings: "Sensory receptors" and "Proprioception" and we would allocate approximately equal time to each of these areas. Since the time available for the symposium was very limited, we were able to select only five speakers for each of the two sessions. We both found it difficult to select such a small number of key speakers from the many international contributors currently active in these fields. Inevitably, we were criticized for having made important omissions in our programme. As a result we decided that one way of at least partially redeeming ourselves, as well as doing the subject more justice in terms of the coverage of topics was to try to publish the proceedings of the meeting. We therefore called for manuscripts from our speakers as well as from other colleagues who attended the symposium and who we had been unable to include on the programme. Another important motive for publication was to commemorate the Ian Boyd Symposium, and grasp the opportunity of having so many distinguished researchers in motor control gathered at one site, at the same time. We therefore asked speakers at the other four symposia to contribute manuscripts as well. The response to our call was sufficient to make this book a reality and we are very grateful to those who gave of their time to contribute manuscripts.

As the manuscripts came in, it soon became clear that they fell within one of several clearly defined areas. It led us to include a number of section headings - "Afferent Mechanisms," "Proprioception," "Reflexes," "Locomotion," "Development," "Cerebellar Mechanisms" and "Comparative Studies." Inevitably, any classification of this kind is somewhat arbitrary and a number of manuscripts covered more than one area. Nevertheless, since this is a book concerned with broad areas of motor control, such headings are, in our view, useful to the reader.

Scrutiny of the range of topics covered reveals that as well as a surprising amount of activity in traditional areas such as "Afferent Mechanisms" and "Reflexes", there is a great deal of work currently going on in "Proprioception," "Locomotion," and "Development." One of the important steps forward in the field reported at the symposium and represented here is the concept of phase of locomotion dependent changes in reflex action of muscle receptors. This integrative approach, involving three different areas, points the way for future progress in the field. Here, too, the importance emerged of some of the comparative studies.

Finally, in the preparation of this volume we have received invaluable help from many friends and colleagues. We would like to thank the London staff of Plenum Press, particularly Joanna Lawrence, for their support, encouragement, and expert advice. We would also like to thank our secretarial and support staff, particularly Elise McCorriston and Florence McGarrity, without whom none of this would have been possible.

William R. Ferrell
Uwe Proske

CONTENTS

AFFERENT MECHANISMS

PROPRIOCEPTION

REFLEXES

LOCOMOTION

DEVELOPMENT

CEREBELLAR MECHANISMS

COMPARATIVE STUDIES

AFFERENT MECHANISMS

ISOLATED MUSCLE SPINDLES, THEIR MOTOR INNERVATION AND CENTRAL CONTROL

M.H. GLADDEN

Muscle Spindle Physiology Group
Institute of Physiology
University of Glasgow
G12 8QQ, Scotland, U.K.

SUMMARY

The facility to isolate muscle spindles continues to provide new insights into their physiology, but the rationale for the heterogeneous composition of γ_s-fusimotor units is still unknown. A fusimotor unit consists of a γ_s- or γ_d-motoneurone and all the intrafusal fibres that neurone innervates, though these fibres are distributed among several spindles. γ_d-motoneurones have one effector, the bag$_1$ fibre, but γ_s-motoneurones have two, bag$_2$ and chain fibres. Boyd's view was that there are two types of fusimotor unit whose principal component was either bag$_2$ or chain fibres. The reasons why this attractively simple hypothesis cannot be accepted now unequivocably are explained.

INTRODUCTION

Although Ian Boyd will be remembered for his early work on the mammalian neuromuscular junction with A.R.Martin, and on the composition of peripheral nerve with M.R.Davey, it was his pioneering work on isolated muscle spindles that aroused most interest. So, almost obligatorily, this became the topic of this contribution to an Ian Boyd Symposium from the laboratory he established in Glasgow. In the 1960's he began to produce films of isolated living muscle spindles which showed clearly for the first time the contractions of intrafusal fibres activated by stimulating their axons in the muscle nerve (Boyd, 1958; 1966; 1970; 1971). To some of us who hitherto had seen only beautiful, but dead histological specimens of spindles, his preparations were fascinating. Here were sensory endings being extended by the contracting fibres, just as one imagined they should. These films also illustrated the distinct differences in mechanical behaviour of the two types of intrafusal fibre then recognised, bag and chain fibres. The facility to observe living muscle spindles provided new insights also into the function of γ-motoneurones, and it was one aspect of the latter, the innervation of spindles by γ_s-motoneurones, which was concerning Boyd when he died in 1987. The background to his problem, and progress since his death will be reviewed here. But first the problem may be put into context by reviewing briefly the far simpler case of the innervation of spindles by dynamic γ-motoneurones.

γ_d-Fusimotor Units

Although up to one half of the motor axons supplying a particular muscle can be solely fusimotor (Boyd and Davey, 1968, Table 7), only one quarter or less of these will have a dynamic action, the rest being static gamma axons. Each γ_d-axon innervates at least 2-3 spindles; single spindles receive 0-4 γ_d -axons, and each axon supplies one or both poles of a bag$_1$ fibre. Contraction of bag$_1$ fibre poles induced by excitation of γ_d -motoneurones stiffens them, so that when the spindle is stretched a greater proportion of the total extension is applied across the primary sensory ending around the central portion of the fibres, producing the familiar dynamic effect, the enhanced sensitivity to stretching first described by Matthews (1962). The magnitude of the effect can be adjusted by change in frequency, being maximal between 75 and 100Hz.

Bag$_1$ fibres were first recognised to be responsible for dynamic effects by methods based on the isolation of spindles (Bessou and Pagès, 1975; Gladden, 1976; Boyd, Gladden, McWilliam and Ward, 1977) And, of course, the mechanical behaviour of bag$_1$ fibres which underlay their ability to influence the dynamic responsiveness of the primary sensory ending could be directly observed in these preparations (Boyd, Gladden and Ward, 1981).

A small proportion of bag$_1$ fibre poles are inaccessible to control, being without any innervation. At the other extreme, some receive as many as three axons; presumably in these cases summation occurs if the contractile apparatus is sub-maximally activated by each input, and summation will also occur if β_d- and γ_d -axons innervate the same bag$_1$ fibres. Here, then, is a homogeneous system in which all the spindles of a muscle supplied by γ_d -axons (and some few are not) can be influenced in qualitatively the same way through their bag$_1$ fibres; the total strength of the effect can be modulated by change in frequency and by recruitment within the γ_d -motoneurone pool.

There is one anatomical inhomogeneity, however, which turns out to be functionally unimportant, so far as is yet known. About 10% of γ_d -axons supply one chain fibre in addition to bag$_1$ fibres. Although chain fibre contraction causes driving (see later), in this instance driving does not occur, as driving is not regularly associated with dynamic effects when γ_d -axons are stimulated (Emonet-Dénand, Laporte, Matthews and Petit, 1977). The contractile properties of this chain fibre might be atypical as a consequence of its special innervation, although they do contract, as movement was observed in one such case (Gladden, 1992). Alternatively, perhaps the pattern of the terminal branches of the Ia axon (see Banks, 1986) is designed so as to quench automatically any input from this chain at a node.

γ_s-Fusimotor Units

Barker, Emonet-Dénand, Laporte, Proske and Stacey (1973) first demonstrated conclusively that γ_s-axons have two effectors, chain fibres and bag fibres. These particular bag fibres were later identified as bag$_2$ rather than bag$_1$ which are innervated by γ_d -axons. When considering single isolated spindles, about one third of γ_s-axons innervate chain fibres only, another third innervate bag$_2$ fibres only and the remainder innervate chain and bag$_2$ fibres together (Bessou and Pagès, 1975; Boyd et al., 1977). Although chain and bag$_2$ fibres can be thus wired up together, the odd thing is that these fibres are by no means equivalent in their action on primary and secondary endings because their mechanical properties are quite different (see Boyd, 1981). Could the CNS have some independent control, albeit imperfect? To find out how the CNS uses intrafusal fibres isolated spindles were exteriorized. Contractions of intrafusal fibres under CNS control could be observed, and activity of the γ_s- and γ_d -motoneurones that controlled them monitored over many hours (Gladden and McWilliam, 1977a and b; Gladden, 1981). Bag$_2$ fibres were recruited regularly by stimulation of areas of CNS separate from loci effective for chains, for example from separate areas of cortex. However, this evidence could never be definitive since the function of all the ramifications of single γ_s motor units could not be monitored due to the technical limitations of the preparation. The possibility could not be excluded, for example, that a γ_s-motoneurone activating a bag$_2$ fibre in the spindle which had been isolated might not activate chains in other, unseen spindles. Nevertheless, Boyd was sufficiently encouraged to investigate the composition of γ_s-fusimotor units by direct and indirect physiological tests to determine whether individual γ_s-fusimotoneurones showed any preference for either chain or bag$_2$ fibres. Some distinct preference would be necessary were the CNS to exert any independent control over the two fibre types.Boyd stimulated functionally single γ_s-axons in ventral root filaments and definitively identified the fibres contracting in one spindle which was isolated. In other spindles of the same muscle the type of fibres innervated was identified with an indirect test (see below and Fig. 1; Boyd and Ward, 1982).

4

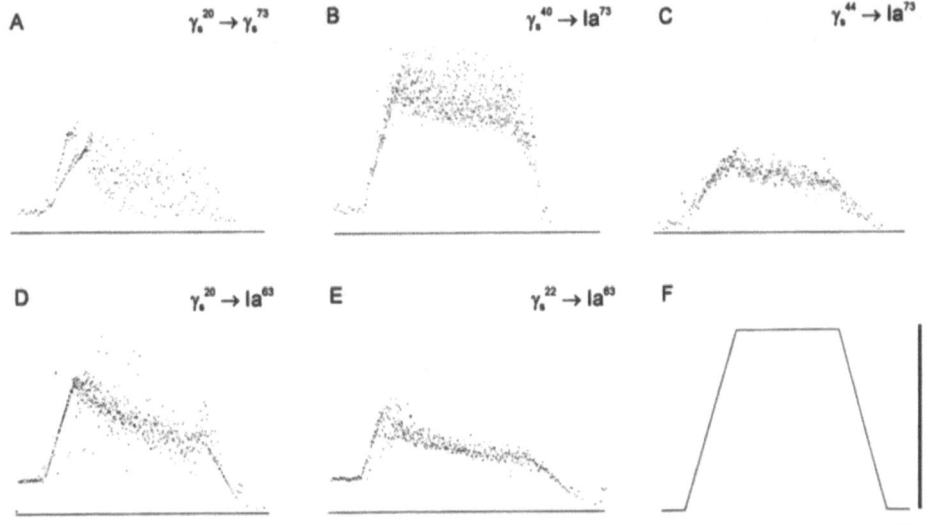

A $\quad\gamma_s^{20}\to\gamma_s^{73}$ B $\quad\gamma_s^{40}\to Ia^{73}$ C $\quad\gamma_s^{44}\to Ia^{73}$

D $\quad\gamma_s^{20}\to Ia^{63}$ E $\quad\gamma_s^{22}\to Ia^{63}$ F

Figure 1. *Examples of the indirect test Boyd used to indicate the fibre types innervated by γ_s-axons.*
A,B,C: responses of one spindle primary ending (Ia^{73}) and D,E of another (Ia^{63}). F: time course of the applied stimulus frequencies; no response followed this exactly. The 1:1 or 1:2 driving at lower frequencies in A,D and E indicated chain innervation by γ_s^{20} and γ_s^{22}. Lack of driving in C indicated bag_2 innervation by γ_s^{44} (note that the non-driving response in C does not follow the rising phase of the frequency ramp in F). In B, γ_s^{40} was known to innervate one chain fibre in addition to the bag_2 fibre, accounting for some driving. Similar driving responses were obtained for γ_s^{20} in one other spindle (three in all) and for γ_s^{22} in two other spindles. Non-driving responses were obtained for γ_s^{44} in four other spindles, and for γ_s^{40} in two other spindles - i.e. γ_s^{20} and γ_s^{22} were Boyd's "static chain" fusimotor units, and γ_s^{40} and γ_s^{44} were "static bag" fusimotor units. Vertical calibration bar: 0-150 impulses/s. Time calibration 1s.

His conclusion was typically clear cut (Boyd, 1986); γ_s-motoneurones had a preference either for bag_2 or for chain fibres; they all fell into two groups, so that a motor unit would express chiefly mechanical properties of either bag_2 or chain fibres. He called them "static bag and static chain γ-motoneurones". Members of the same group would augment effects if combined in individual spindles (Fig. 2), as with γ_d -effects.

All of us who have followed up Boyd's work have had difficulties both with his conclusion and with his indirect test. In explaining these difficulties I am conscious of the fact that he is not here to defend himself. But a defence or apologia is not really necessary; his conclusions were the logical outcome of his experimental evidence, this I can vouch for, as I have re-examined it all from the original data. Boyd was right within the limitations of his experimental conditions, and his evidence could be very persuasive as Figs. 1 and 2 illustrate. It is simply that we have now moved on.

Boyd's indirect test

Isolation of spindles can be very time-consuming, so Boyd needed some indirect test for the innervation of spindles he could not isolate. He "diagnosed" the γ_s-innervation of spindles he did not isolate from the pattern of Ia responses during ramp frequency stimulation at constant length (Boyd, 1986) . Fast twitches characteristic of chain fibres each briefly extend the primary sensory terminals, and "drive" the Ia afferent frequency at the stimulus frequency, or a sub-harmonic of it if only every second or third twitch is effective (Fig. 1A,D,E). Bag_2 fibres, classically, do not drive (Fig. 1C). Matthews pointed out that time-locking due to chain contraction could be missed in a very irregular Ia response. Dickson, Emonet-Dénand, Gladden, Petit and Ward (1993) used cross-correlograms between γ-stimulation pulses and Ia spikes to detect this. However they also uncovered another difficulty. Some primary sensory endings did not have driven responses even though chain fibres must have been contracting (Dickson, Emonet-Dénand, Gladden and Petit 1992). There are, too, the non-classical responses. Some bag[2] fibres can drive at low frequency; some chain fibres do not have propagated potentials and are thought not to drive. Bag[2] fibre driving can be accommodated by excluding driving confined to low frequencies. Non-driving chains are behaving like bag[2] fibres, and, while this is a nuisance experimentally, may be irrelevant to the CNS. Dissatisfaction with the test itself has led Celichowski, Emonet-Dénand, Laporte and Petit (1994) to devise their own.

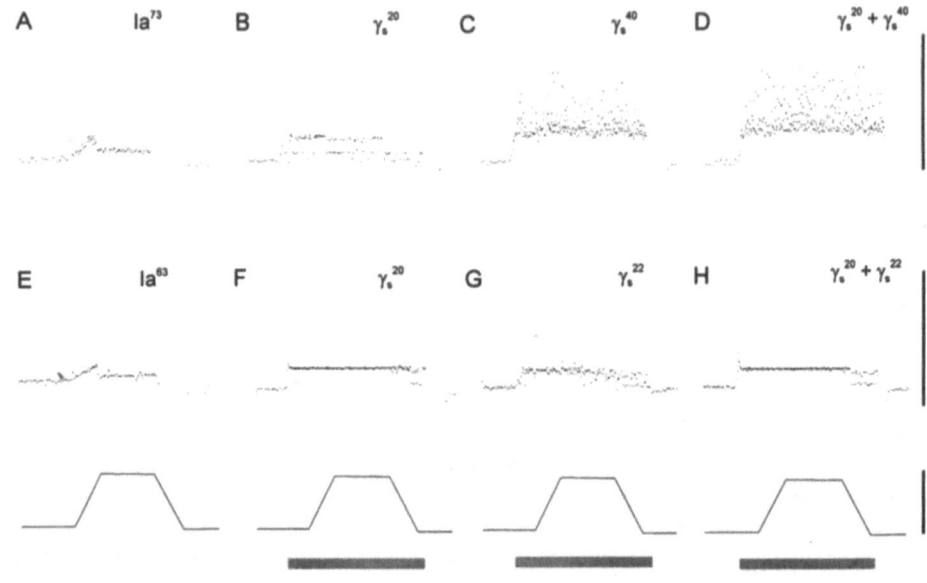

Figure 2. *Combined stimulation of* γ_s*-axons belonging to Boyd's two groups.*

A&E: responses of the two primary endings illustrated in Fig. 1 to standard ramp and hold stretches. The weak driving response of Ia[73] to constant stimulation at 50 impulses/s of γ_s^{20} (B), a "static chain" axon, when combined with γ_s^{40} (C), a "static bag" axon, was abolished (D). Irregularity increased, even though γ_s^{40} innervated a chain fibre in addition to the bag[2]. In a second spindle (Ia[63]) γ_s^{20} had a stonger driving effect (F) which was preserved (H) when combined with the weaker driving effect of γ_s^{22} (G), another "static chain" axon.

Time course of stretches to tenuissimus muscle in lowest traces applies to all records. Duration of stimulation, all at 50 impulses/s shown stippled below stretches. Time scale: 1s; Frequency calibration 0-200 impulses/s. Records for Figs. 1 and 2 from I.A. Boyd's last experiment.

Boyd's division of γ_s-motoneurones

The idea of a separate set of γ_s-motoneurones to control the two static effectors, bag[2] or chains, largely independently is staightforward and logical, but unhappily evidence now is heavily stacked against it. Boyd (1986) had also linked his two γ_s-motoneurone types directly to two patterns of post-synaptic morphology on chain fibres, but Gladden and Sutherland (1989) could not find support for this when all histological data from Boyd's experiments was assembled. Three types of γ_s-motoneurone were possible, but not two. Banks (1991) independently repeated Boyd's work on the same muscle, tenuissimus, but stimulated γ-axons with maintained frequencies, not Boyd's test, and used silver-staining for follow-up histology instead of isolating spindles. He rejected the idea of "types" in favour a continuum related to conduction velocity because he felt there were too many intermediate effects. His fastest conducting γ_s-axons preferentially innervated bag[2], and the slowest were more likely to innervate chains. A tendency for the fastest gamma axons to innervate bag fibres can be detected also in larger muscles (Emonet-Dénand and Gladden, 1993). Dickson *et al.*,(1993) supplemented Boyd's test with the cross-correlogram (see above), and extended testing to the peroneal muscles at long and short lengths within the physiological range. They divided the Ia responses into non-driving and driving, and concluded that there was a component of the γ_s-population which survived the most rigorous test for driving. Most recently Celichowski *et al.,*(1994) used their own indirect test to determine quantitatively the types of fibre innervated by γ_s-axons in peroneus tertius spindles. They concluded that γ_s-motoneurones showed no particular preference for either bag$_2$ or chains.

At present it seems unlikely that Boyd's division of γ_s-motoneurones can survive, even as a convenient shorthand - at least not in the way he described them - motoneurones whose axons innervated either bag$_2$ or chains in every spindle, with some degree of allowable cross innervation in addition. In Boyd's data very few axons did not fit this definition, but in the most recent, very careful study by Celichowski *et al.,*(1994) the proportion of such axons was as high as 36%.

It is unlikely that Boyd himself would have let the matter rest there. He would have pointed out that peroneus tertius could not be regarded as a "typical" muscle. It has a very high proportion of capsules involved in tandem spindles (67%), with 21% of capsules as the b$_2$c type (Scott and Young, 1987). This proportion of tandem-linked capsules is as high as in any of the dorsal neck muscles (41-67%, Richmond and Abrahams, 1975), where it has been postulated that tandem spindles play some specialised functional role (Richmond, Stacey, Bakker and Bakker, 1985).

Central Activation

By contrast with peripheral studies, central studies suggest that the CNS does not treat all γ_s-motoneurones alike. Wand and Schwartz (1985) compared primary and secondary responses in hindlimb flexors, and found that injection of picrotoxin into the substantia nigra removed a static excitation signalled by primary sensory endings, but a residual excitation of secondary sensory endings remained, and was revealed by denervation. They suggested that the L-DOPA antagonist had removed a static excitation mediated through bag$_2$ fibre contraction, leaving largely unaffected a static drive to chain fibres which principally excited secondaries, and that this supported the idea of a dual control of γ_s-motoneurones. Dickson and Gladden (1990; 1992) stimulated the mid-brain just dorsal to the substantia nigra and found an unexpected reciprocal effect on γ_s-motoneurones innervating the same tenuissimus spindles; some were excited and others simultaneously inhibited. In every case encountered, bag$_2$ fibres were recruited and chains inhibited. In one experiment the same effect was elicited from the sensorimotor cortex as well as the midbrain, though not so reliably (Asgari-Khozankalaei and Gladden, 1990). Image analysis of the intrafusal fibre movements confirmed the visual observations.

Although these central studies might appear to endorse Boyd's view, all that can be safely asserted is that they do suggest that the γ_s-motoneurone pool is not a homogeneous entity to the CNS. The selection of γ_s-motoneurones, however, could be according to cell size as might be suggested from Banks' (1991) work, or by some functional criterion so as to activate a group such as the non-driving group of Dickson *et al.*, (1993).

OVERVIEW

By following Boyd's line of enquiry we have reached an enigma, for it is not easy to see how to reconcile the more recent peripheral results with the central studies. Yet the static system is impressive for its capabilities, so to understand how it is used seems a worthwhile goal. The greater number of γ_s-axons supplying muscles in comparison with γ_s-axons suggests they are functionally important. The strength of some static effects, their ability to abolish the Ia length signal and occlude dynamic effects, and the fact that they modulate both afferent outflows from spindles, from primaries and secondaries, makes it unlikely that γ_s-motoneurones are allowed to operate without rigorous central control. A recruitment order for central and reflex activation can be demonstrated in anaesthetised preparations (Dickson and Gladden, 1992; Dickson *et al.*, 1992), and many studies have demonstrated that inhibition from central sources can turn them off, totally or partially.

Possibly we are asking the wrong questions. Our present approach is coloured by the standard tests we apply, as in Figs.1 and 2, with high stimulation frequencies which may be quite outwith the normal physiological range of a particular γ_s-fusimotor unit. New lines of enquiry ought to be based on our understanding of spindle physiology which has evolved since Boyd first began to isolate spindles. For example, it is not generally appreciated that γ_s-motor units with strong driving in several spindles will synchronise the Ia afferent potentials recorded in the dorsal roots, at least to within two or three milliseconds. The effect is potentially powerful, similar to a tendon jerk reflex, but confined to a few spindles. There is also the possibility of plasticity in chain fibres, as myosin expression can vary along their length (Dickson, Emonet-Dénand, Gladden, Petit, Rowlerson and Sutherland, 1991; Yoshimura, Dickson, and Gladden, 1992) and be changed by endurance exercise (Yoshimura, Fujitsuka, Kawakami, Ozawa, Ojala, and Fujitsuka, 1992).

During the IUPS Congress the Boyd Symposium was immediately preceded by a symposium on interneurones during which no speaker made any mention of the gamma system. Unhappily that may reflect the difficulties those of us working in this area have had in the past in reaching a consensus which can be easily incorporated into projected schemes of motor control. Perhaps it is partly a matter of perspective, and what now seem to be inconsistencies and difficulties may be simply manifestations of the versatility of spindle repertoire.

ACKNOWLEDGEMENTS. I.A.Boyd's work on spindle wiring was supported by the MRC. Current work is supported by the Wellcome Trust.

REFERENCES

Asgari-Khozankalaei, A. and Gladden, M.H. (1990) Simultaneous excitation and inhibition of tenuissimus static γ-motoneurones during cortical stimulation in the anaesthetised cat. *Journal of Physiology*, **429**, 9P.

Banks, R. W. (1986) Observations on the primary sensory ending of tenuissimus muscle spindles in the cat. *Cell Tissue Research* , **246**, 309-319.

Banks, R.W. (1991) The distribution of static γ-axons in the tenuissimus muscle of the cat. *Journal of Physiology* **442**, 489-512.

Barker, D., Emonet-Dénand, F, Laporte, Y. Proske, U. and Stacey, M.J. (1973). Morphological identification and intrafusal distribution of the endings of static fusimotor axons in the cat. *Journal of Physiology* **230**, 405-427.

Bessou P AND Pagès B (1975) Cinematographic analysis of contractile events produced in intrafusal muscle fibres by stimulation of static and dynamic fusimotor axons. *Journal of Physiology* **252**, 397-427.

Boyd, I.A. (1958). An isolated mammalian muscle-spindle preparation. *Journal of Physiology* **144**, 11-12P.

Boyd, I.A. (1966). The behaviour of isolated mammalian muscle spindles with intact innervation. *Journal of Physiology* **186**, 109P.

Boyd, I.A. (1970) The Muscle Spindle (film). *Journal of Physiology*, **210**, 23-24P.

Boyd, I.A. (1971) The mammalian muscle spindle - an advanced study (film). *Journal of Physiology*, **214**, 1-2P.

Boyd, I.A. (1981). The action of the three types of intrafusal fibre in isolated cat muscle spindles on the dynamic and length sensitivities of primary and secondary sensory endings. In "Muscle Receptors and Movement" pp. 17-32. (Eds. Taylor, A. and Prochazka, A.). Macmillan, London.

Boyd, I.A. (1986). Two types of static γ axon in cat muscle spindles. *Quarterly Journal of Experimental Physiology* **71**, 307-327.

Boyd, I.A. and Davey, M.R. (1968). *Composition of Peripheral Nerves*. Edinburgh and London: E and S Livingstone.

Boyd, I.A., Gladden, M.H., McWilliam, P.N. and Ward, J. (1977). Control of Dynamic and Static Nuclear Bag Fibres by Gamma and Beta Axons in Isolated Cat Muscle Spindles. *Journal of Physiology* **265**, 133-162.

Boyd, I.A., Gladden, M.H. and Ward, J. (1981). The contribution of mechanical events in the dynamic bag1 intrafusal fibre in isolated cat muscle spindles to the form of the Ia afferent axon discharge. *Journal of Physiology* **317**, 80-81P.

Boyd, I.A. and Ward, J. (1982). The diagnosis of nuclear chain intrafusal fibre activity from the nature of the group Ia and group II afferent discharge of isolated cat muscle spindles. *Journal of Physiology* **329**, 17-18P.

Celichowski, J., Emonet-Dénand, F., Laporte, Y., and Petit, J. (1994) Distribution of static γ-axons in cat peroneus tertius spindles determined by exclusively physiological criteria. *Journal of Neurophysiology*. (in press)

Dickson, M., Emonet-Dénand, F., Gladden, M.H., Petit, J., Rowlerson A. AND Sutherland, F.I. (1991) 'Non-driving' excitation of Ia afferents by static γ-axons in anaesthetised cats. *Journal of Physiology* **438**, 209P.

Dickson, M., Emonet-Dénand, F., Gladden, M.H., Petit, J. (1992) The relation between the ability of static axons to drive Ia afferents during ramp or tonic stimulation and the type of intrafusal fibres innervated. In *Muscle afferents and Spinal control of movement*, ed. Jami, L., Pierrot-Deseilligny,E. and Zytnicki, D., pp. 43-46. Pergamon, Oxford.

Dickson, M., Emonet-Dénand, F., Gladden, M.H., Petit, J. and Ward, J. (1993) Incidence of non-driving excitation of Ia afferents using ramp frequency stimulation of static γ-axons in cat hindlimbs. *Journal of Physiology* **460**, 657-673.

Dickson, M. and Gladden, M.H. (1990) Dynamic and static gamma effects in tenuissimus muscle spindles during stimulation of areas in the mes- and di-encephalon in anaesthetised cats. *Journal of Physiology* **423**, 73P.

Dickson, M. and Gladden, M.H. (1992) Central and reflex recruitment of γ-motoneurones of individual muscle spindles of the tenuissimus muscle in anaesthetised cats. In *Muscle afferents and Spinal control of movement*, ed. Jami, L., Pierrot-Deseilligny,E. and Zytnicki, D., pp. 37-42. Pergamon, Oxford.

Dickson, M., Gladden, M.H., and Yoshimura, A. (1992) γ-Reflex activity in cats anaesthetised with barbiturates. *Journal of Physiology* **459**, 461P.

Emonet-Dénand, F. and Gladden, M.H., (1993) Type of fusimotor excitation induced in individual spindles by their fastest-conducting γ axons. Proceedings of the International Union of Physiological Sciences XVIII (Congress XXXII, Glasgow), 322.6/P.

Emonet-Dénand, F., Laporte Y, Matthews P.B.C.AND Petit J (1977) On the subdivision of static and dynamic fusimotor actions on the primary ending of the cat muscle spindle. *Journal of Physiology* **268**, 827-861.

Gladden, M.H. (1976) Structural features relative to the function of intrafusal muscle fibres in the cat. *Progress in Brain Research,* **44**, 51-59.

Gladden, M.H. (1981) The activity of intrafusal muscle fibres during central simulation in the cat. In: Muscle Receptors and Movement, Eds. Taylor, A. and Prochazka, A. Macmillan, London 109-122.

Gladden, M.H. (1992) Muscle receptors in mammals. In *Comparative aspects of mechanoreceptor systems,* ed. Ito F., *Advances in Comparative and Environmental Physiology,* **10**, 281-302.

Gladden, M.H. and McWilliam, P.N. (1977a) The activity of intrafusal muscle fibres during cortical stimulation in the cat. *Journal of Physiology,* **273**, 28-29P.

Gladden, M.H. and McWilliam, P.N. (1977b) The activity of intrafusal muscle fibres in anaesthetised, decerebrate and spinal cats. *Journal of Physiology,* **273**, 49-50P.

Gladden, M.H. and Sutherland, F.I. (1989) Do cats have three types of static γ axon? *Journal of Physiology,* **414**, 19P.

Matthews, P.B.C. (1962). The differentiation of two types of fusimotor fibre by their effects on the dynamic response of muscle spindle primary endings. *Quarterly Journal of Experimental Physiology* **47**, 324-333.

Richmond, F.J.R. and Abrahams, V.C. (1975) Morphology and distribution of muscle spindles in dorsal muscles of the cat neck. *Journal of Neurophysiology,* **38**, 1322-1339.

Richmond, F.J.R., Stacey, M.J., Bakker, G.J. and Bakker, D.A. (1985) Gaps in spindle physiology: why the tandem spindle? In *The Muscle Spindle* ed. I.A. Boyd and M.H. Gladden, Macmillan, London., pp. 75-81.

Scott, J.J.A. and Young, H. (1987) The number and distribution of muscle spindles and tendon organs in the peroneal muscles of the cat. *Journal of Anatomy,* **151**, 143-155.

Wand, P. and Schwarz, M. (1985) Two types of static fusimotor neurones under separate central control? *Neuroscience Letters,* **58**, 145-149.

Yoshimura A., Dickson M., and Gladden M.H. (1992) Mechanical properties of chain fibres and regional variations in their histo- and immunohistochemical reactivity in tenuissimus muscles of anaesthetised cats. *Journal of Physiology,* **459**, 502P.

Yoshimura A., Fujitsuka C., Kawakami K., Ozawa N., Ojala H. and Fujitsuka N. (1992) Novel myosin isoform in nuclear chain fibres of rat muscle spindles produced in response to endurance swimming. *Journal of Applied Physiology,* **73**, 1925-1931.

RECENT DEVELOPMENTS IN THE PHYSIOLOGY OF THE MAMMALIAN MUSCLE SPINDLE

U. PROSKE

Department of Physiology
Monash University
Clayton, Victoria 3168, Australia

SUMMARY

Experiments have been carried out on passive muscle spindles of the soleus muscle of anaesthetised cats. Use is made of the property of spindles to show after-effects. It is proposed that after a muscle stretch the intrafusal fibres of spindles fall slack, due to their thixotropic property. It is possible to take up the slack selectively in individual intrafusal fibres by stimulating their fusimotor (γ) fibres. It was shown that vibration sensitivity of a slack spindle could be restored after a period of stimulation of γ static (γ_S) but not γ dynamic (γ_D) axons. Similarly the level of resting discharge of the slack spindle could be significantly raised, as could the response to a slow stretch, by stimulating γ_S but not γ_D axons. It is concluded that, at muscle lengths at which spindles show after-effects, responses of passive spindles come from the terminals of the afferent fibre which lie on intrafusal fibres innervated by γ_S axons, the bag$_2$ fibre and perhaps also chain fibres but not the bag$_1$ fibre, innervated by γ_D axons. This conclusion remains to be reconciled with recent observations which suggest that the sensory region of the bag$_1$ fibre is more compliant than that of bag$_2$ and chain fibres.

INTRODUCTION

During the last 40 years, our understanding of the mammalian muscle spindle has taken a number of important steps forward. In the 1950s it was recognition of two kinds of sensory endings, the primary ending and the secondary ending. In the 1960s it was demonstration of two kinds of fusimotor fibres, static and dynamic. The 1970s brought recognition of two kinds of nuclear bag fibres, the dynamic bag$_1$ and the static bag$_2$ fibre. And what have the 1980s brought? There have been a number of new and important observations which have changed our view of the internal functioning of the spindle. For example, we have a better understanding of why there are two different kinds of bag fibres present in spindles and there is now a lot more information available on the pattern of fusimotor innervation.

The 1980s brought the first detailed description of the development of the muscle spindle. It is of interest that the first intrafusal fibre to develop is the bag$_2$ fibre. Then comes the bag$_1$ fibre and finally the chain fibres (Milburn, 1984). The bag$_2$ fibre is the largest intrafusal fibre within the spindle and its mechanical properties are intermediate between those of the bag$_1$ fibre (slow) and the chain fibres (fast). The bag$_2$ fibre appears to be commonly associated with the chain fibres and many static fusimotor axons innervate both bag$_2$ and chain fibres.

For many observations on the pattern of the spindle motor innervation we are indebted to Professor Ian Boyd, after whom the symposium that forms the basis of this book, was named. Boyd pioneered the technique of studying intrafusal contractions in semi-isolated spindles while stimulating identified fusimotor fibres. The method allowed him to associate a particular contraction with the afferent response it initiated. His observations led Boyd to consider that static fusimotor fibres might be separated into two kinds, those predominantly innervating bag_2 fibres and others which supplied chain fibres. The bag_1 fibre appears to be innervated largely, or exclusively by dynamic fusimotor axons.

In the spindle physiology literature, much emphasis has been placed on the bag_1/dynamic fusimotor fibre system. The reason for so much interest is perhaps because of the very dramatic effect dynamic fusimotor stimulation has on the spindle stretch response and because of the significance this may have for servo control theories of muscle contraction. Stepping back and looking at recent developments, the picture is emerging of two intrafusal systems, the bag_1/γ dynamic system and the bag_2: chain/γ static system. Indeed some spindles show bag_2 and chain fibres lying within a separate compartment from bag_1 (Fig. 1, Banks et al., 1982).

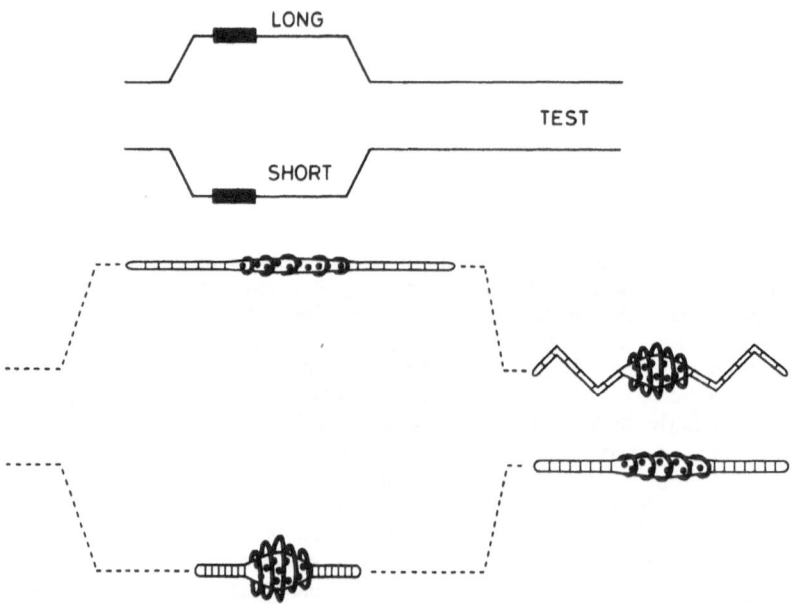

Figure 1. Diagram to illustrate how muscle spindles may fall slack following muscle conditioning. At the top of the figure is shown the conditioning: test sequence. The muscle is either stretched (line moves upwards, `LONG') or shortened (line moves down, `SHORT'), a fusimotor strength contraction carried out (solid bar) and several seconds later the muscle is returned to its initial length and the test stimulus is given (`TEST'). The cartoon below shows how an intrafusal fibre of a muscle spindle may fall slack. Stretching the muscle stretches the intrafusal fibres. To ensure that all stable cross bridges in intrafusal fibres reset at that length, the fusimotor strength contraction is given. On return to the initial length, the intrafusal fibres, stiffened by the presence of stable cross bridges, fall slack. Slack allows the sensory spirals to close. Conditioning by shortening and contracting the muscle does not lead to development of slack. On return to the initial length, the intrafusal fibre lies taut and the sensory spirals are opened out.

If we accept that the bag_1/dynamic fusimotor system is concerned with phase advance and minimising time delays in feedback loops (perhaps also with sensitising the spindle in anticipation of a skilled movement, Prochazka 1989), what is the role of the static system? The present-day view, based largely on the human recordings of Vallbo (1974) is that activity in static fusimotor fibres is concerned with ensuring maintenance of spindle activity during muscle contraction, allowing the central nervous system to continuously monitor progress of the movement as well as providing servo-assistance during its execution.

A recent finding of potentially considerable importance is that some spindles lack a bag_1 fibre. This has been shown by both structural (Price & Dutia, 1987) and physiological means (Price & Dutia, 1987, 1989; Gioux et al., 1991; Scott, 1991). In other words, the information coming from the dynamic system within the spindle is not essential for the afferent feedback from some muscles. Price and Dutia were able to show that spindles which appeared to lack a bag$_1$ fibre showed passive stretch responses and vibration responses that were essentially indistinguishable from those of spindles with bag$_1$ fibres. It raised the question, where in the spindle was the stretch response generated? Up to that time it had generally been assumed that it was associated in some way with the bag$_1$ fibre. The reason was that stimulating the γ dynamic axon, which innervates the bag$_1$ fibre, leads to a large increase in the stretch response of the spindle. By inference, it was assumed that the bag$_1$ fibre was responsible for the stretch response in the passive spindle as well.

The development of the bag$_2$ fibre, before any of the other intrafusal fibres and its ubiquitous distribution hints, I believe, at the importance of this fibre for spindle function. Much of the remainder of this chapter is directed at trying to find out more about the role, within the spindle, of the bag$_2$ fibre.

In the mid 1980s our group introduced a new tool to the study of spindles and their reflex action (Morgan *et al.*, 1984). This was the method of conditioning the muscle and its spindles to put passive spindles into a defined mechanical state. The basis of the method rested on the phenomenon of after-effects (for a review see Proske *et al.*, 1993). We have used it to explore, in particular, the role of the bag$_2$/chain intrafusal system in spindle function.

The underlying idea is to introduce slack in the intrafusal fibres of muscle spindles and then selectively take up the slack, by stimulating identified fusimotor fibres and observing the effects on spindle responses. Slack can be introduced both in intrafusal and extrafusal fibres as a result of the thixotropic property of muscle (Proske *et al.*, 1993). A muscle fibre spontaneously develops stable cross-bridges between actin and myosin, cross-bridges which do not develop any force, but which raise the passive stiffness of the fibre. If stable cross-bridges are allowed to form and the muscle is then passively shortened, the muscle fibres, stiffened by the presence of stable cross-bridges, fall slack. The slack can be removed by a contraction. The cartoon in Fig. 1 illustrates this.

We have now done three different kinds of experiments to explore the role of the various intrafusal fibres, making use of the slack technique. These experiments examine the sensitivity of the passive spindle to vibration, its ability to develop a resting discharge and its response to a slow stretch.

Vibration Sensitivity

In the first experiment we measured the sensitivity of the spindle to locally applied vibration (Morgan *et al.*, 1991). First the spindle is conditioned so that its intrafusal fibres lie slack. This is achieved by contracting the muscle using a nerve stimulus strong enough to recruit fusimotor fibres and therefore to contract both intrafusal and extrafusal fibres, at a length longer than that at which the experiment is to be carried out. On return to the test length spindles will fall slack. In the presence of slack spindles become much less sensitive to mechanical disturbances. Under these conditions gentle palpation of the muscle allows quite precise localization of the spindle, to within a few square millimetres. If a vibrator is placed so that its stylus makes contact with the muscle just at this point it is possible to evoke vibration responses. The real experiment, however, is when after introducing slack in intrafusal fibres, the spindle is conditioned by stimulation of single, identified fusimotor fibres. An example is shown in Fig.2. The response of the slack spindle to vibration without fusimotor conditioning was very weak, just a few impulses at the peak of the vibration pulse. Conditioning with a one second period of γ_D stimulation had little additional effect on the size of the response. However, a similar period of conditioning with a γ_S axon led to a large increase in the vibration response and the afferent became entrained 1:1 to the vibration at its peak amplitude. Stimulating other γ_S axons had little effect on vibration responses.

The conclusions we drew from this experiment were that if after introducing slack in all intrafusal fibres of the spindle, it is selectively taken up in the bag$_1$ fibre by stimulating a γ_D axon, vibration sensitivity was not restored to the level observed when all intrafusal fibres had been made taut by conditioning. Yet stimulating some γ_S especially those which did not have a large chain fibre input, as evidenced by the lack of driving of the afferent response during fusimotor stimulation, fully restored vibration sensitivity in a slack spindle. We concluded that whether the bag$_1$ fibre was slack or taut made little difference to responses and that therefore vibration sensitivity in the passive spindle did not reside in the sensory terminals distributed to the bag$_1$ fibre. Similarly, since γ$_S$ axons which elicited strong driving responses from the spindle were also ineffective in resetting vibration sensitivity, it was concluded that the mechanical state of chain fibres was not a deciding factor either. What seemed to matter was the mechanical state of the bag$_2$ fibre. It might be argued that the bag$_1$ fibre and chain fibres don't fall slack after conditioning, perhaps because they are supported by a network of elastic fibres. So γ stimulation would therefore have no effect. We don't think that is so, at least for γ$_D$, because afferent responses to a slow stretch after hold-long conditioning clearly show a delay in onset representing the time required to take up the slack, a delay which is significantly reduced when slack has been removed by γ$_D$ stimulation or by hold-short conditioning (Fig. 4, see also Gregory *et al.*, 1986).

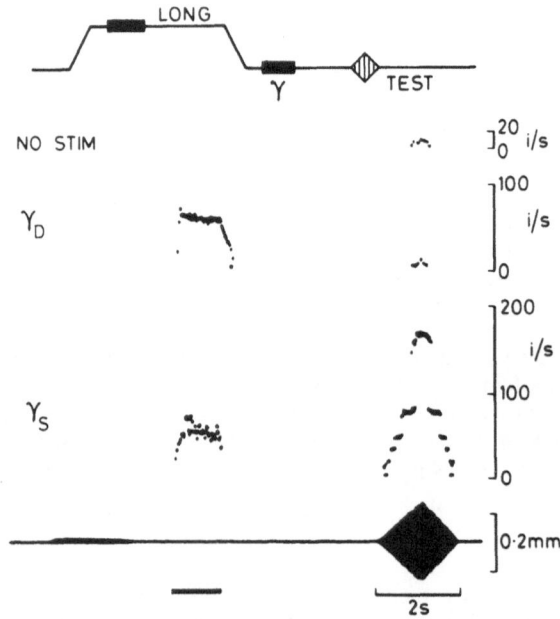

Figure 2. The effect of muscle conditioning on the response of the primary ending of a cat soleus muscle spindle to locally applied vibration. The diagram at the top illustrates the hold-long conditioning, (`LONG'), followed by the test vibration (hatched ◊ `TEST'). The test vibration is preceded by a period of fusimotor stimulation (solid bar, labelled γ). The actual responses are shown below, displayed as instantaneous frequencies. The top trace is hold-long followed by test vibration only, without any intervening γ stimulation. In the second trace, following conditioning, an identified dynamic fusimotor fibre (γ$_D$) is stimulated (150 pps for 1s) before the test vibration is applied. In the third trace static fusimotor stimulation (γ$_S$) precedes the test vibration pulse. The continuous trace at the bottom of the figure is a record of the displacement of the vibrator (vibration frequency 180Hz). Below are shown a time calibration which applies to the displacement and frequency traces and a bar indicating the period of fusimotor stimulation. Stimulating a γ$_S$ axon sensitises the spindle to vibration, stimulating a γ$_D$ has no effect. (Redrawn from Morgan *et al.*, 1991).

Resting Discharge

In a similar kind of experiment we asked, where in the spindle is the resting discharge generated? The experiment consisted of shortening the muscle from a pre-selected length. As a result of the shortening, the spindle fell silent for a number of seconds and then gradually resumed its resting discharge (Fig. 3). Under these conditions it was found that stimulating a γ$_D$ soon after the shortening step did not significantly reduce the time for resumption of the resting discharge. Stimulation of some γ$_S$ axons, on the other hand, was followed by a prompt resumption and the subsequent discharge was maintained at a higher rate (Fig. 3). The conclusion again was that for the recovery of a resting discharge it did not matter whether bag$_1$ was slack or taut. Yet if one or more of the other intrafusal fibres was tightened, this strongly influenced the level of resting discharge and its rate of recovery after a shortening step (Proske *et al.*, 1991).

Stretch Responses

In the third experiment the responses of spindles to a slow stretch were examined after the introduction of slack in intrafusal fibres, with and without a second conditioning stimulus given to an identified fusimotor fibre (Fig. 4). The response to stretch, without fusimotor stimulation, began about one-third of the way through the stretch and consisted of a gradual increase in firing up to a peak level which was then maintained for the remainder of the stretch. At this peak rate, spindle firing was characterised by a very irregular discharge. If a γ_D was stimulated for one second before onset of the stretch, the response to stretch began earlier, but its shape remained much the same. Stimulating a γ_S also produced an earlier onset of the stretch response but this time the whole shape of the response was changed, spindle firing increasing rapidly up to a peak from which it showed little further increase for the remainder of the stretch.

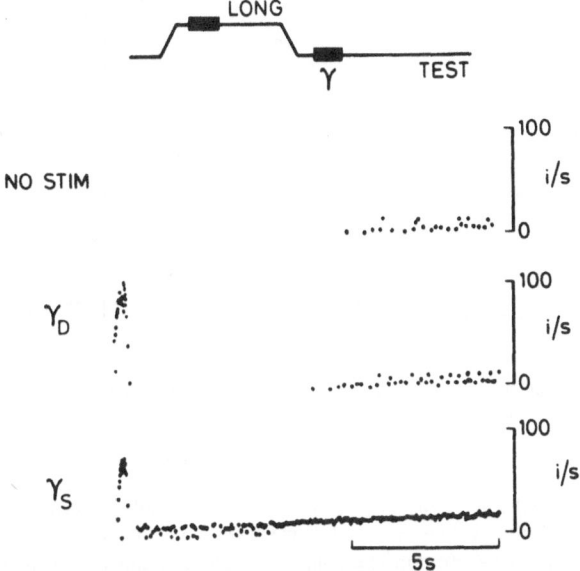

Figure 3. The effect of muscle conditioning on the resumption time of the discharge of a muscle spindle. At the top is shown a diagram illustrating the conditioning sequence as in previous figures. Below are shown instantaneous frequency records of spindle responses. When, following hold-long conditioning, no γ stimulus was given, the resting discharge resumed within about 10 seconds after return to the initial length. If a γ dynamic (γ_D) was stimulated (150 pps for 0.5sec) immediately after return to the initial length, resting discharge resumed slightly earlier. If a γ static (γ_S) was stimulated (150 pps for 0.5sec), resting discharge resumed immediately after the end of stimulation and continued at a higher rate than previously. (Redrawn from Proske *et al.*, 1991).

CONCLUSIONS

From these three experiments we concluded, as had Price and Dutia (1989), that in the passive spindle input to the afferent response from the nerve terminals on the bag$_1$ fibre is not the major determinant of dynamic responsiveness, expressed as the response to stretch and vibration. Our findings have extended this view and established that the bag$_1$ fibre is also not the source of the spindle resting discharge. While using spindle driving by static fusimotor stimulation as a means of identifying chain fibre inputs to the spindle is not always reliable, our data do clearly indicate two classes of γ_S based on their ability to re-set spindle resting discharge, stretch responses and vibration sensitivity. Since the current view is that γ_S axons innervate bag$_2$ fibres, chain fibres or both, (Celichowski *et al.*, 1994) we interpret the fact that non-driving γ_S were best at resetting the spindle, that these axons supplied predominantly bag$_2$ fibres.

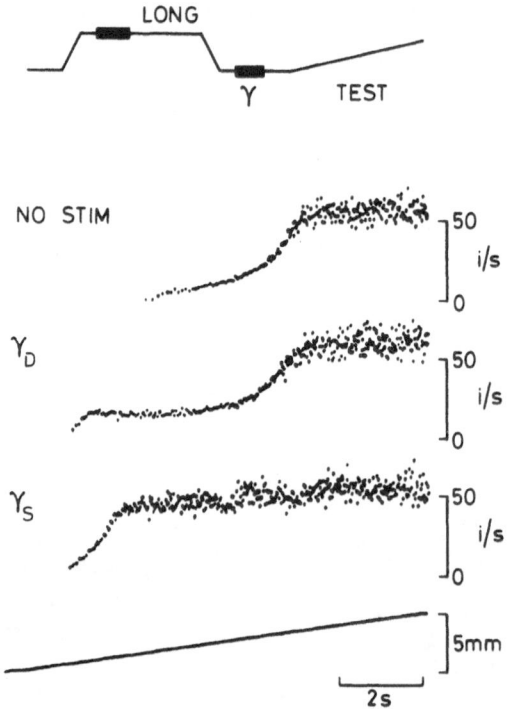

Figure 4. The effect of muscle conditioning on the response of a spindle primary endings to a slow stretch with and without preceding fusimotor stimulation. At the top of the figure is shown, diagrammatically, the conditioning: test sequence, as in previous figures. The actual responses to the stretch are shown as instantaneous frequency traces below. The continuous trace at the bottom of the figure gives the length change of the muscle during the stretch. The top frequency trace shows the stretch response after hold-long conditioning without any subsequent fusimotor stimulation. The middle frequency trace shows the stretch response with an immediately previous period of dynamic fusimotor stimulation (γ_D, 150 pps for 2s). The bottom frequency trace shows the effect of preceding static fusimotor stimulation (γ_S, 100 pps for 2s) on the subsequent stretch response. Length and time calibration at the bottom of the figure apply to the length and frequency traces. Following γ_D stimulation although the response to the stretch starts a little sooner than without any γ conditioning its shape is very similar. After γ_S stimulation the stretch response is quite different, rising rapidly to high rates. (Redrawn from Proske *et al.*, 1992).

We conclude that under the conditions of our experiments, the source of the impulse activity in the passive spindle is predominantly the sensory terminals on the bag$_2$ fibre. However, all of our observations are based on spindle after-effects which depend on the ability of intrafusal fibres to fall slack. It is known that after-effects disappear at long muscle lengths. It might therefore be argued that our conclusions about bag$_2$ fibres are valid only over the range of muscle lengths for which the spindles show after-effects. We have previously measured the length dependence of after-effects (Gregory *et al.*, 1986, 1988). When intrafusal fibres fall slack as a result of hold-long conditioning, the take-up of that slack shows up as a delay in onset of the afferent response to fusimotor stimulation. When the delay was measured for a γ_D, after a 5mm shortening step it was about 150ms at L_M-8 (maximum body length - 8mm, Gregory *et al.*, 1986). If the experiment was carried out at a length 2mm longer, L_M-6, the delay fell to 50ms, while at L_M-4 it was small and disappeared rapidly. At even longer lengths there was only the minimum delay expected from axonal conduction time and time for initiation of the afferent response. It meant that beyond L_M-4, the intrafusal fibre, in this case the bag_1 fibre innervated by γ_D, was no longer falling slack. At this point all other signs of after-effects had gone as well. Our interpretation of these findings was that at long muscle lengths the rising level of passive tension in the muscle exerts lateral compressive forces on the intrafusal fibres leading to rapid removal of any slack produced by muscle conditioning. At these lengths, therefore, it was not possible to do the kinds of experiments that we have described above. We therefore cannot make any firm claims about the intrafusal source of the afferent responses in spindles at long muscle lengths. It is conceivable that at long lengths bag_1 or chain fibres take over from bag_2 as the source of the impulse activity. More likely, however, is that at long lengths inputs from all three kinds of intrafusal fibres contribute to the afferent response. These kinds of predictions should be tested experimentally. If our observations are correct they suggest that intrafusal fibres are able to fall slack and remain so for a significant period over a range of muscle lengths up to about L_M-6. Given that the optimum length for a contraction in soleus is L_M-8 to L_M-10, it suggests that the bag_2 input is likely to be important over much of the working range of the muscle provided, of course, that the spindle remains passive. The situation will change dramatically as soon as fusimotor fibres become active. All of this leads to the conclusion that at short and intermediate muscle lengths the passive mammalian spindle behaves as a single intrafusal fibre spindle, like its reptilian ancestor.

However, before it can be firmly concluded that bag_2 fibres are the principal source of impulse activity some recent new findings on isolated spindles must be considered. Chua and Hunt (1993) observed isolated cat spindles under differential interference microscopy after the sensory endings had been stained with a fluorescent dye. In spindles where the bag_1 and bag_2 fibres could be viewed simultaneously, the sensory terminals on bag_1 were extended 50 to 100 percent more than those on bag_2 in response to either steady or sinusoidal stretch. Movement of the endings on the bag_2 fibre was also delayed relative to those on bag_1 suggesting that the bag_1 terminals were more likely to contribute to dynamic sensitivity. The authors suggested that the difference in extension of sensory endings was due to a difference in compliance of the equatorial, nuclear bag region of bag_1 and bag_2 fibres.

It is, of course, not a simple matter to extrapolate from an isolated spindle to spindles in a whole muscle. Nevertheless the observed difference in compliance of the two kinds of intrafusal fibres is the opposite to what might have been predicted from the conclusion that bag_2 was the main source of the impulse activity. It will be important to correlate the observed extension of the sensory endings in isolated spindles with the pattern of afferent response. In addition it would be interesting to introduce slack in the intrafusal fibres of isolated spindles and see what effect this had on the afferent response. In any case these observations do not fit readily with the findings on whole animals and it remains for additional experiments to help sort out this issue.

In conclusion, there is a growing body of experimental observations that supports the view that in passive spindles the afferent response to muscle stretch, the response to vibration and the resting discharge are all coming from sensory terminals located on the bag_2 fibre. This is likely to hold true over a range of muscle lengths, at least up to the optimum length for a contraction. Recent observations on isolated spindles provide a picture which is not consistent with this view and it will require additional experiments before the question of the source of activity in passive spindles can finally be resolved.

REFERENCES

Banks, R.W., Barker, D. & Stacey, M.J. (1982). Form and distribution of sensory terminals in cat hindlimb muscle spindles. *Philosophical Transactions of the Royal Society of London* **299**, 329-364.

Celichowski, J., Emonet-Dénand, F., Laporte, Y. & Petit, J. (1994). Distribution of static γ axons in cat peroneus tertius spindles determined by exclusively physiological criteria. *Journal of Neurophysiology* **72**, 722-732.

Chua, M. & Hunt, C.C. (1993). Relative compliance of bag_1 and bag_2 sensory regions of the cat muscle spindle. 32nd Internaitonal Congress of Physiological Sciences, Glasgow, **322**, 1P.

Gioux, M., Petit, J. & Proske, U. (1991). Responses of cat muscle spindles which lack a dynamic fusimotor supply. *Journal of Physiology* **432**, 557-571.

Gregory, J.E., Morgan, D.L. & Proske, U. (1986). After-effects in the responses of cat muscle spindles. *Journal of Neurophysiology* **56**, 451-461.

Gregory, J.E., Morgan, D.L. & Proske, U. (1988). Responses of muscle spindles depend on their history of activation and movement. In: "Transduction and cellular mechanisms in sensory receptors". *Progress in Brain Research* eds. W. Hamann & A. Iggo, **74,** 85-90.

Morgan, D.L., Prochazka, A. & Proske, U. (1984). The after-effects of stretch and fusimotor stimulation on the responses of primary endings of cat muscle spindles. *Journal of Physiology* **356,** 465-477.

Milburn, A. (1984). Stages in development of cat muscle spindles. *Journal of Embryology and Experimental Morphology* **82,** 177-216.

Morgan, D.L., Proske, U. & Gregory, J.E. (1991). Responses of primary endings of cat muscle spindles to locally applied vibration. *Experimental Brain Research* **87,** 530-536.

Price, R.F. & Dutia, M.B. (1987). Properties of cat neck muscle spindles and their excitation by succinylcholine. *Experimental Brain Research* **68,** 619-630.

Price, R.F. & Dutia, M.B. (1989). Physiological properties of tandem muscle spindles in neck and hind-limb muscles. *Progress in Brain Research* eds. J.H.J. Allum & M. Hulliger **80,** 47-56.

Prochazka, A. (1989). Sensorimotor gain control: A basic strategy of motor systems? *Progress in Neurobiology* **33,** 281-307.

Proske, U., Gregory, J.E. & Morgan, D.L. (1991). Where in the muscle spindle is the resting discharge generated? *Experimental Physiology* **76,** 777-785.

Proske, U., Morgan, D.L. & Gregory, J.E. (1993). Thixotropy in skeletal muscle and in muscle spindles: A review. *Progress in Neurobiology* **41,** 705-721.

Scott, J.J.A. (1991). Responses of Ia afferent axons from muscle spindles lacking a bag_1 intrafusal muscle fibre. *Brain Research* **543,** 97-101.

Vallbo, Å, B. (1974). Human muscle spindle discharge during isometric voluntary contraction. Amplitude reactions between spindle frequency and torque. *Acta Physiologica Scandinavica* **90,** 319-336.

THE HUMAN MUSCLE SPINDLE AND ITS FUSIMOTOR CONTROL

D. BURKE and S.C. GANDEVIA

Prince of Wales Medical Research Institute and
University of New South Wales
Sydney 2031, Australia

SUMMARY

It is now 25 years since Hagbarth and Vallbo (1968) described the technique of microneurography, with the first direct recordings of the activity of muscle afferents in human subjects. It is appropriate to take stock of what has been learnt about fusimotor control of human muscle spindles since then in order to define the course of future experiments.

Muscle spindle discharge is used in microneurographic recordings to infer the action of the fusimotor system. This Chapter considers some of the fusimotor and non-fusimotor mechanisms which can influence the behaviour of human muscle spindle afferents. Voluntary contractions under isometric (or lengthening) conditions activate the fusimotor system such that overall spindle discharge increases. Evidence is presented that this can lead to considerable reflex "support" to motoneuronal output. This contribution may diminish during muscle fatigue and be absent during voluntary shortening contractions of sufficient speed.

Further studies will continue to focus on possible flexibility in the activation of fusimotor neurones produced by reflex and voluntary drives. Little data exist on possible abnormalities in fusimotor-induced spindle feedback or in reflexes acting on fusimotor neurones in patients with motor disorders.

Non-Fusimotor Effects on Spindle Responsiveness

Excised human muscle spindles behave much as would be expected from the cat (Poppele and Kennedy, 1974; Newsom Davis, 1975), but a spindle isolated from its environment gives limited insight into normal behaviour: under natural conditions its response will vary with potentially conflicting extrafusal influences. There is a wide range of variability in behaviour of human spindles, affecting background discharge, responsiveness to stretch and ease of activation in a voluntary contraction of the receptor-bearing muscle (e.g. Burke, 1981; Edin and Vallbo, 1990). Human spindles, though usually larger in absolute terms than cat spindles, are relatively small when the size of the extrafusal muscle is considered (for review, see Burke and Gandevia, 1990). Human spindles insert more often onto perimysium than do cat spindles, and there are more connective tissue connections between spindle and extrafusal muscle. These anatomical differences would render the human spindle effectively "in-series" with some muscle fibres, and such behaviour has been documented (Burke, Aniss and Gandevia, 1987).

Meyer-Lohmann, Riebold and Robrecht (1974) emphasized that the responsiveness to static stretch of cat spindles varies with the location of the spindles in the muscle, presumably reflecting their different exposure to the length change applied to the tendon. Given the greater size of human muscle, it is reasonable to expect that spindle location could

Neural Control of Movement, Edited by W.R. Ferrell
and U. Proske, Plenum Press, New York, 1995

be an even greater factor in human subjects, perhaps accounting for much of the variability of behaviour seen within a single spindle pool. Our current studies suggest that spindle location and musculo-tendinous architecture are particularly important in determining the passive behaviour of human spindle endings in complex muscles with a number of tendons (Gandevia, Hales, Wilson and Burke, unpublished).

Another major factor contributing to the behaviour of the spindle is its length and fusimotor history. Spindle "after effects" have been well described in the cat (e.g. Brown, Goodwin and Matthews, 1969; Morgan, Prochazka and Proske, 1984; Gregory, Morgan and Proske, 1987; Gregory *et al* ., 1990) and observed in human recordings (e.g. Edin and Vallbo, 1988; Macefield, Hagbarth, Gorman, Gandevia and Burke, 1991; Ribot-Ciscar, Tardy-Gervet, Vedel and Roll, 1991). One example, using a protocol developed by Proske and colleagues is shown in Figure 1. Particularly for muscles which are relaxed for long periods and which do not undergo extreme length changes these effects will act to sensitize the spindles such that stretch (and fusimotor drive) will produce a stronger spindle output. This could have important reflex and kinaesthetic consequences (Gregory *et al* ., 1987, 1990).

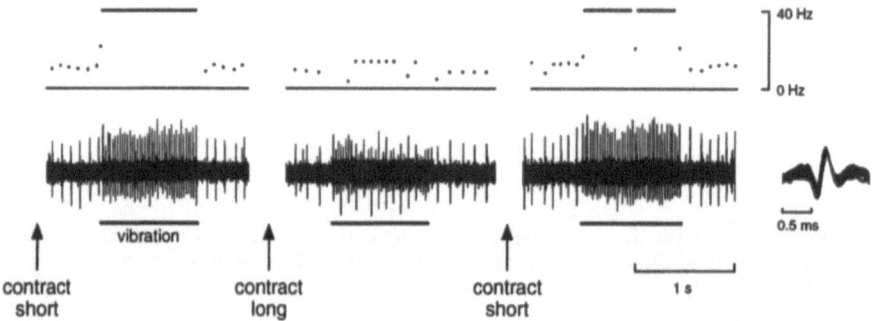

Figure 1. Spindle afferent in tibialis anterior. Vibration sensitivity tested three times (40Hz, 0.25mm) at the same intermediate length. Following contraction at a shorter length the spindle was sensitized such that vibration (indicated by the horizontal bar) produced one-to-one driving of the discharge. Following contractions at a longer length the spindle was less responsive to the test vibration. Superimposed spindle spikes are shown at right (Gandevia, Burke, Cordo and Hales, unpublished observations).

The ability of individual spindles to respond not only to fusimotor drive but also to mechanical events peculiar to their precise location in the muscle makes the discharge of some spindle endings a poor reflection of the change in length applied to a muscle tendon. Spindle endings are so sensitive that they encode the arterial pulsation within a muscle (McKeon and Burke, 1981), the body disturbance produced by respiration (Gandevia and Burke, 1985), and the twitch contraction of single nearby motor units (McKeon and Burke, 1983). An individual spindle afferent carries a unique message related to its receptor's environment. Is this important or is it merely noise, the price to be paid for maintaining high sensitivity, something that can be removed by pooling the activity of many afferents?

Binder and Stuart (1980) postulated that a spindle would have greatest reflex feedback onto motor units that were in the same region of muscle and were thereby more capable of affecting the discharge of that spindle. This hypothesis has since been modified and elaborated (see Windhorst, Hamm and Stuart, 1989 for review). Accordingly, evidence has been sought for "reflex partitioning" involving low-threshold motoneurones in human tibialis anterior, but without success (McKeon, Gandevia and Burke, 1984). At least in this muscle the findings indicated that reflex actions were spread across low-threshold motoneurones in the pool without regard to topography.

With an analogous rationale, we have sought evidence that a single spindle can produce a significant reflex effect on the discharge of homonymous motoneurones (Gandevia, Burke and McKeon, 1984) or can evoke a clear sensation when activated in isolation (Macefield, Gandevia and Burke, 1990), but again without success. In both cases the *population* of spindle endings have clear reflex and perceptual consequences, respectively. The latter result contrasts with the observation that microstimulation of single cutaneous and joint afferents can evoke appropriate sensations (Macefield *et al* ., 1990). The over-riding impression is that the discharge of a spindle is important, not by itself, but by virtue of its contribution to the population response.

What Does the Nervous System Extract from the Spindle Discharge?

The contribution of a spindle to the population response can take two forms. First, it can contribute to the strength of the input and, secondly, it can contribute by signalling deviations from the background discharge which, if consistent across the population, would evoke reflex and perceptual consequences.

The original formulation of the servo theory required that the strength of spindle feedback could drive the discharge of motoneurones (Merton, 1953), a view contested by Matthews on the grounds that there was insufficient gain in the reflex pathway (see Matthews, 1972). Vallbo (1971) demonstrated that the contraction-associated increase in spindle firing followed rather than preceded the onset of EMG when human subjects performed rapid isometric voluntary contractions. These data contributed to the idea that spindle feedback provides "servo assistance" (or "supportive excitation") to a contraction in which the major drive on the α-motoneurone pool comes directly from the descending voluntary command.

The extent of reflex support to a voluntary contraction has been estimated by recording the maximal discharge rates of motor units before and after a partial block of the peroneal nerve with local anaesthetic (Hagbarth, Kunesch, Nordin and Wallin, 1986) and by recording the discharge rates of a motor axons proximal to a complete nerve block of the ulnar or peroneal nerves (respectively, Gandevia, Macefield, Burke and McKenzie, 1990; Macefield, Gandevia, Bigland-Ritchie, Gorman and Burke, 1993). All three studies indicate that the discharge rates of motoneurones in maximal voluntary efforts are reduced by about one-third when afferent feedback is removed. In addition, Macefield *et al*.,(1993) found that there was a similar reduction in discharge when subjects performed submaximal and minimal efforts, indicating that afferent feedback contributes significantly to force generation, no matter what the strength of contraction (Fig. 2).

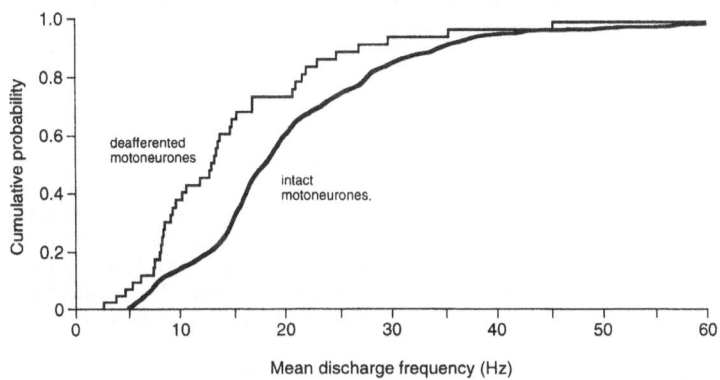

Figure 2. Data for normal motor units in tibialis anterior (solid curve at right) and for motor axons recorded proximal to a complete peroneal nerve block (thin curve at left). Results plotted as cumulative probability histograms. Discharge frequencies during voluntary contractions ranging from minimal to maximal levels are lower when the muscle is paralysed and homonymous and heteronymous feedback from it is blocked (from Macefield *et al*., 1993, with permission).

If spindles maintain a background discharge, they can respond to muscle shortening (as well as lengthening). During the course of a voluntary movement, there are inevitably small irregularities in performance of the movement, due to the changing mechanical properties of the limb and any load on it, and perhaps to irregularities in the motor programme. Spindle discharge can reflect these departures from the intended movement (Vallbo, 1973; Burke, Hagbarth and Löfstedt, 1978a,b; Hagbarth and Young, 1979; see also Al-Falahe, Nagaoka and Vallbo, 1990), and the resultant change in supportive excitation to the contracting muscle may be translated into a reflex decrease or increase in EMG through an unloading or a "stretch" reflex, respectively (Burke *et al*., 1978a,b; Hagbarth and Young, 1979).

Efficacy of Fusimotor Drive

Much of what has been said above depends on the maintenance of an adequate spindle discharge during a contraction. Muscle shortening, whether overt or "internal", will necessitate an increase in fusimotor drive sufficient to offset the shortening. It is therefore appropriate to examine the extent to which fusimotor activity alters spindle discharge in natural or near-natural tasks in human subjects.

In previous reviews, evidence has been presented that when subjects are relaxed there is little if any fusimotor drive directed to non-contracting muscles, and that this remains so when subjects engage in mental or motor tasks that do not disturb the receptor-bearing muscle (Vallbo, Hagbarth, Torebjörk and Wallin, 1979; Burke, 1981). It is not even possible to train subjects to modulate spindle activity without contracting or otherwise disturbing the receptor-bearing muscle (Gandevia and Burke, 1985). However, recent studies indicate that the inability to alter spindle discharge does not imply complete fusimotor quiescence (Ribot, Roll and Vedel, 1986; Aniss, Diener, Hore, Burke and Gandevia, 1990; Gandevia, Wilson, Cordo and Burke, 1993). Without producing EMG or detectable muscle stretch, electrically evoked volleys in low-threshold cutaneous afferents can alter muscle spindle discharge from the pretibial muscles of standing subjects (Aniss *et al*., 1990) and from the forearm extensor muscles of subjects not engaged in any specific motor task (Gandevia *et al*., 1993). Thus, although we have been unable to influence spindle discharge selectively by natural activation of descending motor systems (or by stimulating them directly - e.g. the corticospinal system: Rothwell, Gandevia and Burke, 1990; vestibulo-fugal systems: Burke, McKeon and Westerman, 1980), the discharge of some spindle endings can be altered selectively through a fusimotor reflex involving low-threshold cutaneous afferents. At least in the forearm muscles, this reflex action seems to involve dynamic fusimotor neurones (Gandevia *et al*., 1993).

Leaving aside reflex actions (which have been demonstrable for a minority of afferents), the only reliable method for activating human muscle spindle endings through the fusimotor system is a deliberate voluntary contraction of the receptor-bearing muscle (Vallbo *et al*., 1979). The contraction-associated effect involves γ motoneurones (Burke, Hagbarth and Skuse, 1979) and possibly β motoneurones (Rothwell *et al*., 1990), probably of the static type but possibly also of dynamic type (Vallbo, 1973, 1974). The fusimotor drive is directed largely, if not exclusively, to the contracting muscle, and seems to be graded in proportion to effort (Vallbo, 1974). However, not all spindle endings increase their discharge when the receptor-bearing muscle contracts, particularly if the contraction is weak (Burke, Hagbarth and Skuse, 1978c; Edin and Vallbo, 1990). The latency to spindle acceleration may be long in slow contractions, and the initial response from an active ending may be a decrease in discharge before the acceleration. These findings have led to the suggestion that the voluntarily driven increase in fusimotor drive is not distributed equally to all spindles in the muscle, and that there may be an orderly recruitment of fusimotor neurones as contraction strength increases (Burke *et al*., 1978c; *cf*. Edin and Vallbo, 1990). Alternatively, the thresholds for activation of spindle endings by voluntary effort could represent differences in the extent to which different endings are unloaded by the extrafusal contraction. The suggestion that spindle recruitment in a contraction occurs at a reproducible level of force (effort) has not been supported for upper-limb spindle endings (Edin and Vallbo, 1990). Whether this results from a difference in the organization of the fusimotor outflows for the upper and lower limbs or merely to a difference in test paradigm remains to be established.

In an isometric contraction, spindle discharge does not remain steady even when subjects maintain a steady force level (despite fatigue) or a steady effort (Macefield, Hagbarth, Gorman, Gandevia and Burke, 1991). In prolonged contractions, spindle discharge declines in the face of the increasing EMG required to maintain a constant force. After contraction for 30 s, discharge frequencies are only 70% of those early in the contraction. Thus, the degree of reflex support from homonymous spindles must decrease as the contraction progresses, unless there are compensatory changes in central gain. This conclusion immediately places a further limitation on the extent to which the muscle spindle/fusimotor mechanism operates as a simple servo to support a contraction.

Only when voluntary isotonic movements are slow or when the contracting muscle must move a load is the increase in fusimotor drive sufficient to maintain spindle discharge in the face of shortening (Burke *et al*., 1978a,b; Al-Falahe *et al*., 1990). Similar findings have been presented for voluntary alternating movements in the cat (Prochazka, Stephens and

Wand, 1979). Furthermore, Prochazka *et al.*, have presented evidence that, in the rapid cyclical movements associated with locomotion, the "subthreshold" fusimotor inputs do not act to sensitize the spindle so that it is more ready to respond to a perturbation.

These findings must prompt reassessment of the role of the fusimotor system in the control of movement. Presumably fusimotor drive can be important only by affecting muscle spindle discharge. Only in isometric contractions or lengthening contractions can a significant increase in spindle feedback be expected. In concentric contractions, the speed of movement and the load on the contracting muscle will determine whether spindle discharge is maintained. Spindles in the antagonist behave as if being passively stretched, unless the antagonist is contracted (Burke *et al*., 1978b; Roll and Vedel, 1982; Logigian, Wiegner and Young, 1989).

Delicate manipulation involves near-isometric contractions of the active muscle, commonly with co-contraction of antagonists and with movement proceeding as much by relaxation of the lengthening muscle as by increased contraction of the shortening. With the lower limbs, isometric, eccentric and slow concentric contractions are common when assuming the upright posture, when lifting an object from the ground, and during locomotion. Under all these circumstances, muscle spindles can provide significant reflex support. For the muscles that act on the hand, fusimotor activity can maintain or increase spindle activity during slow deliberate movements, perhaps those used when learning a task. It is worth noting, however, the discharge of muscle spindles in a voluntary task does not change as an initially naive subject gains skill (Al-Falahe and Vallbo, 1988; Vallbo and Al-Falahe, 1990), suggesting that the pattern of fusimotor drive does not alter as the spindle feedback becomes less critical to performing the task.

The Fusimotor System in Disorders of Motor Control

Much less is known about fusimotor function in patients, although it has been established that the question of overactivity of the fusimotor system is not primarily responsible for disturbances of muscle tone (Burke, 1983). In retrospect this question seems somewhat naive, dependent on much the same logic as that responsible for the original servo hypothesis, that normal movement could be driven through the fusimotor system.

As emphasized in the preceding sections, the fusimotor system is relatively quiescent at rest but is brought into activity during movement, whether by descending drives or reflexly by the afferent feedback generated by movement. Thus a search for fusimotor dysfunction should focus on the movement disorder rather than the genesis of an abnormality of tone in a resting patient. More appropriate experimental questions would be whether defective supraspinal control of fusimotor activity generates inappropriate afferent feedback for a particular movement or whether the abnormality of movement itself results in abnormal feedback.

ACKNOWLEDGEMENT. The authors' work is supported by the National Health and Medical Research Council of Australia.

REFERENCES

Al-Falahe, N.A., Nagoaka, M. and Vallbo, Å.B. (1990). Response profiles of human muscle afferents during active finger movements. *Brain* **113,** 325-346

Al-Falahe, N.A. and Vallbo, Å.B. (1988). Role of the human fusimotor system in a motor adaptation task. *Journal of Physiology* **401,** 77-95

Aniss, A.M., Diener, H.-C., Hore, J., Burke, D. and Gandevia, S.C. (1990). Reflex influences on muscle spindles in human pretibial muscles during standing. *Journal of Neurophysiology* **64,** 671-679

Binder, M. D. and Stuart, D. G. (1980). Motor unit - muscle receptors interactions: Design features of the neuromuscular control system. In *Spinal and Supraspinal Mechanisms of Voluntary Motor Control and Locomotion. Progress in Clinical Neurophysiology,* Vol. 8, edited by J.E. Desmedt, pp. 72-98. Karger: Basel

Brown, M.C., Goodwin, G.M. and Matthews, P.B.C. (1969). After-effects of fusimotor stimulation on the response of muscle spindle primary afferent endings. *Journal of Physiology* **205,** 677-694

Burke, D. (1981). The activity of human muscle spindle endings in normal motor behavior. *International Review of Physiology* **25,** 91-126

Burke, D. (1983). Critical examination of the case for or against fusimotor involvement in disorders of muscle tone. In: *Motor Control Mechanisms in Health and Disease, Advances in Neurology, Vol. 39,* edited by J.E. Desmedt, pp. 133-150. Raven Press: New York

Burke, D., Aniss, A.M. and Gandevia, S.C. (1987). In-parallel and in-series behavior of human muscle spindle endings. *Journal of Neurophysiology* **58**, 417-426

Burke, D. and Gandevia, S.C. (1990). The peripheral motor system. In: *The Human Nervous System*, edited by G. Paxinos, pp. 125-145. Academic Press: New York

Burke, D., Hagbarth, K.-E. and Löfstedt, L. (1978a). Muscle spindle responses in man to changes in load during accurate position maintenance. *Journal of Physiology* **276**, 159-164

Burke, D., Hagbarth, K.-E. and Löfstedt, L. (1978b). Muscle spindle activity in man during shortening and lengthening contractions. *Journal of Physiology* **277**, 131-142

Burke, D., Hagbarth, K.-E. and Skuse, N.F. (1978c). Recruitment order of human spindle endings in isometric voluntary contractions. *Journal of Physiology* **285**, 101-112

Burke, D., Hagbarth, K.-E. and Skuse, N.F. (1979). Voluntary activation of spindle endings in human muscles temporarily paralysed by nerve pressure. *Journal of Physiology* **287**, 329-336

Burke, D., McKeon, B. and Westerman, R.A. (1980). Induced changes in the thresholds for voluntary activation of human spindle endings. *Journal of Physiology* **302**, 171-181

Edin, B.B. and Vallbo, Å.B. (1988). Stretch sensitization of human muscle spindles. *Journal of Physiology* **400**, 101-111

Edin, B.B. and Vallbo, Å.B. (1990). Muscle afferent responses to isometric contraction and relaxation in humans. *Journal of Neurophysiology* **63**, 1307-1313

Gandevia, S.C. and Burke, D. (1985). Effect of training on voluntary activation of human fusimotor neurons. *Journal of Neurophysiology* **54**, 1422-1429

Gandevia, S.C., Burke, D. and McKeon, B. (1984). Coupling between human muscle spindle endings and motor units assessed using spike-triggered averaging. *Neuroscience Letters* **71**, 181-186

Gandevia, S.C., Macefield, G., Burke, D. and McKenzie, D.K. (1990). Voluntary activation of human motor axons in the absence of muscle afferent feedback: the control of the deafferented hand. *Brain* **113**, 1563-1581

Gandevia, S.C., Wilson, L., Cordo, P. and Burke, D. (1993). Fusimotor reflexes in relaxed forearm muscle evoked by afferents from the hand. *Proceedings of the International Union of Physiological Sciences* 322.27

Gregory, J.E., Morgan, D.L. and Proske, U. (1987). Changes in size of the stretch reflex of cat and man attributed to after effects in muscle spindles. *Journal of Neurophysiology* **58**, 628-640

Gregory, J.E., Mark, R.F., Morgan, D.L., Patak, A., Polus, B. and Proske, U. (1990). Effects of muscle history on the stretch reflex in cat and man. *Journal of Physiology* **424**, 93-107

Hagbarth, K.-E., Kunesch, E.J., Nordin, M., Schmidt, R.F. and Wallin, E.U. (1986). Gamma loop contributing to maximal voluntary contractions in man. *Journal of Physiology* **380**, 575-591

Hagbarth, K.-E. and Vallbo, Å.B (1968). Discharge characteristics of human muscle afferents during muscle stretch and contraction. *Experimental Neurology* **22**, 674-694

Hagbarth, K.-E. and Young, R.R. (1979). Participation of the stretch reflex in human physiological tremor. *Brain* **102**, 509-526

Logigian, E.L., Wiegner, A.W. and Young, R.R. (1989). Alpha-gamma co-silence in human wrist flexor muscles during slow visually-guided passive wrist flexion. *Neuroscience Research Communications* **4**, 117-124

Macefield, G., Gandevia, S.C. and Burke, D. (1990). Perceptual responses to microstimulation of single afferents innervating the joints, muscles and skin of the human hand. *Journal of Physiology* **429**, 113-129

Macefield, G., Hagbarth, K.-E., Gorman, R.B., Gandevia, S.C. and Burke, D. (1991). Decline in spindle support to α motoneurones during sustained voluntary contractions. *Journal of Physiology* **440**, 497-512

Macefield, V.G., Gandevia, S.C., Bigland-Ritchie, B., Gorman, R.B. and Burke, D. (1993). The firing rates of human motoneurones voluntarily activated in the absence of muscle afferent feedback. *Journal of Physiology* **471**, 429-443

Matthews, P.B.C. (1972). *Mammalian Muscle Receptors and their Central Actions*. Edward Arnold: London

Merton, P.A. (1953). Speculations on the servo control of movement. In *The Spinal Cord. Ciba Foundation Symposium*. Edited by G.E.W. Wolstenholme, J.A. Malcolm and J.A.B. Gray, pp. 247-260. Churchill: London

Meyer-Lohmann, J., Riebold, W. and Robrecht, D. (1974). Mechanical influence of the extrafusal muscle on the static behaviour of deefferented primary muscle spindle endings in cat. *Pflügers Archiv* **352**, 267-278

McKeon, B. and Burke, D. (1981). Component of muscle spindle discharge related to arterial pulse. *Journal of Neurophysiology* **46**, 788-796

McKeon, B. and Burke, D. (1983). Muscle spindle discharge in response to contraction of single motor units. *Journal of Neurophysiology* **49**, 291-302

McKeon, B., Gandevia, S. and Burke, D. (1984). Absence of somatotopic projection of muscle afferents onto motoneurons of same muscle. *Journal of Neurophysiology* **51**, 185-194

Morgan, D.L., Prochazka, A. and Proske, U. (1984). The after-effects of stretch and fusimotor stimulation on the responses of primary endings of cat muscle spindles. *Journal of Physiology* **356**, 465-477

Newsom Davis, J. (1975). The response to stretch of human intercostal muscle spindles studied *in vitro*. *Journal of Physiology* **249**, 561-579

Poppele, R.E. and Kennedy, W.R. (1974). Comparison between behaviour of human and cat muscle spindles recorded *in vitro*. *Brain Research* **75**, 316-319

Prochazka, A., Stephens, J.A. and Wand, P. (1979). Muscle spindle discharge in normal and obstructed movements. *Journal of Physiology* **287**, 57-66

Ribot, E., Roll, J.-P. and Vedel, J.-P. (1986). Efferent discharges recorded from single skeletomotor and fusimotor fibres in man. *Journal of Physiology* **375**, 251-268

Ribot-Ciscar, E., Tardy-Gervet, M.F., Vedel, J.-P., and Roll, J.-P. (1991). Post-contraction changes in muscle spindle resting discharge and stretch sensitivity. *Experimental Brain Research* **86**, 673-678

Roll, J.-P. and Vedel, J.-P. (1982). Kinesthetic role of muscle afferents in man studied by tendon vibration and microneurography. *Experimental Brain Research* **47**, 177-190

Rothwell, J.C., Gandevia, S.C. and Burke, D. (1990). Activation of fusimotor neurones by motor cortical stimulation in human subjects. *Journal of Physiology* **430**, 105-117

Vallbo, Å.B. (1971). Muscle spindle response at the onset of isometric voluntary contractions in man. Time difference between fusimotor and skeletomotor effects. *Journal of Physiology* **218**, 405-431

Vallbo, Å.B. (1973). Muscle spindle afferent discharge from resting and contracting muscles in normal human subjects. In: *New Developments in Electromyography and Clinical Neurophysiology. Volume 3.* Edited by J.E. Desmedt, pp. 251-262. Karger: Basel

Vallbo, Å.B. (1974). Human muscle spindle discharge during isometric voluntary contractions. Amplitude relations between spindle frequency and torque. *Acta Physiologica Scandinavica* **90**, 319-336

Vallbo, Å.B. and Al-Falahe, N.A. (1990). Human muscle spindle response in a motor learning task. *Journal of Physiology* **421**, 553-568

Vallbo, Å.B., Hagbarth, K.-E., Torebjörk, H.E. and Wallin, B.G. (1979). Somatosensory, proprioceptive, and sympathetic activity in human peripheral nerves. *Physiological Reviews* **59**, 919-957

Windhorst, U., Hamm, T.M. and Stuart, D. (1989). On the function of muscle and reflex partitioning. *Behavioral Brain Sciences* **12**, 629-681

MECHANISMS UNDERLYING THE EXCITATION OF MUSCLE AFFERENTS BY SUXAMETHONIUM

M.B. DUTIA

Department of Physiology
Medical School
Teviot Place
Edinburgh EH8 9AG
U.K.

SUMMARY

Topical application or systemic administration of suxamethonium (succinylcholine, SCh) has proved to be a useful technique for the relatively rapid, reversible excitation of muscle spindles, independently of their gamma fusimotor innervation. Current evidence indicates that the powerful excitation of primary Ia afferents is due to the contracture of the bag_1 and bag_2 intrafusal muscle fibres in the presence of SCh. The effects of SCh on long chain fibres are presently unknown. Secondary endings are excited much less powerfully than primary endings, and there is debate about the mechanisms responsible. Secondaries may be excited as a result of the hyperkalaemic "side-effects" of systemic SCh administration or by depolarisation of collateral sensory terminals they may have on the bag intrafusal fibres. It is likely that all spindle afferents (as well as other non-spindle afferents) are affected by the increase in serum potassium levels, while only those spindle afferents with substantial collateral terminals on bag fibres are affected by their contraction. The application of SCh as a means of assessing the contribution of individual intrafusal fibres to the afferent discharge of different spindle endings is briefly reviewed.

INTRODUCTION

The observations of Gladden (1976) and Boyd (1985) on isolated visualised spindles established clearly that acetylcholine (ACh) and its analogue suxamethonium (succinylcholine, SCh) induced a sustained contracture of the nuclear bag intrafusal muscle fibres, while the nuclear chain fibres like extrafusal fibres were paralysed and did not contract. In the context of the possible mechanisms by which SCh excited muscle spindle afferents, these experiments established unequivocally that intrafusal muscle fibre contraction did occur. Previously Granit, Skoglund and Thesleff (1953) had suggested that SCh may have a direct depolarising action on the afferent nerve terminals, and Douglas and Gray (1953) and others had described the excitation of cutaneous afferents by ACh, presumably by such a direct action. Subsequently Smith (1963) in considering the elevation of plasma potassium levels following SCh administration, remarked that the hyperkalaemic effects were large enough to contribute to the excitation of spindle afferents. A third possible mechanism, that the depolarised membranes of the intrafusal muscle cells in the presence of SCh might electrotonically excite the sensory nerve terminals in contact with them, had been proposed by Brinling and Smith (1960).

There is now no doubt that SCh-induced contraction of the the nuclear bag fibres is the

Neural Control of Movement, Edited by W.R. Ferrell
and U. Proske, Plenum Press, New York, 1995

major cause of the excitation of spindle primary afferents (although it should be noted that the effects of SCh on long chain fibres, if any, have yet to be determined). Nevertheless the relative importance of other possible mechnisms, particularly for secondary endings which may have 75 - 100% of their sensory terminals on the nuclear chain fibres (Banks *et al.,* 1982), is still a matter for debate. This brief review attempts to bring together various lines of evidence relevant to this debate, which is of some current interest given the usefulness of SCh as an experimental tool in the classification of muscle afferents (Price and Dutia 1989).

EFFECTS OF SCH ON MUSCLE SPINDLE PRIMARY AND SECONDARY ENDINGS

Rack and Westbury (1966) were the first to record the effects of SCh on the responsiveness to muscle stretching of primary and secondary spindle afferents. They showed that following an intravenous injection of SCh the response of primary endings was very similar to that during dynamic fusimotor stimulation, and concluded that SCh excited primary afferents through the contraction of intrafusal fibres in a similar way to dynamic gamma axons. They confirmed the earlier finding of Fehr (1965) that secondary endings were excited much less powerfully than primary endings, and noted that this corresponded with the weak actions of dynamic fusimotor axons on them, but did not speculate on the possible mechanism of excitation of secondary endings. Fehr's elegant study had compared the effects of SCh injection on the resting discharge of pairs of primary and secondary afferents that innervated the same spindle, and had shown that the excitation of secondary endings was not only less marked but also had a longer latency (15-135 seconds longer than that for primary endings), reached a maximum an average of 63 seconds after the maximum effect on primary endings, and declined much more slowly (Fig. 1). Fehr suggested that SCh caused the contraction of both nuclear bag and nuclear chain fibres so exciting both types of sensory endings, and proposed that the differences in the time-course of excitation were due to differences in the responsiveness of the nuclear bag and nuclear chain fibres to the drug. While the direct observations in the isolated spindle eliminated the possibility that the chain fibres were involved, the proposal of Rack and Westbury was proved partly correct in that SCh did indeed cause the the bag$_1$ fibre to contract in a similar way to dynamic fusimotor axons. The effects of the contraction of the bag$_2$ fibre were probably also observed by Rack and Westbury as brief periods immediately after the SCh injection when the spindle

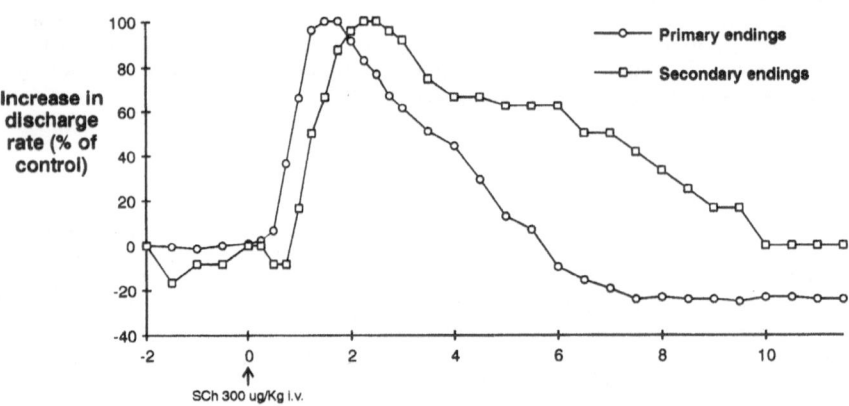

Time after SCh injection (minutes)

Figure 1. Timecourse of the effects of SCh on four simultaneously recorded pairs of primary (circles) and secondary (squares) afferents from an isolated segment of the tenuissimus muscle (data from Fehr, 1965, Fig. 1). Each point is the average change in discharge rate expressed as a percentage of the control discharge rate, at the maximal physiological length of the muscle. Primary endings increased their discharge from an average of 28 imp/sec to a peak of 80 imp/sec, while the secondaries' discharge rate increased from 46 imp/sec to 52 imp/sec. Note the longer latency from SCh injection to the onset of excitation of the secondary endings, the later peak in their activation and the much slower recovery to control compared to primary endings. See Fehr (1965) for further details.

response was similar to that during combined static and dynamic gamma stimulation, but they interpreted this as possibly a damaging effect of larger doses of SCh.

The first study following the experimental demonstration of the SCh-induced contraction of the nuclear bag fibres was carried out by Dutia (1980) who used slow intra-arterial infusions of SCh rather than single injections, in an effort to observe the effects of the recruitment of each of the two types of bag fibre on the afferent responsiveness to stretching. During an infusion of SCh at $100\mu g/kg/minute$, most primary endings went through three stages of excitation in order. In the first stage, which began soon after the start of infusion, the discharge rate of the endings was gradually facilitated but without any increase in the responsiveness to stretching. At the onset of the second stage the dynamic sensitivity of the primary afferents increased dramatically, and this was interpreted as the point at which the the the bag_1 nuclear bag fibre was recruited as the intramuscular levels of SCh increased above its threshold. With continued infusion the primary afferents continued to respond with a very vigourous dynamic response and a high dynamic index until, at the onset of the third stage, there was an increase in the steady discharge rate ("biasing", Boyd 1981, 1985) and in the position response. The dynamic index was reduced and the afferent response resembled that during strong combined dynamic and static fusimotor stimulation. This was interpreted as the point at which the higher-threshold bag_2 fibre had also been recruited, so that both bag_1 and bag_2 fibres were in contraction. These changes wore off in reverse order when the SCh infusion was terminated.

Thus two discrete events corresponding to the expected effects of the contraction of the bag_1 and bag_2 nuclear bag fibres were recognizable during slow SCh infusion in soleus primary afferents; however SCh also had an early, facilitatory action which was not accompanied by any indication of intrafusal fibre contraction. Secondary endings typically showed only a very gradual facilitation of their discharge rate, similar to the early facilitatory effects on primary endings (Dutia 1977, 1980). In some afferents with conduction velocities between 60-80m/sec (i.e. intermediate between the Group Ia and Group II ranges) discrete increases in either the dynamic sensitivity or the position sensitivity were seen, similar to but much smaller in amplitude than the analogous effects on primary endings. These were termed "intermediate" sensory afferents, which differed from typical secondary endings in that they appeared to have functionally significant collateral terminals on the nuclear bag fibres and were therefore also influenced by their recruitment. However, as in primary endings, the discrete events which changed these afferents' response to stretching, and which were interpreted as indicating the onset of bag_1 or bag_2 fibre contraction, were always preceeded by a period of gradual facilitation.

Subsequent studies have focussed largely on using the SCh-induced contraction of the nuclear bag fibres as an experimental technique with which to activate and classify spindle afferents, either in situations where the conventional conduction velocity measurements were not feasible or in addition to these and other measurements (e.g. Amassian and Eberle 1986, Wadell et al.,1991, Nishimura, Johnston and Munson 1993). Thus Price and Dutia (1987, 1989) analysed the effects of SCh on spindle afferents from the dorsal neck muscles. Afferents from bag_2-chain (b_2c) capsules of tandem muscle spindles were shown to respond differently during slow i.a. SCh infusion than those from "normal" b_1b_2c spindles, so enabling them to be identified and characterised electrophysiologically. b_2c afferents differed from b_1b_2c afferents in not showing an increase in dynamic sensitivity when activated by SCh, reflecting their lack of a bag_1 intrafusal fibre (Richmond and Abrahams 1975, Richmond et al.,1986). The usefulness of this technique was confirmed in parallel experiments on tenuissimus spindles, where the diagnosis of a spindle afferent as either b_2c or b_1b_2c type was subsequently confirmed histologically (Price and Dutia 1989). Scott (1991) analysed the effects of SCh on peroneus tertius muscle spindles and also distinguished b_1b_2c, b_2c and secondary afferent response types based on their responses to stretching. Taylor et al.,(1992a,b) applied a statistical analysis to the effects of SCh on populations of spindle afferents from the jaw-closing muscles and the gastrocnemius muscles of the cat. They subdivided the population of spindle afferents from each muscle into four groups corresponding to b_1b_2c, b_2c, b_1c and secondary (c) types, on the assumption that the increase in dynamic response of an afferent in the presence of SCh indicated the strength of functional input from the bag_1 fibre and the increase in steady firing (biasing) indicated the strength of input from the bag_2 fibre. In these studies the effects of SCh were measured 1.5 minutes following a single i.v. injection, enabling the semi-automated analysis of a large number of afferents.

SCh Effects on Spindle Afferents Related to Nuclear Bag Fibre Contraction

Activation by SCh is a ready means of causing the maximal contraction of the nuclear bag fibres independently of the gamma fusimotor axons that innervate them, and it is of interest to compare the effects of activating the bag_1 and bag_2 fibres in this way in the different muscles examined so far. In soleus muscle spindle primary afferents bag_1 fibre contraction was indicated by a very large increase in dynamic response (Rack and Westbury 1966, Dutia 1980). Similar effects were seen in peroneus tertius by Scott (1991) and in gastrocnemius afferents by Taylor *et al.*, (1992b). In neck afferents, however, the peak dynamic response rarely reached frequencies greater than 150 imp/sec, in contrast to peak frequencies of 250-350 imp/sec in soleus spindles. Presumably the peak discharge rates of neck spindle primary afferents are limited by their smaller diameter in relation to hindlimb afferents (Richmond and Abrahams 1975, 1979; Richmond, Anstee, Sherwin and Abrahams 1976, Richmond, Bakker, Bakker and Stacey 1986). In jaw-opening muscles the majority of afferents showed peak frequencies of less than 200 imp/sec, while some reached peak values of over 300 imp/sec when activated by SCh (Taylor *et al.*, 1992a). Functionally it may be important to note that in afferents with a relatively low upper limitation in discharge frequency, the effects of static fusimotor activity (particularly the biassing associated with bag_2 contraction) will tend to dominate even when any dynamic co-activation is maximal.

In soleus spindles, bag_2 fibre contraction was indicated by an increase in both biassing and the position sensitivity of the primary afferent (e.g. Dutia 1980). In contrast neck b_1b_2c and b_2c afferents showed only a marked biassing and an increase in variability of the interspike intervals, as did tenuissimus spindles activated by SCh (Price and Dutia 1987, 1989). This latter pattern is in agreement with the direct observations of Boyd, Murphy and Moss (1985) on the effects of bag_2 fibre contraction on the length sensititivity of primary endings. Scott (1991) also reported that in peroneus tertius primary afferents there was an increase in bias corresponding to bag_2 fibre contraction, but found that in one primary afferent out of seven the position sensitivity also increased. The description by Rack and Westbury (1966) of an increase in dynamic as well as static sensitivity following doses of 200-1000µg/Kg SCh, and the data in their Fig. 3, suggest that they also observed not only biassing but also an increase in position response related to bag_2 fibre contraction. The population data from gastrocnemius (Taylor *et al.*, 1992b) is not directly comparable, but the high values of "static difference" (the difference between discharge rates before the stretch and 0.5 sec after the end of the dynamic phase) of 100-150 imp/sec in many cases, suggests that in these spindles also bag_2 fibre contraction increased both biassing and the position response in many primary afferents. Thus the common feature of bag_2 fibre contraction appears to be a strong biassing of the primary afferent discharge, and this is accompanied in some spindle types by an increase in position sensitivity. The reasons for the apparent differences in the effects of bag_2 contraction on primary position sensitivity in different spindles are not clear.

SCh Effects on Spindle Afferents not Related to Nuclear Bag Fibre Contraction

The gradual facilitatory effects of SCh described above have been observed in soleus, tenuissimus, neck and peroneus tertius spindle primary and secondary afferents (Dutia 1980, Price and Dutia 1987, Scott 1991). In jaw spindle b_1b_2c afferents, Durbaba *et al.*, (1991) have reported that the "static difference" and dynamic index increased together following an intravenous injection of SCh, reaching a peak 1 minute after the injection, while the initial discharge rate peaked earlier at 30 sec and then declined. While the analogous data is not presented for gastrocnemius spindles in their subsequent study (Taylor *et al.*. 1992b), the mean increase in initial discharge after SCh also appears to peak earlier than the increases in peak frequency and static index (their Fig. 1; but note that in this case the mean values are from a sample which presumably included both primary and secondary afferents, tending to smear any differences in the timecourse of activation (Fehr, 1965), and that the values were measured 1.5 minutes after SCh injection for every afferent).

The mechanisms responsible for the general facilitatory effects on spindle primary and secondary afferents are still not fully resolved. While Dutia (1980) followed earlier workers and proposed that they were due to either a direct or indirect depolarising action of SCh on the sensory nerve terminals, it has recently been suggested that the SCh-induced contraction of the nuclear bag fibres could also explain the excitation of all secondary endings (Taylor *et al.*,1994). This suggestion follows on the histological demonstration of Banks *et al.*,(1982) that although secondary sensory endings innervate mainly the nuclear chain fibres, the majority of secondary afferents also have collataral sensory terminals on the nuclear bag fibres and may therefore be expected to be influenced by their contraction. The extent to which these collateral terminals provide effective input to the main afferent axon is likely to be very variable. In visualised tenuissimus spindles, selective contraction of the bag_2 fibre by gamma stimulation was shown to affect the majority of secondaries tested to some extent, indicating that these collateral terminals do provide a functional input to the afferent axon (Boyd, Sutherland and Ward 1985). In contrast however bag_1 fibre contraction evoked by dynamic gamma stimulation only rarely influenced secondary endings despite the sensory innervation of this fibre also by secondary afferent collaterals, and Banks *et al.*, (1982) suggested that the collateral terminals on bag_1 fibres though morphologically extensive in some cases may not be functionally significant.

In either case, a number of arguments can be put forward against the possibility that the excitation of secondary endings is due entirely to bag fibre contraction as suggested by Taylor *et al.*,(1994). The systematic differences in the timecourse of excitation of primary and secondary endings following SCh injection (Fehr 1965; Fig. 1) make it unlikely that the two processes are dependent on the same mechanism. From the observations of Rack and Westbury (1966) that a strong static effect (now interpreted as the contraction of the bag_2 fibre) was seen only for the first minute or less following an i.v. injection, it can be expected that any input from bag_2 terminals to a secondary afferent should excite it strongly only for this time. By contrast during a slow infusion bag_2 contraction appears to occur as a late, discrete and recognizable step, and any input from bag_2 collaterals to secondary endings should be strongly increased at this point. Such an effect is seen only in "intermediate" afferents (Dutia 1980). Significantly, in the most direct experiments reported so far in which bag_2 fibre contraction was seen to excite secondary endings, Boyd (1981) obtained such effects in four afferents, all with high conduction velocities. Three of the four afferents showed increases in dynamic and static sensitivities when the bag_2 fibre was made to contract, and therefore strongly resembled "intermediate" afferents rather than typical secondaries in which these sensitivities are reduced by SCh. The fact that the majority of secondary endings do not show effects related to the timecourse of bag_2 fibre contraction indicates that either the collateral sensory terminals are not functionally significant in most secondary endings, or that the additional input they provide is too small to be apparent above the facilitation caused by SCh itself.

Of the three routes suggested so far by which SCh might depolarise the sensory nerve terminals themselves, the direct action of the drug on the nerve membrane (Granit *et al.*, 1953) and an electrotonic spread of depolarisation from the intrafusal fibres (Brinling and Smith 1960) appear unlikely since there is no evidence for such effects in isolated frog muscle spindles (Ottoson 1961). Recent experiments *in vitro* have also shown that carbachol has no effect on large-diameter mechanoceptive cutaneous afferents, although smaller diameter C-fibres were strongly excited (Steen and Reeh, 1993). There is on the other hand ample evidence for a significant elevation of serum potassium levels following even single injections of SCh. The hyperkalaemia following SCh administration is recognized as an important side-effect with major possible complications in the clinical use of SCh as a muscle relaxant (Roth and Wuthrich 1969, Laurence 1987; for a review see Yentis 1990). Numerous studies have shown increases in serum potassium concentration of 0.4 - 2 mmol/litre, representing a change of about 10-50% above normal levels, within one to three minutes following a single injection (Stevenson 1960, Yentis 1990, Van der Bijl and Roelofse 1993). Since these levels of serum potassium represent an equilibrium between the rate of release from the depolarised end-plates of skeletal muscle cells and the rate of excretion, the intramuscular potassium levels particularly in the endplate regions are likely to be higher. Indeed in limbs with compromised venous drainage serum potassium levels may reach much higher levels after SCh administration (Yentis 1990).

Elevation of external potassium has been shown to facilitate the afferent discharge of primary and secondary endings in isolated spindles, in a manner very similar to that seen *in vivo* (Kidd and Vaillant 1974). In a brief report Prochazka and Somjen (1986) reported that infusion of KCl weakly excited only 5 out of 11 spindles and one tendon organ *in vivo*, and their conclusion that serum potassium did not influence spindle excitability has been cited by Taylor *et al.*,(1994) as an argument against any indirect effects of SCh. However in this study the elevated serum potassium levels were maintained for various rather short periods

(0.5 - 3 min) and it is not clear whether the positive results which *were* obtained were seen only with an appropriately long exposure. It may be of interest to repeat this study more systematically to determine if close intra-arterial administration of KCl mimics the facilitatory effects of SCh *in vivo*, although it has to be borne in mind that the relationship between serum potassium levels and intramuscular levels of extracellular potassium is complex. The timecourse of the development of serum hyperkalaemia over a period of a few minutes after SCh administration and its slow decline afterwards, is compatible with the timecourse of the facilitatory effects observed in both primary and secondary spindle afferents. The small facilitatory effects of SCh on other non-spindle afferents are also most easily explained as consequences of the hyperkalaemia, and this explanation is supported by the fact that such effects can be observed *in vivo* but not *in vitro* (for a related discussion see Steen & Reeh 1993).

CONCLUSION

Administration of SCh is a useful technique for inducing the maximal, reversible contraction of the nuclear bag intrafusal fibres and strongly exciting muscle spindle primary afferents, either globally by systemic injection and infusion or in a particular group of muscles by topical application or restricted intra-arterial infusion. This technique has been used to assess the contribution of the sensory nerve terminals on the bag fibres to the primary afferent discharge, and in studies of the reflex function of spindle afferent inputs. However the interpretation of the effects obtained depends on an understanding of the mechanisms by which SCh acts on the spindle and its action on other afferents. The current evidence indicates that the predominant effects on primary b_1b_2c and b_2c afferents are due to the strong contraction of the bag_1 and bag_2 intrafusal fibres. However SCh administration, even as a single injection, has a significant systemic "side-effect" in causing the elevation of serum potassium levels, as a consequence of its depolarising blockade of extrafusal muscle cells. The gradual facilitatory effects of SCh on spindle primary and secondary afferents as well as other non-spindle afferents are probably due to this hyperkalemic "side-effect" of SCh administration.

REFERENCES

Amassian, V.E. & Eberle, L. (1986). Muscle spindle projections to feline thalamic neurones identified by local arterial injection of succinylcholine. *Journal of Physiology* **371,** 43P.

Banks, R.W., Barker, D. and Stacey, M.J. (1982). Form and distribution of sensory terminals in cat hindlimb muscle spindles. *Philosophical Transactions of the Royal Society B* **99,** 329-364.

Boyd, I.A. (1981). The action of the three types of intrafusal muscle fibre in isolated cat muscle spindles on the dynamic length sensitivities of primary and secondary sensory endings. In *Muscle Receptors and Movement,* ed. Taylor A. and Prochazka, A., pp. 17-32. Macmillan, London.

Boyd, I.A. (1985). Review. In *The Muscle Spindle,* ed. Boyd, I.A. & Gladden, M.H., pp. 129-150. Macmillan, London.

Boyd, I.A., Murphy, P.R. & Moss, V.A. (1985). Analysis of primary and secondary afferent responses to stretch during activation of the dynamic bag_1 or the static bag_2 fibre separately in cat muscle spindles. In *The Muscle Spindle,* ed. Boyd, I.A. and Gladden, M.H., pp. 153-158. Macmillan, London.

Boyd, I.A., Sutherland, F. & Ward, J. (1985). The origin of the increase in the length sensitivity of secondary sensory endings produced by some fusimotor axons. In *The Muscle Spindle,* ed. Boyd, I.A. & Gladden, M.H., pp. 153-158. Macmillan, London.

Brinling, J.C. & Smith, C.M. (1960). A characterisation of the stimulation of mammalian muscle spindles by succinylcholine. *Journal of Pharmacology and Experimental Therapeutics* **129,** 56-60.

Douglas, W.W. & Gray, J.A.B. (1953). The excitant action of ACh and other substances on cutaneous sensory pathways and its prevention by hexamethonium and d-tubocurarine. *Journal of Physiology* **119,** 118-128.

Durbaba, R., Rodgers, J.F. & Taylor, A. (1991). Time course of the effects of succinylcholine on spindles of the jaw muscles of the anaesthetised cat. *Journal of Physiology* **435,** 58P.

Dutia, M.B. (1977). Activation of cat spindle secondary sensory endings by intravenous infusion of suxamethonium. *Journal of Physiology* **273,** 89-90P.

Dutia, M.B. (1980). Activation of cat muscle spindle primary, secondary and intermediate sensory endings by suxamethonium. *Journal of Physiology* **304,** 315-330.

Dutia, M.B. & Price, R.F. (1990). Response to stretching of identified b2c spindle afferents in the anaesthetised cat. *Journal of Physiology* **420,** 101P.

Fehr, H.U. (1965). Activation by succinylcholine of primary and secondary endings of the same de-efferented muscle spindle during static stretch. *Journal of Physiology* **178,** 98-110.

Gladden, M.H. (1976). Structural features relative to the function of intrafusal muscle fibres in the cat. *Progress in Brain Research* **44,** 51-59.

Granit, R., Skoglund, S. and Thesleff, S. (1953). Activation of muscle spindles by succinylcholine and decamethonium. The effects of curare. *Acta physiologica scandinavica* **28,** 134-151.

Kidd, G.L. & Vaillant, C.H. (1974). The interaction of K+ and stretching as stimuli for primary muscle spindle endings in the cat. *Journal of Anatomy* **119,** 196.

Laurence. A.S. (1987). Myalgia and biochemical changes following intermittent suxamethonium administration. *Anaesthesia* **42,** 503-510.

Nishimura H., Johnson, R.D. & Munson, J.B. (1993). Rescue of neuronal function by cross-regeneration of cutaneous afferents into muscle in cats. *Journal of Neurophysiology* **70,** 213-222.

Ottoson, D. (1961). The effect of acetylcholine and related substances on the isolated muscle spindle. *Acta Physiologica Scandinavica* **53,** 276-287.

Price, R.F. & Dutia, M.B. (1987). Properties of cat neck muscle spindle afferents and their excitation by succinylcholine. *Experimental Brain Research* **68,** 619-630.

Price, R.F. & Dutia, M.B. (1989). Physiological properties of tandem muscle spindles in neck and hind-limb muscles. *Progress in Brain Research* **80,** 47-56.

Prochazka, A. & Somjen, G.G. (1986). Insensitivity of cat muscle spindles to hyperkalaemia in the physiological range. *Journal of Physiology* **372,** 26P.

Rack, P.M.H. & Westbury, D.R. (1966). The effects o suxamethonium and acetylcholine on the behaviour of cat muscle spindles during dynamic stretching and during fusimotor stimulation. *Journal of Physiology* **186,** 698-713.

Richmond, F.J.R. & Abrahams, V.C. (1975). Morphology and distribution of muscle spindles in dorsal muscles of the cat neck. *Journal of Neurophysiology* **38,** 1322-1339.

Richmond, F.J.R. & Abrahams, V.C. (1979). Physiological properties of muscle spindles in the dorsal neck muscles of the cat. *Journal of Neurophysiology* **42,** 604-617.

Richmond, F.J.R., Anstee, G.C.B., Sherwin, E.A. & Abrahams, V.C. (1976). Motor and sensory fibres of neck muscle nerves in the cat. *Canadian Journal of Physiology and Pharmacology* **54,** 294-304.

Richmond, F.J.R., Bakker, G.J., Bakker, D.A. & Stacey, M.J. (1986). The innervation of tandem muscle spindles in the cat neck. *Journal of Comparative Neurology* **245,** 483-497.

Roth, F. & Wuthrich H. (1969). The clinical importance of hyperkalaemia following suxamethonium administration. *British Journal of Anaesthesia* **41,** 311.

Scott, J.J.A. (1991). Responses of Ia afferent axons from muscle spindles lacking a bag_1 intrafusal muscle fibre. *Brain Research* **543,** 97-101.

Smith, C.M. (1963). Neuromuscular pharmacology: drugs and muscle spindles. *Annual Review of Pharmacology* **3,** 223-242.

Steen, K.H. & Reeh, P.W. (1993). Actions of cholinergic agonists and antagonists on sensory nerve endings in rat skin in vitro. *Journal of Neurophysiology* **70,** 397-405.

Stevenson, D.E. (1960). Changes in the blood electrolytes of anaesthetised dogs caused by suxamethonium. *British Journal of Anaesthesia* **32,** 364-371.

Taylor, A., Durbaba, R. & Rodgers, J.F. (1992a). The classification of afferents from muscle spindles of the jaw-closing muscles of the cat. *Journal of Physiology* **456,** 609-628.

Taylor, A., Rodgers J.F., Fowle, A.J. & Durbaba, R. (1992b). The effect of succinylcholine on cat gastrocnemius muscle spindle afferents of different types. *Journal of Physiology* **456,** 629-644.

Taylor, A., Durbaba, R. & Rodgers, J.F. (1994). The site of action of succinylcholine on muscle spindle afferents in the anaesthetised cat. *Journal of Physiology,* in press (Proceedings of the Bristol Meeting, February 1994)..

van der Bijl, P. & Roelofse, J.A. (1993). Serum potassium and sodium levels following intravenous suxamethonium in pediatric dental anaesthesia. *Journal of Oral and Maxillofacial Surgery* **51,** 875-878.

Wadell, I., Johansson, H., Sjolander, P., Sojka, P., Djupsjobacka, M & Niechaj, A. (1991). Fusimotor reflexes influencing secondary muscle spindle afferents from flexor and extensor muscles in the hind-limb of the cat. *Journal de physiologie* **85,** 223-234.

Yentis, S.M. (1990). Suxamethonium and hyperkalaemia. *Anaesthesia and Intensive Care* **18,** 92-101.

QUANTITATIVE ASPECTS OF THE USE OF SUCCINYLCHOLINE IN THE CLASSIFICATION OF MUSCLE SPINDLE AFFERENTS.

A. TAYLOR, R. DURBABA and J.F. RODGERS

Sherrington School of Physiology
UMDS, St. Thomas' Hospital Campus
London, SE1 7EH, U.K.

SUMMARY

Testing with succinylcholine (SCh) has been shown to give valuable information about the nature of the contacts of muscle spindle afferents on the three different intrafusal muscle fibre types. The important features of the methods to be used in order to allow quantitative interpretation of the test are reviewed. From large populations of gastrocnemius and jaw muscle afferents in the cat it was concluded that only two measurements of the ramp and hold stretch responses were needed to detect significant endings on bag_1 and bag_2 fibres. These were the increments in dynamic stretch response and in initial frequency caused by the SCh. Evidence is reviewed to support the view that essentially all the effects of SCh, under the dosage conditions used, are due to contraction of the bag fibres. The resulting predictions regarding patterns of afferent terminations on intrafusal fibres are compared with the conclusions from independent histological observations. From studies of jaw muscle afferents it is concluded that the SCh classification can give new insights into factors which influence central connectivity patterns.

INTRODUCTION

The time-honoured method of classifying muscle spindle afferents as belonging to primary or to secondary sensory endings according to conduction velocity has served very well, but can now be seen to have certain limitations. Separation according to conduction velocity was based on studies of limb muscle nerves, in which peaks in the fibre diameter spectra are well defined. In other situations, notably the cranial and the axial muscles, which have short nerves and no clear bimodal distribution of spindle afferent fibre diameters, methods have had to be devised based on afferent discharge properties. Cooper (1961) was able to show that in the decerebrate animal, in which there is ongoing fusimotor discharge, primary afferents are more dynamically sensitive than secondaries. Matthews (1963) using ramp and hold stretches showed the existence of a positive correlation between dynamic index and conduction velocity for hindlimb muscles, but the general usefulness of this test is limited because the separation of primaries from secondaries is only clear in the presence of adequate dynamic fusimotor activation. Rack and Westbury (1966) were able to show that the same effect could be achieved by exposure to succinylcholine (SCh), which was believed to enhance the dynamic sensitivity of primary afferents by contraction of bag fibres. After the recognition of the two different bag fibre types (Ovalle and Smith, 1972) and the attribution of dynamic effects to bag_1 activation and static effects to bag_2 activation, Gladden (1976) showed that both types were contracted by acetylcholine. Thus it followed that SCh (thought to act in the same way as acetylcholine) should cause static effects due to bag$_2$ fibres as well dynamic effects due to bag$_1$ fibres. Dutia (1980) made a study of soleus spindles from this viewpoint and layed the foundations for further work in which it was possible to recognise primary afferents which lacked terminals on bag$_1$ fibres in hindlimb and neck muscles (Price and Dutia, 1987; 1989).

In the above work the response to SCh was used in an essentially qualitative way and no quantitative criteria were available for making objective decisions regarding spindle afferent types more generally. Two new studies were therefore carried out recently on spindles of jaw elevator muscles and of hindlimb muscles in the cat to find statistically based measures, which could be used for this purpose (Taylor, Durbaba and Rodgers,1992; Taylor, Rodgers, Fowle and Durbaba,1992). The outcome of this work was the conclusion that, provided a suitably standardised technique was used, it was possible to estimate separately the strength of the influence upon each afferent of the bag$_1$ and the bag$_2$ intrafusal fibres. This permitted categorisation of afferents as b$_1$c, b$_1$b$_2$c, b$_2$c or c according to the combinations of intrafusal fibre influence on each unit. In the case of the hindlimb spindles, the numerical distribution of units amongst these categories was consistent with the best available histological data (Banks, Barker and Stacey,1982). Subsequently, the value of this approach to spindle afferent classification has been emphasised in a number of reports (Rodgers, Fowle, Durbaba and Taylor,1993; Taylor, Durbaba and Rodgers,1993; Taylor, Rodgers, Fowle and Durbaba,1992), some of which are in abstract form (Fowle, Taylor, Rodgers and Durbaba,1992; Rodgers, Durbaba, Fowle and Taylor,1993; Rodgers, Taylor, Durbaba and Fowle,1992; Rodgers, Taylor, Fowle and Durbaba,1992; Taylor, Durbaba, Rodgers and Fowle,1993; Taylor, Rodgers, Fowle and Durbaba,1992). The object of the present account is to set out some of the practical considerations which should be observed in using the method and to show how it can give new insights into various aspects of muscle spindle physiology.

METHODS

Muscle spindles must either be de-efferentated or must have fusimotor drive eliminated by deep anaesthesia in order to secure reliable control conditions. It is convenient to give the SCh as a single intravenous injection of 200μg kg^{-1} rather than as a prolonged infusion either intravenous or intra-arterial. The IV route is preferred because this is expected to give a much better defined concentration in the blood reaching the muscle than would be the case with intra-arterial injection. In the latter case the actual concentration must depend not only on the speed of the infusion and its concentration, but also on the blood flow to the muscle and the exact position of the catheter tip in relation to arterial branches. There seems to be no advantage in using a prolonged IV infusion, because this will produce a slowly rising concentration and it is not clear at what stage the effect on the spindle response should be measured. In the hindlimb muscle experiments it was found that the response rose more quickly and was larger and more repeatable if the muscle blood flow was enhanced by a period of 30s stimulation via the muscle nerve at 2 x threshold at 10Hz, immediately before the SCh injection. Recovery from a dose of SCh is complete in 30min. The stimulus adopted for testing the spindle responses in triceps surae was a ramp and hold stretch of 5mm in 1s, held for 1.5s and shortening in 1s, repeated continuously every 6s. Spindle discharge was assessed from cycle histograms of 50 bins each of 100ms, starting 0.5s before each ramp stretch. The SCh injection was given after recording 5 control stretch cycles and the recording then continued for 5mins. The computer system employed (CED 1401 interface and Spike 2 software) recorded afferent discharges as the times of occurrence of each spike and continuously displayed instantaneous frequency plots from up to six single units recorded simultaneously, from which the reliability of triggering could easily be checked.

Interpretation

The data from the ramp and hold stretch responses were assessed by means of three direct measurements, the initial frequency (IF) averaged from the 0.5s preceeding the onset of stretch, the peak frequency (PF) and the static index (SI) being the frequency 0.5s after the peak. From these were calculated the dynamic difference (DD) = PF - IF, the dynamic index (DI) = PF - SI and the static difference (SD) = SI - IF. From distribution histograms of these measures it was evident that it was the changes in them due to SCh (indicated by the prefix Δ) which best represented the effects of the drug rather than the absolute values and of the various possibilities it was ΔDD and ΔIF which most clearly suggested the existence of distinct subclasses of behaviour. In fact, both these measures are distributed bimodally, but vary independently, so implying the existence of four subclasses of afferent behaviour. Since bag_1 (b_1) fibre contraction principally determines the dynamic response to stretch, one may take ΔDD to indicate the strength of bag_1 influence on each afferent. On the other hand the effect of bag_2 fibre contraction is principally to increase the IF, so one may take ΔIF to indicate the strength of bag_2 (b_2) influence on each afferent. Since it is agreed that all afferents terminate upon chain (c) fibres, the four subclasses referred to above may be designated as b_1c, b_1b_2c, b_2c and c according to the combinations of intrafusal fibre influencing each afferent. In order to attribute units to one or other of the four groups according to values of ΔIF and ΔDD it was necessary to choose optimal dividing values in the distributions and this was done by fitting normal curves to the distributions of the logarithmically transformed variables. As the data base has expanded since the initial reports some anomalous results have come to light. These principally concern units which have not shown their maximum rise in DD or in IF during the period in which the measurement was being made. One problem was that some units very strongly affected by SCh could have the dynamic component partly occluded by the bag_2 effect at the height of the response (Rodgers, Taylor, Fowle and Durbaba, 1993). Another concern was the occurrence of collapse of the discharge during the ramp stretch in some strongly affected units ("depolarisation block" of Price and Dutia, 1989). These problems are now eliminated by using the maximum values of ΔIF and ΔDD, rather than the mean over a longer period. The hindlimb muscle spindle data shown in this paper are obtained in this way. Any increase in IF greater than 1 impulse.s^{-1} was taken to indicate the presence a bag_2 effect, whilst the best value of ΔDD for dividing significant from non-significant bag_1 effects was chosen as 71.4 impulses.s^{-1} from the results of curve fitting.

RESULTS

Figure 1 shows scatter plots of ΔIF and ΔDD against conduction velocity (CV) for 545 gastrocnemius afferents. In A there is a weak positive correlation between ΔIF and CV. It is evident that there are some units with virtually no increment in IF and these are concentrated in the low CV range. They correspond to the secondary afferents with sensory endings restricted to chain fibres. In Fig. 1 B at first sight there appears to be a positive correlation between ΔDD and CV also, however closer inspection shows that the units actually fell into two clusters. One forms a horizontal band of points with all values of CV except the highest, but with small positive or negative values of ΔDD. The other, with CV in the primary range, has a wide scatter of values of ΔDD. Within each group there is no correlation of ΔDD with CV. It is very clear from Fig. 1 B that all the units with high values of ΔDD (b_1c and b_1b_2c units) are in the primary range of CV. Some high CV units have only small values of ΔDD, but have significant values of ΔIF. They correspond to b_2c primary afferent units described by Price and Dutia (1989).

The quantitative SCh classification was first worked out on jaw muscle spindles in the cat, a situation in which conduction velocity measurements are of very little help in separating primary from secondary afferents. One unexpected feature of the results was the high proportion of b_2c afferents. It was not clear whether these represented the occurrence of an unusually high proportion of primary afferents innervating spindle capsules lacking b_1 fibres or whether there were many secondary afferents with much more influence from b_2 fibres than expected. Another possible method of separating secondary afferents depends on measuring control values of dynamic index (cDI). From hindlimb studies it transpires that, while primaries may have high or low values of cDI, secondaries never have high values. This is clearly confirmed by the plots of Fig. 2 in which the distributions of values of cDI are shown for the subgroups defined by SCh testing. Fig. 2 A shows that values of cDI for b_1b_2c units are distributed approximately symmetrically above and below a value of 27.0 impulses s^{-1}. In the present data the best dividing point for CV to separate high and low populations was found by curve fitting to be 64.1m.s^{-1}. The low CV b_2c group (secondaries with some bag_2 connection) have values of cDI all below 27.0 impulses s^{-1} (Fig. 2 C). Remarkably, nearly all of the fast b_2c group (primaries with no significant b_1 conection) also have cDI values below 27.0 impulses s^{-1} (Fig. 2 B). The obvious difference between units in Fig. 2 A and Fig. 2 B is the presence or absence respectively of a significant bag_1 fibre input. It follows therefore that high values of cDI, when they occur, must be due to bag_1 fibre effects. It is nevertheless true that a substantial number of b_1b_2c primaries have low values of cDI and only show their capacity for dynamic behaviour when the bag_1 fibres are activated by SCh or by dynamic fusimotor stimulation. Evidently, the bag_1 fibre contact is a necessary but not a sufficient condition for high cDI in the passive state. Possibly there is some other anatomical feature related to the pattern of afferent branching which is important. A previous study of spindles in peroneus muscles (Banks, Ellaway and Scott, 1980) had failed to reveal a group of afferents with low dynamic sensitivity corresponding to b_2c primaries. The likely reason for this is now seen to be that though units of this group do all have low cDI values, many of the b_1b_2c primaries are in the same situation. A particularly striking feature of Fig. 2 is that the distributions of values of cDI in the b_2c primary and secondary groups (B and C) are very similar. This implies that, given terminations on the same intrafusal fibres, afferents are indistinguishable in this respect despite their widely different conduction velocities. This favours the view that there is no fundamental biophysical difference between the transducing properties of primary and secondary spindle afferents.

Returning to the jaw muscle spindle afferents, it is evident from Fig. 2 D that the distribution of cDI values for the whole b_2c group (CV unknown) is essentially the same as

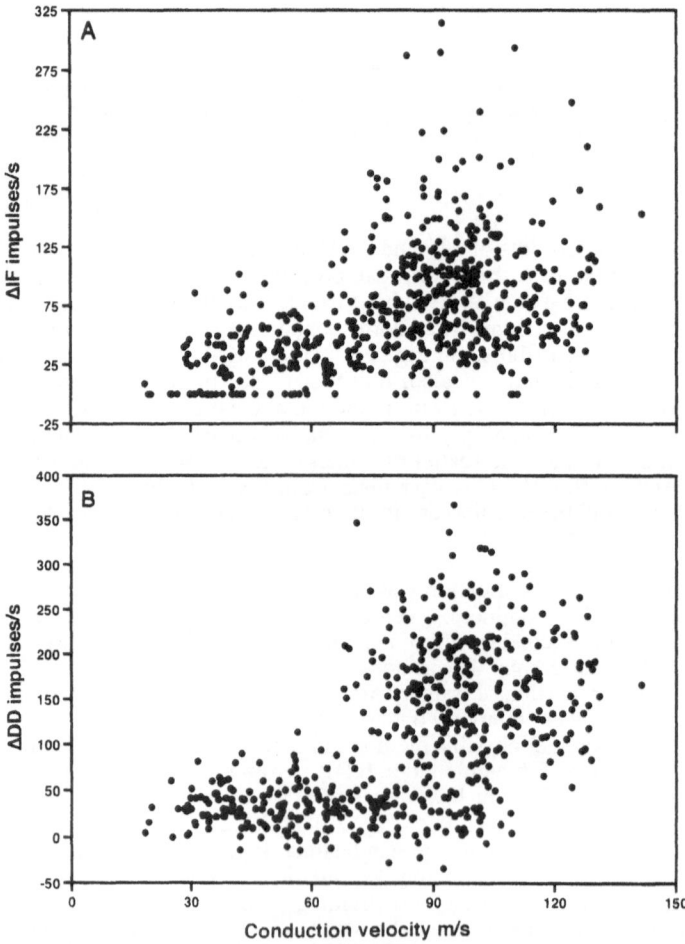

Figure 1. Scatter plots for 545 gastrocnemius spindles afferents of all types to show the relationship of the responses to SCh to conduction velocity. In A the influence of the bag_2 fibre is estimated by ΔIF. In B the influence of the bag_1 fibre is esimated by the value of ΔDD.

for B and C, but with a few additional high values. Unfortunately this does not enable us to decide the division of these jaw b_2c units between primary and secondary types. The only available evidence on this question arises from a recent histological study by Sciote (1993) in which very few tandem spindle capsules were found in cat jaw muscles and these are the usual source of b_2c primary afferents.

DISCUSSION

Some direct evidence linking the diagnosis of a b_2c type SCh response of primaries with histologically proved b_2c capsules in tenuissimus was obtained by Price and Dutia (1989), but this has not been attempted with larger muscles like triceps. However, our data have shown very satisfactory correspondence between the proportions of the different sub-types of spindle afferent as diagnosed with SCh and those expected from independent histological studies (Banks *et al.*, 1982) as seen in Table 1. The table shows the percentages of the types of intrafusal innervation estimated separately for primary and secondary afferents by the two methods. The agreement in the case of primaries is very complete and the distributions are not significantly different by the χ^2 test. There are, on the other hand, considerable differences in the case of the secondaries. The SCh test appears greatly to underestimate the number of b_1b_2c secondaries and to overestimate the b_2c type relative to the histological observations. The explanation may be that, though microscopical examination revealed as many as 67.8% of secondaries with some terminations on b_1 fibres, these endings are small in area and may have very little functional effect. Again the excess of c type by the SCh test relative to histology may arise because some of the terminations reported on b_2 fibres may be too small to exert a detectable effect. Some confidence in the finding of approximately 20% of c type secondaries comes from the finding that these afferents all had very low conduction velocities. This is precisely what would have been predicted, since Banks *et al.*, (1982) found that those afferents ending only on chain fibres had the smallest diameters.

Potential Difficulties with the SCh Method of Classification

It has been suggested that some of the effect of SCh on spindle afferents might be exerted directly on their sensory terminals either through receptor sites or through the liberation of potassium ions from the muscle (Dutia and Ferrell, 1980). In the earlier work, when secondary endings were thought to be restricted to chain fibres, the response of secondary endings to SCh was explained in this way. However, now that secondary afferents are known to terminate frequently on bag_2 fibres in addition, it has been important to re-examine the matter. If there were to be a significant action other than via the contraction of intrafusal bag fibres this would cast some doubt on the interpretations described above. Some recent work on secondary afferents argued strongly against the need to invoke any mechanism other than contraction of bag fibres (Taylor, Morgan, Gregory and Proske, 1993). In the presence of SCh, secondaries were able to maintain a discharge at rates of shortening five times greater than normally. Also the excitation due to SCh showed a length dependence similar to that for fusimotor stimulation. Effects were completely blocked by gallamine and the time course of the subsequent recovery was very similar to that for the recovery of fusimotor responses. The fact that some secondary endings are completely unaffected and that these have the lowest conduction velocity afferents also strongly favours the idea that only those afferents with endings on bag fibres respond to the drug. A general effect on muscle mechanoreceptors is also ruled out by the unresponsiveness of Golgi tendon organs, while cutaneous afferents also are unaffected at these doses (Taylor, Durbaba and Rodgers,1994). The possibility that some of the action of SCh is exerted through potassium ions also has not proved to be supportable. First, the single IV SCh dose produces only a 0.8mmol/l rise in arterial potassium, while Prochazka and Somjen, 1986) found that levels as great as 6 to 11mmol/l caused only occasional small effects on spindles. We have also recently shown that even large rises of free potassium ions produced intramuscularly by tetanic stimulation have no excitatory effect (Taylor *et al.*, 1994). From all these lines of evidence therefore we conclude that with the dosage of SCh employed in this work essentially all its effects on spindle afferents can be attributed to contraction of intrafusal bag fibres and the interpretations made on this basis are safe.

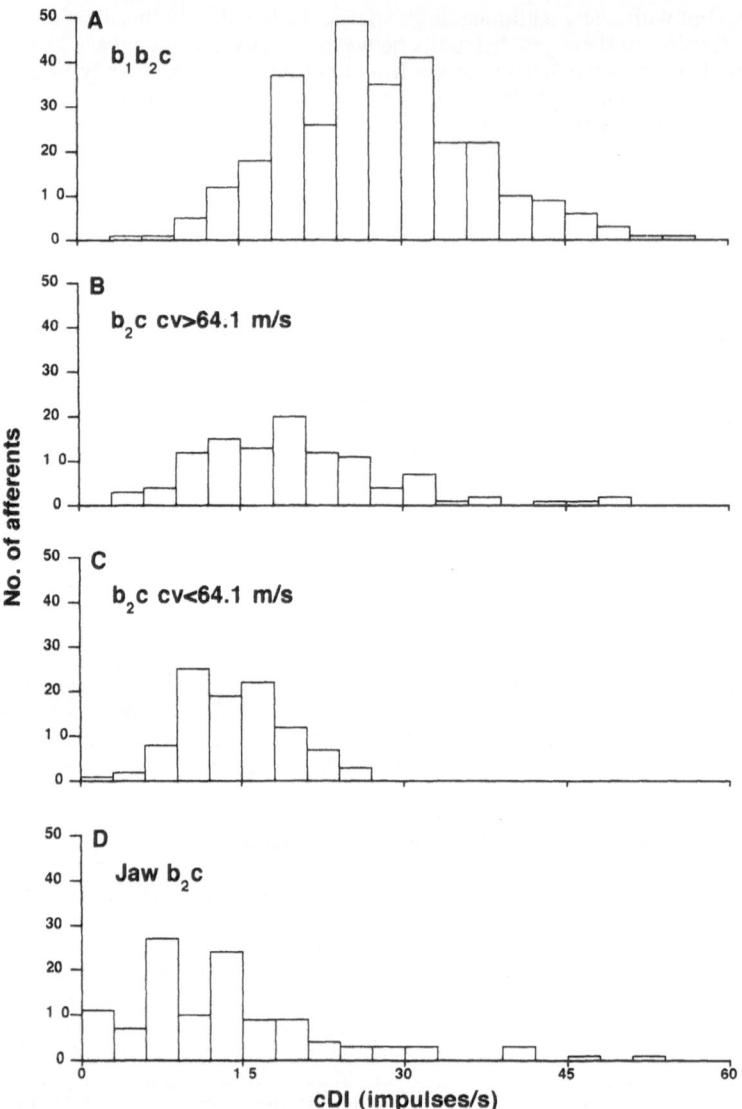

Figure 2. Histograms to show the distribution of values of control dynamic index (cDI) in the various subgroups of gastrocnemius muscle spindle afferents as defined by SCh testing and conduction velocity (A, B and C). In D is shown the distribution of b_2c afferents from jaw muscles.

Applications of the Classification Method

Study of the responses to SCh emphasises the very wide range of properties displayed by spindle afferents which are dependent on variations in the patterns of distribution of endings on the intrafusal muscle fibres. One consequence of this is that care must be taken when using changes in spindle responses to stretch to deduce the underlying fusimotor activity. The simplest example is that a b_2c primary afferent cannot be used to look for dynamic effects, but other consequences have also been demonstrated (Taylor *et al.*,1992).

Another interesting question is the relation between the spindle afferent type and its central projection. If the variation between afferent types has any functional significance then we may expect to find some correlations with strength and distribution of central synaptic connections. An initial study of this has been made in the case of the jaw muscle spindles by estimating the size of the extracellular focal synaptic potentials due to individual

Table 1. Proportions of the four types of innervation of muscle spindles by primary and secondary afferents estimated by succinylcholine testing (SCh) and by the histological studies of Banks *et al.* (1982)

	PRIMARY UNITS		SECONDARY UNITS	
	SCh	Histology	SCh	Histology
b_1c	1.2	1.6	0	5.1
b_1b_2c	71.3	75.3	8.0	67.8
b_2c	26.5	23.1	72.3	20.8
c	1.0	0	19.7	6.3

afferents by spike-triggered averaging (Taylor *et al.*, 1993). Previous work has clearly established that primary afferents make much stronger connections to motoneurones than do secondaries and so it might have been expected that the central projection strength should be related to the strength of the influence upon afferents of the intrafusal bag$_1$ fibre. In the event however it proved to be ΔIF, the measure of bag$_2$ influence which was positively correlated with focal synaptic potential amplitude. The reason for this is not yet clear, but it may be a developmental consequence of the fact that bag$_2$ fibres develop first and are innervated first by the growing Ia afferents, which are also making their central connections at that stage. The observations are currently being extended to hindlimb muscle studies. If they can be confirmed in that situation it will add greatly to the impetus to understand the functional relevance of the evolution of the three intrafusal fibre types in the control of movement.

ACKNOWLEDGEMENT. We acknowledge support from the St Thomas' Hospital Research Endowments, Action Research and the Medical Research Council for research described above.

REFERENCES

Banks, R. W., Barker, D. and Stacey, M. J. (1982). Form and distribution of sensory terminals in cat hindlimb muscle spindles. *Philosophical Transactions of the Royal Society B.* **99**,329-364

Banks, R. W., Ellaway, P. H. and Scott, J. J. (1980). Responses of de-efferented muscle spindles of peroneus brevis and tertius muscles in the cat. *Journal of Physiology* **310**,53P

Cooper, S. (1961). The responses of the primary and secondary endings with intact motor innervation during applied stretch. *Quarterly Journal of Experimental Physiology.* **46**,389-392

Dutia, M. B. and Ferrell, W. R. (1980). The effect of suxamethonium on the response to stretch of Golgi tendon organs in the cat. *Journal of Physiology.* **306**,511-518

Fowle, A. J., Taylor, A., Rodgers, J. F. and Durbaba, R. (1992). Mesencephalic and diencephalic areas for fusimotor control in the anaesthetised cat. *Journal of Physiology* **446**,230P

Gladden, M. H. (1976). Structural features relative to the function of intrafusal muscle fibres in the cat. *Progress in Brain Research.* **44**,51-59

Matthews, P. B. C. (1963). The response of de-efferented muscle spindle receptors to stretching at different velocities. *Journal of Physiology.* **168**,660-678

Ovalle, W. K. and Smith, R. S. (1972). Histochemical identification of three types of intrafusal muscle fibres in the cat and monkey based on the myosin ATPase reaction. *Canadian Journal of Physiological and Pharmacological Sciences* **50**,195-202

Price, R. F. and Dutia, M. B. (1987). Properties of cat neck muscle spindle afferents and their excitation by succinylcholine. *Experimental Brain Research.* **68**,619-630

Price, R. F. and Dutia, M. B. (1989). Physiological properties of tandem muscle spindles in neck and hindlimb muscles. *Progress in Brain Research.* **80**,47-56

Prochazka, A. and Somjen, G. G. (1986). Insensitivity of cat muscle spindles to hyperkalaemia in the physiological range. *Journal of Physiology.* **372**,26P

Rack, P. M. H. and Westbury, D. R. (1966). The effects of suxamethonium and acetylcholine on the behaviour of cat muscle spindles during dynamic stretching and during fusimotor stimulation. *Journal of Physiology.* **186**,698-713

Rodgers, J. F., Durbaba, R., Fowle, A. J. and Taylor, A. (1993). Flexor and extensor muscle fusimotor activation from midbrain stimulation in the anaesthetized cat. *Journal of Physiology.* **459**,462P

Rodgers, J. F., Fowle, A. J., Durbaba, R. and Taylor, A. (1993). Sine versus ramp stretches for characterising fusimotor actions on muscle spindles in the anaesthetized cat. *Journal of Physiology* **467**, 298P

Rodgers, J. F., Taylor, A., Durbaba, R. and Fowle, A. J. (1992). The value of dynamic index in assessing muscle spindle afferent properties in the anaesthetized cat. *Journal of Physiology* **446**,22P

Rodgers, J. F., Taylor, A., Fowle, A. J. and Durbaba, R. (1992). Differences in the time course of effects of succinylcholine on different muscle spindle afferent types in the anaesthetized cat. *Journal of Physiology.* **446**,565P

Rodgers, J. F., Taylor, A., Fowle, A. J. and Durbaba, R. (1993). The occlusion of Ia muscle spindle afferent dynamic sensitivity by bag_2 activation in the anaesthetized cat. *Journal of Physiology* **459**,218P

Sciote, J. J. (1993). Fibre type distribution in the muscle spindles of the cat jaw-elevator muscles. *Archives of Oral Biology* **38**,685-688

Taylor, A., Durbaba, R. and Rodgers, J. F. (1992). The classification of afferents from muscle spindles of the jaw-closing muscles of the cat. *Journal of Physiology* **456**,609-628

Taylor, A., Durbaba, R. and Rodgers, J. F. (1993). Projection of cat jaw muscle spindle afferents related to intrafusal fibre influence. *Journal of Physiology* **465**,647-660

Taylor, A., Durbaba, R. and Rodgers, J. F. (1994). The site of action of succinylcholine on muscle spindle afferents in the anaesthetised cat. *Journal of Physiology* **476**, 26P

Taylor, A., Durbaba, R., Rodgers, J. F. and Fowle, A. J. (1993). Reciprocal actions of midbrain stimulation on static and dynamic fusimotor neurones of the hindlimb in anaesthetised cats. *Journal of Physiology* **473**, 15P

Taylor, A., Morgan, D. L., Gregory, J. E. and Proske, U. (1993). The mode of action of succinycholine on secondary muscle spindle endings of soleus in the anaesthetized cat. *Journal of Physiology* **459**,217P

Taylor, A., Rodgers, J. F., Fowle, A. J. and Durbaba, R. (1992). Conduction velocity and bag_1 influence on gastrocnemius muscle spindle afferents in the anaesthetized cat. *Journal of Physiology* **446**,21P

Taylor, A., Rodgers, J. F., Fowle, A. J. and Durbaba, R. (1992). The effect of succinylcholine on cat gastrocnemius muscle spindle afferents of different type. *Journal of Physiology.* **456**,629-644

Taylor, A., Rodgers, J. F., Fowle, A. J. and Durbaba, R. (1992). Some problems in the interpretation of spindle afferent recordings. In *Muscle Afferents and Spinal Control of Movement.*, ed. Jami, L., Pierrot-Deseilligny, E. and Zytnicki, D., pp 105-111. Pergamon Press Ltd., Oxford

TENDON ORGAN DISCHARGES AND THEIR CENTRAL EFFECTS DURING MUSCLE CONTRACTIONS

L. JAMI, D. ZYTNICKI and G. HORCHOLLE-BOSSAVIT

CNRS URA 1448
Université René Descartes
45 rue des Saints-Pères
75270 Paris Cedex 06, France

SUMMARY

This review compares the ensemble discharges of the tendon organ population of a muscle during sustained contractions with the effects exerted by such contractions on two of the spinal targets of Ib afferent input, motoneurones and cells of origin of the dorsal spinocerebellar tract. The message carried by the total Ib afferent traffic provides elaborate information about the moment to moment variations of muscle force. This information does not appear directed to motoneurones since intracellular recordings in homonymous and synergic motor pools demonstrated a quick suppression of the classical negative feedback from tendon organs, due to presynaptic inhibition of Ib afferents. In dorsal spinocerebellar tract neurones, effects were not uniform. Declining inhibitions, similar to those observed in motoneurones, were recorded in about half of the contraction-sensitive neurones. Other neurones displayed excitation, either transient or persistent throughout the duration of sustained contractions. These neurones could represent the origin of a pathway carrying the message from the tendon organ population to the cerebellum and cerebral cortex.

INTRODUCTION

The sensory equipment of mammalian skeletal muscles includes contraction sensors, the Golgi tendon organs, innervated by Ib afferent fibres. The tendon organs of the cat hindlimb muscles are usually located at musculo-tendinous or musculo-aponeurotic junctions rather than within tendons. In a typical tendon organ, the sensory terminals are laying within an encapsulated fascicle of collagen bundles that fuse with the muscle tendon or aponeurosis at one end, and with the individual tendons of a few muscle fibres at the other end. The contraction of these fibres is the specific stimulus of the tendon organ (Houk and Henneman 1967), and, as they belong to several motor units, the receptor can signal the activity of each unit (Fig 1A). Systematic investigations have further shown that a particular motor unit can activate several tendon organs (Figure 1B). In various muscles of the cat hindlimb the ratios of motor unit to tendon organ numbers are in a range of 1.9 to 9.1 (see Jami 1992). Taken together, these data strongly suggest that the contraction of every motor unit in a muscle is monitored by at least one tendon organ.

In the group of motor units acting on a particular tendon organ, the range of mechanical properties is the same as in the total population of the muscle. This is illustrated in Figure 1A, showing the range of tetanic forces of 47 motor units of the cat peroneus longus muscle (i.e., 50% of the total population). Of these units, 15 (open squares in Fig. 1A) were found to activate the same tendon organ and their forces ranged between 1.2 and 75gf, in close parallel to the 1.2-112gf range of the 32 other units which did not act on that particular tendon organ.

The low threshold of tendon organs for contractile force and the consistency of their responses to the slightest contractions (see, e.g., Gregory and Proske 1979, Fukami 1981) would suggest that they monitor active muscle force, possibly working as dynamometers. If this were the case, some clear correlation should appear between the discharge frequency of a tendon organ and the force developed by the contraction of its activating motor units. But observations made on individual tendon organs from various muscles of the cat hindlimb did not confirm this assumption (see references in Jami 1992). Of two motor units acting on the same receptor, the weaker often produces the higher discharge frequency, and motor units activating several tendon organs often elicit different responses from each receptor (see, e.g., Crago, Houk and Rymer 1982; Horcholle-Bossavit, Jami, Petit, Vejsada and Zytnicki 1990). Moreover, the concurrent contractions of two motor units acting on the same tendon organ

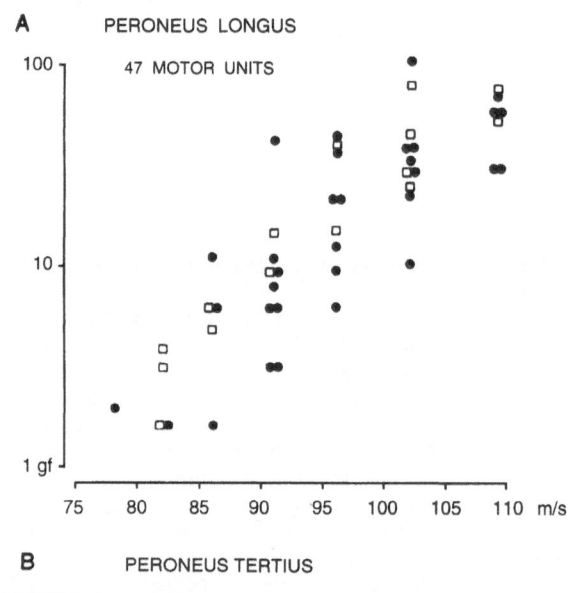

A PERONEUS LONGUS

B PERONEUS TERTIUS

MOTOR UNIT TYPE and conduction velocity (m/s)	TENDON ORGAN AFFERENT and conduction velocity (m/s)									
	1 92	2 88	3 92	4 94	5 96	6 100	7 88	8 96	9 96	10 96
FR 92 a	■	■	■							
FR 92 b	■	■		■		■				
FR 90				■	■		■	■	■	
FR 96					■	■				
S 73								■		
S 78			■							
FF 92					■					
FF 96							■	■	■	■

Figure 1. A: Tetanic forces (expressed in gram-force units) of 47 peroneus longus motor units examined in a single experiment plotted on a logarithmic scale against their axonal conduction velocities. Each symbol represents a motor unit. The open squares represent 15 motor units acting on the same tendon organ (Modified from Jami and Petit 1976). **B:** Actions of 8 motor units on 10 tendon organs examined in a single experiment. Each filled square represents the activation of a tendon organ by a motor unit. The ensemble response of afferents 1-6 to simultaneous contraction of units a and b is illustrated in Fig. 2.

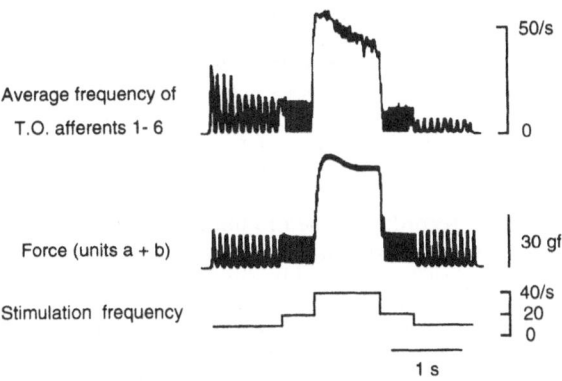

Figure 2.. Instantaneous average frequencygram of the discharges of afferents 1-6 in Fig. 1B in response to the contraction of units a+b stimulated with the "step" pattern. To allow clear visualization of the variations in average discharge frequency, the points in the frequencygram have been joined by continuous lines. See further description in text. (Modified from Horcholle-Bossavit *et al.* 1989)

usually evoke responses at lower frequencies than might be expected from the addition of the responses to each unit contracting separately.

However frustrating for the experimenters, the lack of a clear relation between the force of a contraction and the discharge frequency of an individual tendon organ does not mean that the central nervous system cannot receive useful information on muscle force from these receptors. In their early study, Houk and Henneman (1967, see also Reinking, Stephens and Stuart 1975, Crago *et al.*,1982) suggested that functionally significant messages about a muscle contraction could be encoded in the *pooled discharges* of the tendon organ population of this muscle. Recently, we were able to test this hypothesis in peroneus tertius, a relatively small muscle of the cat leg, containing an average population of 10 tendon organs (Scott and Young 1987). In experiments where 9-11 Ib fibres, representing virtually all the Ib afferents from the muscle, had been isolated for recording in dorsal root filaments, we could therefore observe the total Ib-afferent traffic generated by the contractions of different motor units. These units were activated by stimulating, in isolation or in combination, single motor axons prepared in ventral root filaments. The distribution of the actions of 8 motor units on 10 tendon organs examined in such an experiment is shown in Figure 1B. Each tendon organ was excited by at least one motor unit and each unit was found to act on at least one tendon organ.

Units a and b of Fig 1B developed similar forces and acted on 6 and 4 tendon organs, respectively, eliciting different rates of discharge from individual receptors (Horcholle-Bossavit *et al.*,1990). Various methods were used to assess the ensemble discharge of the activated tendon organs during a period of contraction, either by simply summing up the counts of impulses discharged by each receptor, or by calculating the instantaneous frequency of the pooled discharges or the averaged discharge frequency of all the active tendon organs (see Fig. 2; this procedure is fully described in Horcholle-Bossavit *et al.*,1990). All the methods gave similar results. In the example shown here, motor units a and b were stimulated together with a special pattern in which frequency started at 10/s and was increased by successive steps to 20 and 40/s before returning, also by steps, to 20 and 10/s. This pattern allowed observation in a single run of responses to twitches and to unfused tetani with two different plateau levels (see the force record in Fig. 2). The figure shows that the averaged frequency of the six tendon organs (representing about 50% of the muscle population) activated by units a and b faithfully reflected the variations in the force profile, even though it did not always indicate the actual strength of contractions (compare, e.g., the initial and final series of 10/s twitches). This observation suggests that the message carried by the *total Ib afferent traffic* from a muscle provides better information about the dynamic components of contractile force than about its static level. However, estimation of contraction strength by the central nervous system might be achieved through co-processing of information provided by various mechanoreceptors, not only tendon organs, but also spindles, and non-specific muscle receptors, plus cutaneous and joint receptors, all of which are known to discharge during muscle contraction.

At any rate, observations of the total Ib-afferent traffic from a muscle suggested that the *targets of Ib input* might receive elaborate records of muscle force variations. Experiments were therefore designed to investigate the effects of such messages on two of the main targets of Ib input in the spinal cord, the homonymous and synergic motoneurone and the cells of origin of the dorsal spinocerebellar tract (DSCT) in Clarke's column. Intracellular records were made in anaesthetized cats, first in lumbar motoneurones and second in DSCT neurones, during subtotal unfused contractions of an ankle extensor muscle, gastrocnemius medialis (GM).

A typical example of the records obtained in more than 80% of 156 homonymous and synergic motoneurones is shown in Figure 3 (Zytnicki, Horcholle-Bossavit, Lafleur, Lamy and Jami 1990). Here, contraction was elicited by stimulating the distal end of a cut branch of the GM nerve with the "step" pattern. Comparison of figures 2 and 3 immediately shows that, during a sustained GM contraction, the variations in membrane potential of triceps surae motoneurones did not reflect the force variations as encoded in the ensemble discharge of tendon organs. Inhibitory potentials appeared in the motoneurone at the onset of contraction but quickly decreased and disappeared until the 20-40/s step in stimulation frequency produced an abrupt increase in force. Inhibitory potentials then reappeared but only to decline again in spite of the maintained force plateau. A priori, because tendon organs are activated by contraction and Ib afferents are known to produce di- or trisynaptic inhibition in homonymous and synergic motoneurones (Laporte and Lloyd 1952; Eccles, Eccles and Lundberg 1957), a contraction-induced inhibition could be ascribed to Ib afferents. But here, the rapid decrease of contraction-induced inhibitory potentials, contrasting with the profile of tendon organ population discharges, raised the question whether Ib afferents were indeed responsible for the inhibition recorded in motoneurones.

It was then verified that repetitive electrical stimulation of Ib afferents by themselves could produce similarly decreasing inhibition (see Figs 5 and 6 in Zytnicki *et al.*,1990). The resemblance between the effects elicited in triceps surae motoneurones by GM contraction and by repetitive electrical stimulation of Ib afferents in the GM nerve supported the assumption that the contraction-induced inhibition was due to the action of contraction-induced Ib input (subsequent observations suggested an additional contribution of group II afferents, see Lafleur, Zytnicki, Horcholle-Bossavit and Jami, 1993a,b). Moreover, as electrical stimulation was applied at a fixed strength, producing a constant Ib input throughout the stimulation sequence, the decline of the inhibition elicited by this stimulation necessarily had a central origin, which supported the view that the decline of contraction-induced inhibition also might depend on a central cause.

A second question was the identification of this cause. The contribution of presynaptic inhibition was suggested by data from a different series of experiments. Intra-axonal recordings were made in the intraspinal portion of identified Ib afferents from GM during contractions of this muscle similar to those produced in the experiments where motoneurones had been recorded. Primary afferent depolarizations, allowing diagnosis of presynaptic inhibition, were seen to develop in these fibres with a time course closely following the development of contractile force (Lafleur, Zytnicki, Horcholle-Bossavit and Jami 1992). Another support for the assumption that presynaptic inhibition could account for the decline in Ib inhibition during contraction was obtained from a computer model study. A simple but realistic model of the disynaptic Ib afferent pathway to motoneurones was designed, incorporating as much as possible of the known functional properties of the natural network components, as determined in electrophysiological experiments. Declining inhibitory potentials appeared in the motoneurone-like output stage when functions simulating primary afferent depolarizations of Ib afferents, similar to those recorded experimentally, were introduced in the network (Zytnicki and L'Hôte 1993).

Altogether, these observations pointed to the conclusion that the effects of Ib afferents on their motoneuronal targets are quickly filtered out during contraction. This means that a mechanism is available in the central nervous system, by which *the negative feedback from tendon organs can be suppressed.* Such a suppression could be useful during sustained contractions, because it would facilitate the recruitment of homonymous and synergic motor units when the effort has to be maintained or increased. It could also help the expression of a recently demonstrated short excitatory pathway from Ib afferents to motoneurones. Flexibility of Ib afferent effects on extensor motoneurones has been observed on preparations displaying fictive locomotion pattern : disynaptic inhibitory action gives way to a disynaptic excitatory action during the extension phase of the locomotor cycle (McCrea, Shefchyk and Pearson

1993). As presynaptic inhibition is under supraspinal control, the suppressive mechanism may not be systematically turned on whenever contraction occurs. In fictive locomotion again, it was shown that presynaptic inhibition of primary afferents is modulated, in phase with the locomotor cycle, by the central pattern generator (see Cabelguen, Dubuc, Gossard and Rossignol 1992).

A third, and so far unanswered, question regards the reappearance of inhibitory potentials in the motoneurone upon rapid force increase at the 20-40/s step (Fig. 3). The force step elicits an increase in the ensemble discharge of tendon organs (Fig. 2) which for a short while overcomes the effects of presynaptic inhibition. Differences in time courses of pre- and postsynaptic Ib inhibitions might account for this breakthrough of the negative feedback from tendon organs to motoneurones. Excitability changes in Ib interneurones, known to receive a wide variety of input from segmental and supraspinal sources (Harrison and Jankowska 1985), would be expected to play a major role in the efficacy of Ib inhibitory pathway.

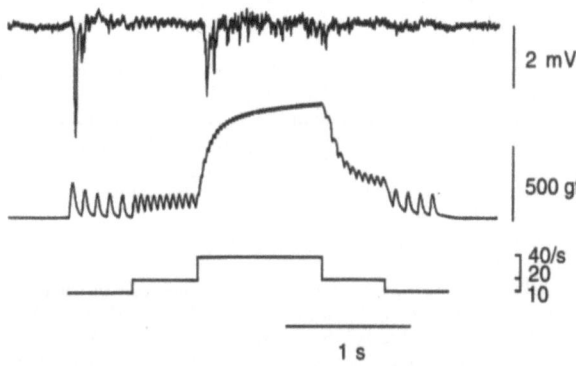

Figure 3. Response of a Gastrocnemius lateralis-Soleus motoneurone (four responses averaged) to GM contraction produced by stimulating a cut nerve branch with the "step" pattern (duration of the 10/s step was shorter here than in Fig. 2). Upper trace, motoneurone membrane potential; middle trace, contractile force ; lower trace, stimulation frequency.(Modified from Zytnicki *et al.*,1990)

The DSCT neurones in Clarke's column are known to receive Ib afferent input by at least two pathways. The first one produces monosynaptic excitation (Lundberg and Oscarsson 1956) while the second elicits disynaptic inhibition via the same interneurones that mediate Ib inhibition of motoneurones (Hongo, Jankowska, Ohno, Sasaki, Yamashita and Yoshida 1983). Investigation of DSCT neurones during triceps surae tetanic contractions showed non-uniform effects (Zytnicki, Lafleur and Kouchtir 1991). Of 145 cells examined, only 77 (53%) "saw" the contractions and four patterns could be distinguished : 1) In 38 DSCT neurones, the effect was the same as in homonymous and synergic motoneurones, that is, a declining inhibition. This could be expected because presynaptic inhibition of Ib afferent fibres during contractions (Lafleur *et al.*, 1992) is likely to affect the transmission of Ib afferent information to both targets, motoneurones and Clarke's column cells, reached via the same interneurone. 2) Excitations, decreasing with a time course similar to that of inhibition decline in motoneurones, were observed in 15 DSCT neurones. The excitation might be due to the monosynaptic connections between Ib afferents and DSCT cells while the decrease could depend on the presynaptic inhibition of group I afferent terminals projecting directly onto DSCT cells, as reported by Jankowska and Padel (1984). 3) Mixed excitations and inhibitions were observed in 13 neurones. 4) But more interestingly, in a final group of 11 DSCT neurones, excitation persisted throughout the contraction, suggesting that some of the excitatory terminals of group I afferents in Clarke's column were not affected by presynaptic inhibition. This would imply a non-uniform distribution of presynaptic inhibition among all the terminal branches of an individual group I afferent, as suggested by recent observations (Eguibar, Quevedo, Jimenez and Rudomin 1993). At any rate, the DSCT neurones in which excitation persisted could represent the origin of a pathway carrying the elaborate information from tendon organ populations to the cerebellum and also to the cerebral cortex (McIntyre, Proske and Rawson 1984, 1985).

CONCLUSION

The observations reviewed in this paper point to variability and flexibility in the effects of input from tendon organs at the spinal level. Filtering by presynaptic inhibition and/or modulation of excitability in Ib interneurones entail deep transformations of the messages carried by Ib afferents. How information concerning contraction parameters are retrieved at higher levels, probably by collating messages from several sources, muscular and extramuscular, including corollary discharges, remains to be elucidated.

REFERENCES

Cabelguen, J.-M., Dubuc, R., Gossard, J.-P. and Rossignol, S. (1992). Presynaptic mechanisms during locomotion in the Cat. in *Muscle Afferents and Spinal Control of Movement*, eds. Jami, L., Pierrot-Deseilligny, E. and Zytnicki, D. pp 453-458, Pergamon Press, Oxford.

Crago, P.E., Houk, J.C. and Rymer, W.Z. (1982). Sampling of total muscle force by tendon organs. *Journal of Neurophysiology* **47**, 1069-1083.

Eccles, J.C., Eccles, R.M. and Lundberg, A. (1957). Synaptic actions on motoneurones caused by impulses in Golgi tendon organ afferents. *Journal of Physiology* **138**, 227-252.

Eguibar, J.R., Quevedo, J.N., Jimenez, I. and Rudomin, P. (1993). Differential control exerted by the motor cortex on the synaptic effectiveness of two intraspinal branches of the same group I afferent fiber. *American Society for Neurosciences*, 23rd Annual Meeting Abstract 588.6.

Fukami, Y. (1981). Responses of isolated Golgi tendon organs of the cat to muscle contraction and electrical stimulation. *Journal of Physiology* **318** , 429-443.

Gregory, J.E. and Proske, U. (1979). The responses of Golgi tendon organs to stimulation of different combinations of motor units. *Journal of Physiology* **295** , 251-262.

Harrison, P.J. and Jankowska, E. (1985). Sources of input to interneurones mediating group I non-reciprocal inhibition of motoneurones in the cat. *Journal of Physiology* **361**, 379-401.

Hongo, T., Jankowska, E., Ohno, T., Sasaki, S., Yamashita, M. and Yoshida, K. (1983). The same interneurones mediate inhibition of dorsal spinocerebellar tract cells and lumbar motoneurones in the cat. *Journal of Physiology* **342** , 151-180.

Horcholle-Bossavit, G., Jami, L., Petit, J., Vejsada, R. and Zytnicki, D. (1989). Encoding of muscle contractile tension by Golgi tendon organs. in *Stance and Motion : Facts and Concepts*. eds. Gurfinkel, V., Ioffe, M., Massion, J. and Roll, J.-P., pp 1-10, Plenum, New-York.

Horcholle-Bossavit, G., Jami, L., Petit, J., Vejsada, R. and Zytnicki, D. (1990). Ensemble discharge from Golgi tendon organs of the cat peroneus tertius muscle. *Journal of Neurophysiology* **64** , 813-821.

Houk, J.C. and Henneman, E. (1967). Responses of Golgi tendon organs to active contraction of the soleus muscle of the cat. *Journal of Neurophysiology* **30** , 466-481.

Jami, L. (1992). Golgi tendon organs in Mammalian skeletal muscle : functional properties and central actions. *Physiological Reviews* **72** , 623-666.

Jami, L. and Petit, J. (1976). Heterogeneity of motor units activating single Golgi tendon organs in cat leg muscles. *Experimental Brain Research* **24** , 485-493.

Jankowska, E. and Padel, Y. (1984). On the origin of presynaptic depolarization of group I muscle afferents in Clarke's column in the cat. *Brain Research* **295** , 195-201.

Lafleur, J., Zytnicki, D., Horcholle-Bossavit, G. and Jami, L. (1992). Depolarization of Ib afferent axons in the cat spinal cord during homonymous muscle contractions. *Journal of Physiology* **445** , 345-354.

Lafleur, J., Zytnicki, D., Horcholle-Bossavit, G. and Jami, L. (1993a). Declining inhibition in ipsi- and contralateral lumbar motoneurones during contractions of an ankle extensor in the cat. *Journal of Neurophysiology* **70** , 1797-1804.

Lafleur, J., Zytnicki, D., Horcholle-Bossavit, G. and Jami, L. (1993b). Declining inhibition elicited in cat lumbar motoneurones by repetitive stimulation of group II muscle afferents. *Journal of Neurophysiology* **70**, 1805-1810.

Laporte, Y. and Lloyd, D.P.C. (1952). Nature and significance of the reflex connections established by large afferent fibers of muscular origin. *American Journal of Physiology* **169** , 609-621.

Lundberg, A. and Oscarsson, O. (1956). Functional organization of the dorsal spinocerebellar tract in the cat. IV. Synaptic connections of afferents from Golgi tendon organs and muscle spindles. *Acta Physiologica Scandinavica* **38** , 53-75.

McCrea, D., Shefchyk, S. and Pearson, K. (1993). Activation of Golgi tendon organ afferents produces disynaptic excitation and not inhibition of synergists during fictive locomotion. Symposium 028, IUPS Meeting Glasgow, Abstract 31/4.

McIntyre, A.K., Proske, U. and Rawson, J.A. (1984). Cortical projection of afferent information from tendon organs in the cat. *Journal of Physiology* **354** , 395-406.

McIntyre, A.K., Proske, U. and Rawson, J.A. (1985). Pathway to the cerebral cortex for impulses from tendon organs in the cat's hind limb. *Journal of Physiology* **369** , 115-126.

Reinking, R.M., Stephens, J.A. and Stuart, D.G. (1975). The tendon organs of cat medial gastrocnemius : significance of motor unit type and size for the activation of Ib afferents. *Journal of Physiology* **250**, 491-512.

Scott, J.J.A. and Young, H. (1987). The number and distribution of muscle spindles and tendon organs in the peroneal muscles of the cat. *Journal of Anatomy* **151** , 143-155.

Zytnicki, D., Lafleur, J., Horcholle-Bossavit, G., Lamy, F. and Jami, L. (1990). Reduction of Ib autogenetic inhibition in motoneurones during contractions of an ankle extensor muscle in the cat. *Journal of Neurophysiology* **64** , 1380-1389.

Zytnicki, D., Lafleur, J. and Kouchtir, N. (1991). Effects of ankle extensor muscle contractions on dorsal spinocerebellar tract cells in anaesthetized cats. *European Journal of Neuroscience*, suppl 4, p 39.

Zytnicki, D. and L'Hote, G. (1993). Neuromimetic model of a neuronal filter. *Biological Cybernetics* **70**, 115-121.

PROPRIOCEPTION

MUSCLE, CUTANEOUS AND JOINT RECEPTORS IN KINAESTHESIA

D. IAN McCLOSKEY

Prince of Wales Medical Research Institute
High Street, Randwick 2031
Sydney NSW, Australia

SUMMARY

What was the "muscular sense" to Sherrington was later the "joint sense", when it became the conventional wisdom that muscles are "insentient". These terms betray a view, still common in the field, that the basis of kinaesthetic sensibility must be either joint receptors or muscle receptors, but not both. However, the demonstrations that established that muscle receptors do have a role in kinaesthesia were not demonstrations that joint receptors play no part. Indeed, there have been specific demonstrations of kinaesthetic roles for joint receptors. It is clear that both muscle and joint receptors can be involved in giving specific kinaesthetic sensations, and that there is probably considerable redundancy between them. Furthermore, some cutaneous receptors give signals that could be the basis of quite good kinaesthetic performance, although there is not yet strong evidence of their being used for this. Apart from specific kinaesthetic roles for each receptor class, there is evidence of mutual facilitation between the submodalities. Sometimes this facilitation appears to be of a non-specific kind as seen for example in the dependence of kinaesthesia in the fingers upon cutaneous inputs.

INTRODUCTION

Sherrington referred to the sense of limb movement and position as *'the muscular sense'* (Sherrington, 1900), and clearly believed that the discharges from muscle receptors reached consciousness. Physiologists and neurologists who came after him, however, did not share this view. They believed that proprioceptive sensibility could be attributed entirely to the discharges of sensory nerve endings in and around the capsules and ligaments of joints. This belief was based, in part, on the apparent suitability of joint receptors for this role as revealed by electrophysiological recordings said to be of their behaviour (although it later became apparent that some recordings were made, mistakenly, from intramuscular receptors). The belief was reinforced by the apparent unsuitability of intramuscular receptors for conscious proprioception. While stretch-sensitive receptors, such as muscle spindles, in muscles operating around a joint might have seemed appropriate to provide a signal from which the central nervous system could compute joint position or velocity of movement, these receptors could also be made to discharge by fusimotor (gamma) activation without any change of joint position. It seemed that such discharges could not signal joint position unambiguously. Furthermore, intramuscular receptors were thought not to project to the cerebral cortex. For these and other reasons intramuscular receptors were believed to be 'private' to the muscles and to deal only with reflex adjustments at a subconscious level.

Neural Control of Movement, Edited by W.R. Ferrell
and U. Proske, Plenum Press, New York, 1995

SOURCES OF PROPRIOCEPTIVE SIGNALS

The classes of afferent fibre that are candidates for subserving position and movement senses are those from the muscles and tendons, the joint capsules and ligaments, and the skin.

Muscle Receptors

So entrenched was the view of the 'insentience' of muscles that, when a cortical projection from muscles was first discovered (Amassian and Berlin, 1958; Oscarsson and Rosén, 1963; Phillips, Powell and Wiesendanger, 1971), it was taken to indicate that access of sensory information to the cortex need not involve conscious awareness of its message.

Subsequently, however, it was shown in a variety of experiments on normal subjects that proprioceptive signals based on intramuscular receptors can be perceived. These experiments have been reviewed elsewhere (e.g. Goodwin, McCloskey and Matthews, 1972; Goodwin, 1976; Matthews, 1977, 1988, McCloskey, 1978) but briefly including the following:

(i)   High frequency (100 Hz) Vibration, applied transcutaneously over the tendon of muscle, elicits the illusory sensation that the joint at which that muscle operates is moving in the direction that would normally stretch the muscle (Goodwin, McCloskey and Matthews, 1972). There are no illusory movements when the vibration is applied over the joint directly, without directly engaging the muscle. As it is known from animal experiments and from microneurographic studies in human subjects that vibration of this kind powerfully excites muscle spindle primary endings, this observation has been taken to indicate a proprioceptive role for such afferents (Goodwin, McCloskey and Matthews, 1972: Figure 1).

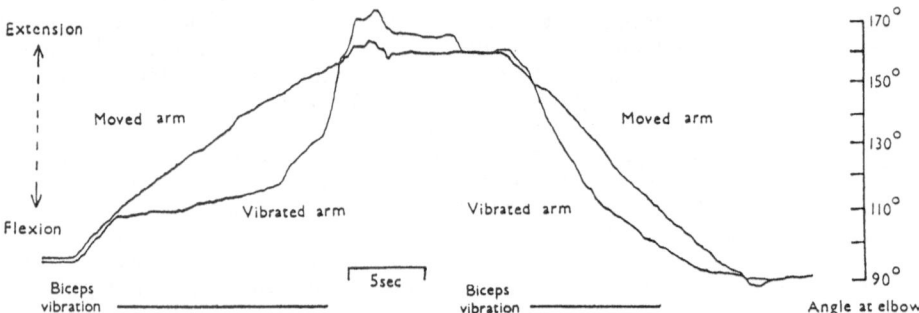

Figure 1. The effect of vibration applied to an arm that the subject was using to make a voluntary movement. The left arm was moved by the experimenter to provide a reference and the subject was asked to track it with his right arm. During the periods indicated vibration was applied to the biceps of the right arm which was the one which was being moved voluntarily. This caused the subject to position the vibrated arm so that it was unduly flexed with regard to the reference arm, that is so that its vibrated muscle was unduly short. This occurred irrespective of whether the vibrated arm was being moved into flexion or extension, although the effect was more dramatic when the arm was being moved into extension. The arm was moving in the vertical plane with the upper arm lying horizontal so that the biceps muscle will have been contracting throughout. From Goodwin, McCloskey and Matthews (1972).

(ii)   Anaesthesia of the joints and skin of a finger or thumb, or the whole hand, is a procedure which does not include the muscles which flex and extend the digits and which lie in the forearm away from the anaesthetized area. However, this procedure does not abolish a subject's ability to detect flexion or extension movements imposed on a digit, an ability which must therefore be ascribed to discharges of sensory receptors in the unaffected muscles.

(iii)   A simpler experiment is to pull upon an exposed tendon in a conscious subject so as to stretch its muscle; this can be done while immobilizing the joint which the muscle normally moves. This simple experiment has sometimes failed to give this result and the subject is said to get no sensation from tendon pulling (Gelfan and Carter, 1967; Moberg, 1972, 1983). However, such experiments were carried out on patients during surgery and it is possible that they were not sufficiently at ease and attentive to perceive rather subtle sensations. When the experiment is performed under laboratory conditions, and in other circumstances on surgical patients, however, the subject reports that the joint which is usually operated by the lengthened muscle seems to move, and to move in the direction which would normally stretch the muscle (McCloskey *et al.,* 1983a; Moberg, 1983). In one of these studies (Moberg, 1983), the proprioceptive sensation was reported only when the muscle was pulled near to the end of its physiological range. In the other, finely graded proprioceptive sensations were evoked, and the detection thresholds of these corresponded quite well to the thresholds for detection of displacements imposed on the same joint prior to the tendon-pulling experiments (McCloskey *et al.,* 1983: Figure 2).

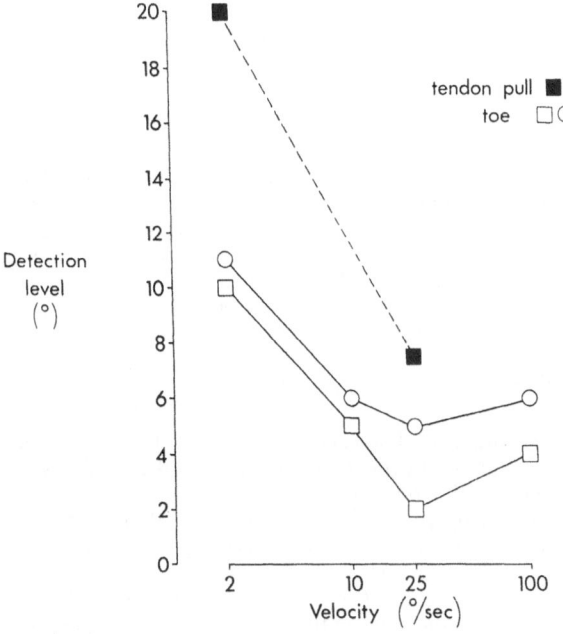

Figure 2. This shows the relation between 70% detection levels of joint rotation and the angular velocity of rotation, for the interphalangeal joint of the big toe. The open symbols show the relation for the intact toe: circles show data for imposed ramp displacements into dorsiflexion and squares for plantar flexion. The filled squares show data obtained subsequently in the experiment on the subject's exposed, transected extensor hallucis longus tendon. Each point represents the smallest angle (or equivalent angle) of imposed displacement, at the velocity shown, for which the subject correctly nominated the direction of joint rotation in seven or more of ten trials. Velocity is plotted on a log scale. In the experiment on the exposed tendon, ramp displacements at 2.5 mm/s (equivalent to 25°/s) were tested in two directions: the 'pulls', detected as plantar flexions, are plotted; 'pushes' at the same velocity, even when 2 mm (20°) in amplitude, were not detected. McCloskey *et al.,*(1983) reproduced with permission.

The intramuscular mechanoreceptors most likely to subserve position and movement sense are muscle spindles. Importantly, high frequency vibration applied to human muscle tendons, a stimulus known to excite muscle spindle afferents (Burke *et al.,* 1976; Roll and Vedel, 1982), evokes powerful illusions of joint movement and altered joint position (Goodwin *et al..* 1972; McCloskey *et al.,* 1983). However, implicit in any conclusion that muscle spindles contribute to perception is the requirement that the central nervous system informs itself internally of the level of fusimotor (gamma) drive, so that the proprioceptively significant element of spindle discharge can be computed.

Joint Receptors

Joint afferents were once considered so critical for detection of limb movement that the term *joint position sense* seemed clinically appropriate. However, with the appreciation that muscle spindle afferents subserve this sensation (see above) the potential kinaesthetic role for joint afferents was given less emphasis. Nevertheless, both psychophysical and electrophysiological studies in human subjects has reaffirmed a possible kinaesthetic role for some specialized joint receptors.

The hand can be positioned so that the distal interphalangeal joint of the middle finger is effectively disengaged from its muscular attachments (Gandevia and McCloskey, 1976; see later). When this is done, applied movements within the normal range cannot be detected at all if the digital nerves of the finger are anaesthetized (Goodwin et al.,1972). This implies that joint and cutaneous receptors subserve the residual kinaesthetic sensation, but it does not distinguish between these two afferent species. This distinction was made when subjects, with the hand postured to eliminate a contribution from muscle afferents, were far less effective in detecting movements applied by a motor when the joint was anaesthetized by injection of a small volume of lignocaine than when the joint was unanaesthetized but distended by a volume of a plasma expander (Ferrell, Gandevia and McCloskey, 1987).

For other joints in the body, a similar role for joint receptors has not yet been found. Indeed, anaesthesia of the knee joint does not impair the detection of extremely slow movements (~1°/min; Clark et al.,1979); nor does replacement of diseased joints impair movement detection (see McCloskey, 1978). In the former studies, muscle afferents were still able to provide kinaesthetic signals, and in the latter, disease may have already eliminated useful kinaesthetic input from joint receptors.

Microneurography has corroborated a possible kinaesthetic role for some digital joint receptors. Although the majority of slowly adapting receptors discharged at the end of an angular range and often responded to more than one axis of movement (e.g. flexion/extension, adduction/abduction), some discharged with movement through the mid range (Burke, Gandevia and Macefield, 1988).

Cutaneous Receptors

Despite the obvious cutaneous displacements and distortions produced by joint movement, the role of the skin in detection of movements is less well established than that of the other afferent groups. It is known that both slowly and rapidly adapting receptors in the skin of the human hand discharge during passive (e.g. Knibestöl, 1975; Burke et al., 1988; Edin and Abbs, 1991) and active movements of the fingers (Hulliger et al.,1979). The discharge of the slowly adapting receptors increases towards the extremes of flexion and extension, much as noted for joint receptors (Hulliger et al.,1979; Burke et al.,1988). Recordings from the radial nerve of afferents innervating the skin on the dorsal aspect of the hand have emphasized the potential for cutaneous receptors, some centimetres proximal to the joint, to encode passive and active movements (Edin and Abbs, 1991). Clearly, the nervous system would need some decoding of cutaneous signals for specification of individual joint movement, and presumably this would require reference to other kinaesthetic inputs.

MULTI COMPONENT PERFORMANCE

One method through which insight has been gained into how the components of proprioception perform individually, and combine together, has involved an anatomical peculiarity to 'disconnect' muscles from their attachment at a joint. If all the joints of the index, ring and little fingers are extended, and then held extended while the middle finger is flexed maximally at its first interphalangeal joint, the terminal joint of the middle finger becomes impossible to move by voluntary muscular effort. This phenomenon has been known to anatomists for very many years. The flexor and extensor of the joint attach also to adjacent fingers and to more proximal joints of the middle finger and, in the posture described, are held by these attachments at lengths which are inappropriate for their action on the terminal joint of the middle finger. Nevertheless, it remains easy to impose movements on the joint. Because the muscles cannot operate the joint, imposed movements do not pull upon the muscles, and proprioceptive acuity can be tested again when all receptor species (intramuscular, joint and cutaneous) can be stimulated.

After injection of local anaesthetic around the digital nerves of the middle finger, afferents from joint and cutaneous receptors are blocked, leaving only the receptors in the long flexor and extensor muscles available for proprioception (Gandevia and McCloskey, 1976; Gandevia *et al.*,1983). Injection of local anaesthetic into the joint capsule will impair joint receptors but spare cutaneous (and muscle) receptors (Ferrell *et al.*,1987). Fig. 3 illustrates some of the findings using these experimental procedures.

Proprioceptive acuity with all receptors available is superior to that when intramuscular receptors cannot contribute or when only they can contribute (Gandevia and McCloskey, 1976: Gandevia *et al.*,1983). By itself, anaesthesia of joint receptors does little to blunt acuity provided that intramuscular receptors can be stimulated (i.e. muscles 'connected') but causes significant deficits when they cannot (i.e. muscles 'disconnected') (Ferrell *et al.*,1987). This implies some redundancy of proprioceptive function between these two receptor classes. It also indicates that the cutaneous receptors in the fingers can provide only poor proprioceptive acuity, as proprioceptive performance is poor when only cutaneous receptors are available. When *both* cutaneous and joint receptors are anaesthetised, proprioceptive performance deteriorates (Gandevia and McCloskey, 1976; Gandevia *et al.*,1983). As cutaneous receptors alone probably have a limited *specific* role in proprioceptive performance, this probably indicates an important non-specific background *facilitatory* role of cutaneous receptors (i.e. facilitating inputs from specific proprioceptive inputs), at least for the fingers.

PROPRIOCEPTION IN EVERYDAY TASKS

There seems little doubt that most, or all, movements depend critically upon proprioceptive inputs. For slower movements, these may well be crucial *during* the task, while for faster ones proprioceptive input probably define start and end points, and provide *post-hoc* knowledge of results.

Simple daily tasks that seem to have a clear proprioceptive component are executed with variable accuracy. It has been shown, for example, that human subjects have a remarkable capacity to discriminate the thicknesses of objects held between finger and thumb (John, Goodwin and Darian-Smith, 1989) in circumstances in which possible contributions from non proprioceptive inputs have been carefully excluded as the basis of discrimination. From this, it might be concluded that proprioception is a sense of high acuity. However, this contrasts with the variable performance within and between subjects in simply approximating the tips of the index fingers of the two hands without looking. From this, it might be concluded that proprioception is a sense of relatively low acuity when compared, for example, to vision or hearing.

It needs to be borne in mind that tasks with a strong dependence on proprioception may not depend solely on proprioception, and may use proprioceptive cues in various ways. Touching the index fingers together without vision, for example, requires not only accuracy of signalling of the position of each finger tip independently (an accuracy that in turn depends upon accuracy of signals about positions of several joints), but a 'matching' of the calibrations of these two signals centrally, together with an accuracy of motor control on each side to achieve the movements necessary to bring the two composite position signals into register. The potential for small errors to summate in such a multi-component system is large.

The inherent instability of the inverted pendulum, which the human body becomes during normal standing, results in small sways. Many, or most, of these are reflexly corrected, and we remain unaware of them. Recent work has shown that even when vision and vestibular inputs are excluded from participation in such reflex correction, and when in addition the feet and ankles are anaesthetized by local ischaemia, reflex corrections are still able to keep the amplitudes of such small sways close to normal (Fitzpatrick, Rogers and McCloskey, in press). It is concluded that these reflex corrections are based on the discharges of sensory receptors (probably muscle spindles) in the muscles operating about the ankles.

Not all swaying of the body is unperceived, as is clear from observation of one's own sway. Essentially simple experiments reveal the capacities of various sensory systems to provide for such perceptions. For example, a subject can be strapped firmly in the standing position to an upright, and then the whole visual environment around him moved slowly by rotation about an axis co-linear with the ankles. This reveals what the visual system allows us to detect. Alternatively, the subject can be blindfolded and strapped to an immobile upright while balancing an upright mass equivalent to their own by operating a plate with axis of rotation at the ankles. The supported mass has the same mass and centre of gravity as the person him/herself. However, when it sways, the ankles rotate, but the person does not sway, and so no signal related to the sway is available from the vestibular or visual systems.

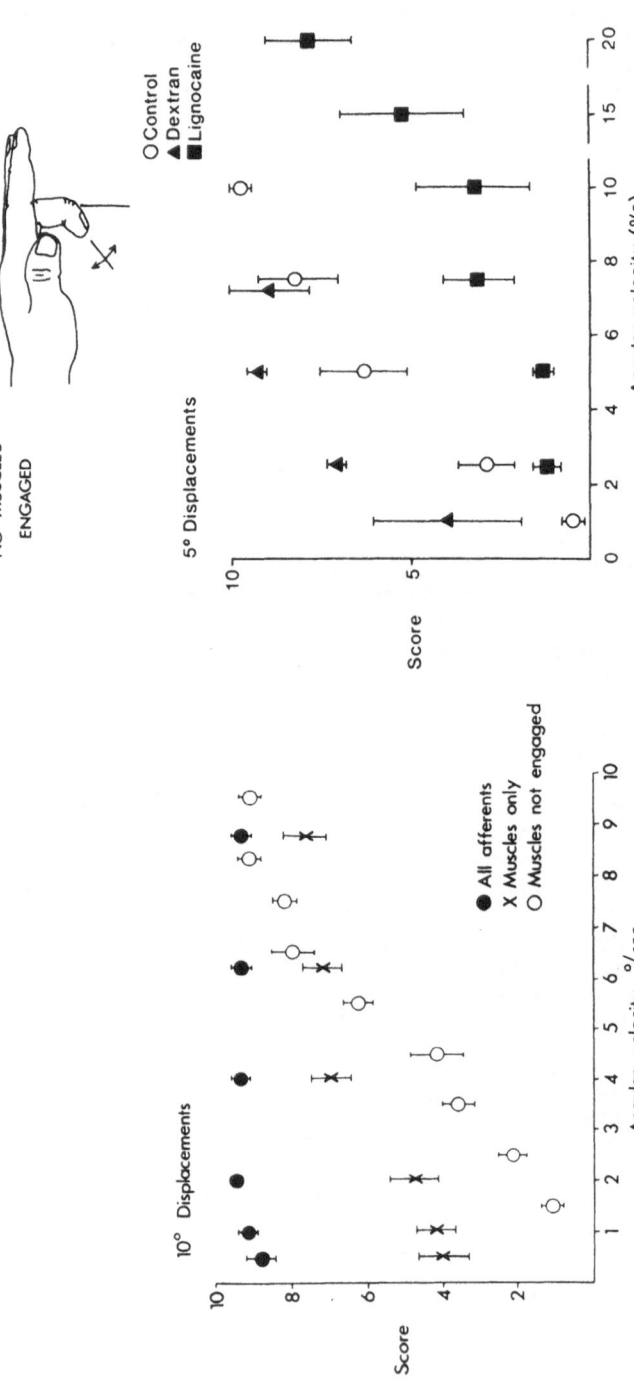

Figure 3. *Left*: Detections made by eight normal subjects of 10° displacements imposed on the terminal joint of the middle finger at various angular velocities. Each subject was presented with sets of ten flexions and ten extensions, randomly mixed, and the 'score' of detections out of ten for each was recorded: detection required correct nomination of the direction of imposed movement. This Figure shows the mean (± SEM) detection scores for subjects with all proprioceptive machinery intact (filled circles); with only muscle afferents available to contribute to kinaesthesia (open circles) Gandevia *et al.* (1983) reproduced with permission.

Right: Detection scores for subjects in whom sets of ten flexion and ten extension, randomly mixed, were imposed on the terminal joint of middle finger while the hand was postured so as to disengage effective muscular action on joint (see inset). Detection required correct nomination of direction. Two scores were obtained for each of three subjects, and means ± SEM are plotted. Circles: performance when joint and/or cutaneous afferents were available; triangles: enhancement of performance resulting from expanding joint capsule by injection of a dextran solution; squares: reduced performance resulting from anaesthesia of joint capsular receptors by intraarticular injection of lignocaine. In this condition only cutaneous afferents were available to provide proprioceptive information. Based on data in Ferrell *et al.* (1987).

Sways such as this are detected by lower limb proprioceptors. Or, finally, the subject can be blindfolded and the body is strapped to the upright of an 'L' shaped frame, and the feet to its horizontal element. The whole frame rotates about the axis of the ankles. When this frame moves, the subject inclines forwards or backwards, but the ankles do not bend, and vision is excluded. Detection of this sway depends upon vestibular signals. Using such devices, it has recently been shown that lower limb proprioceptors are alone sufficient to give a detection performance equal to that achieved during normal standing when all sensory inputs are available. Vision is similarly effective at higher velocities of sway, but allows only a poorer performance for slow sways. In contrast, however, the vestibular system alone gives extremely poor performance for detecting sways - indeed, the thresholds for detection with this system lie well outside the velocities and amplitudes of all normal sways (Fitzpatrick and McCloskey, in press). These results emphasize the centrally important role of proprioception in this everyday detection task.

CONCLUSION

Afferents from muscles, joints and perhaps from skin can contribute to the detection of the movement. Certainly each of these afferent classes will discharge in response to applied movement. Based on their documented responsiveness to very small changes in length and the proprioceptive illusions produced by their relatively selective activation, muscle spindle afferents play a major role in kinaesthesia. Given that subjects often inadvertently contract their muscles during tasks requiring proprioception, cues from muscle receptors are likely to be further improved.

REFERENCES

Amassian, V.E. and Berlin L. (1958). Early cortical projection of Group I afferents in the forelimb muscle nerves of cat. *Journal of Physiology* **143**, 61P.

Burke,D., Gandevia, S.C. and Macefield, G. (1988). Responses to passive movement of receptors in joint, skin and muscle of the human hand. *Journal of Physiology* 402:347-61.

Burke, D., Hagbarth, K-E., Löfstedt, L., and Wallin, B.G. (1976) The responses of human muscle spindle endings to vibration of non-contracting muscles. *Journal of Physiology* **261**, 673-94.

Clark, F.J., Horch, K.W. and Bach, S.M. and Larson, G.F. (1979). Contributions of cutaneous and joint receptors to static knee position sense in man. *Journal of Neurophysiology* **42**, 877-88.

Edin, B.B. and Abbs, J.H. (1991). Finger movement responses of cutaneous mechanoreceptors in the dorsal skin of the human hand. *Journal of Neurophysiology* **65**, 657-70.

Ferrell, W.R., Gandevia, S.C. and McCloskey, D.I. (1987). The role of joint receptors in human kinaesthesia when intramuscular receptors cannot contribute. *Journal of Physiology* **386**, 63-71.

Gandevia, S.C, Hall, L A. and McCloskey, D.I. and Potter, E.K. (1983). Proprioceptive sensation at the terminal joint of the middle finger. *Journal of Physiology* **355**, 507-17.

Gandevia, S.C. and McCloskey, D.I. (1976). Joint sense, and their combination as position sense measured at the distal interphalangeal joint of the middle finger. *Journal of Physiology* **260**, 387-407.

Gelfan, S. and Carter, S. (1967). Muscle sense in man. *Experimental Neurology* **18**, 496-73.

Goodwin, G. M. (1976). The sense of limb position and movement. *Exercise Sport Science Review* **4**, 87-124.

Goodwin, G. M., and McCloskey, D. I. and Matthews, P.B.C. (1972). The contribution of muscle afferent to kinaesthesia shown by vibration induced illusions of movement and by the effects of paralysing joint afferents. *Brain* **95**, 705-48.

Hulliger, M., Nordh, E. and Thelin, A-E. and Vallbo, Å.B. (1979). The responses of afferent fibres from the glabrous skin of the hand during voluntary finger movements in man. *Journal of Physiology* **291**, 233-49.

John, K.T., Goodwin, A.W. and Darian-Smith, I. (1989). Tactile discrimination of thickness. *Experimental Brain Research* **78**, 62-8.

Knibestöl, M. (1975). Stimulus response functions of slowly adapting mechanoreceptors in the human glabrous skin area. *Journal of Physiology* **243**, 63-80.

McCloskey, D.I. (1978) Kinaesthetic sensibility. *Physiological Reviews* **58**, 763-820.

McCloskey, D.I., Cross, M.J., Honner. R. and Potter, E. K. (1983). Sensory effects of pulling or vibrating exposed tendons in man. *Brain* **106**, 21-37.

Matthews, P.B.C. Muscle afferents and kinaesthesia. (1977). *British Medical Bulletin* **33**, 137-42.

Matthews, P.B.C. (1988). Proprioceptors and their contribution to somatosensory mapping: complex messages require complex processing. *Canadian Journal of Physiology and Pharmacology* **66**, 403-38.

Moberg, E. (1972) Fingers were made before forks. *Hand* **4**, 201-6.

Moberg, E. (1983) The role of cutaneous afferents in position sense, kinaesthesia and motor function of the hand. *Brain* **106**, 1-19.

Oscarsson, O. and Rosen, I. (1963). Projection to cerebral cortex of large muscle-spindle afferents in forelimb nerves of the cat. *Journal of Physiology* **169**, 924-45.

Phillips, C. G., Powell, T.P.S. and Wiesendanger, M. (1971). Projection from low threshold muscle afferents of hand and forearm to area 3a of baboon's cortex. *Journal of Physiology* **217**, 419-46.

Roll, J.P. and Vedel, J.P. (1982). Kinaesthetic role of muscle afferents in man, studied by tendon vibration and microneurography. *Experimental Brain Research* **47**, 177-90.

Sherrington, C.S. (1900) The muscular sense. In *Text-Book of Physiology*, vol. 2, ed. Schäfer, E.A., pp1002-25, Young J. Pentland, Edinburgh.

CONTRIBUTION OF JOINT AFFERENTS TO PROPRIOCEPTION AND MOTOR CONTROL

W.R. FERRELL

Institute of Physiology
University of Glasgow
G12 8QQ, Scotland, U.K.

SUMMARY

Diseased joints are often deformed, but the underlying mechanisms are unclear. Previous research indicates that group II joint afferents influence the excitability of both α and γ motoneurones of muscles acting at the joint and thus their tone. It is therefore possible that joint disease could disrupt proprioceptive feedback, thereby altering the balance of forces acting at the joint and producing deformity. Recent observations have shown that patients with rheumatoid arthritis affecting the proximal interphalangeal joint of the index finger have ~50% decrement in position sense with a clear flexion bias in their judgements of finger position. Subjects with the hypermobility syndrome (double-jointedness) also show disturbed position sense even though this group have much less joint damage than patients with rheumatoid arthritis. These findings suggest that proprioceptive feedback from group II articular afferents is altered in joint diseases. Animal experiments performed in cats decerebrated under halothane anaesthesia have shown that the excitatory influence of group II joint afferents on both α and γ motoneurones is reduced and often abolished by conditioning stimuli from group IV afferents from the same joint. Thus, the activation of joint nociceptors, which almost invariably accompanies joint disease, could significantly alter reflex effects mediated via joint proprioceptors, once again resulting in altered muscle tone. Joint deformity could therefore arise by two complimentary mechanisms - altered proprioceptive feedback and enhanced nociceptive discharge from the affected joint.

INTRODUCTION

Arthritic conditions such as rheumatoid arthritis (RA) often result in deformity of the affected joints, but the underlying mechanisms are poorly understood. Degenerative changes occurring in these joints lead to biomechanical instability, but the consistent pattern of deformities suggests that other factors may play a role. One factor which has received little attention is the extent to which the tone of muscles acting at the joint might be affected by the disease process. Wasting of the intrinsic hand muscles is a common feature which accompanies deformity of finger joints, but whether the balance of flexor and extensor tone of these muscles is affected is unknown. Previous work has shown that group II afferents arising from low threshold joint mechanoreceptors influence the excitability of both α motoneurones (Baxendale, Ferrell and Wood 1987) and γ motoneurones of muscles acting at the knee joint (Baxendale, Davey, Ellaway and Ferrell 1992). As chronic joint disease causes extensive damage to the synovial tissues containing these receptors, it is possible that such disease disrupts proprioceptive feedback, thereby altering the balance of forces acting at the joint and producing deformity. A contributory factor in the genesis of joint deformity arises

Neural Control of Movement, Edited by W.R. Ferrell
and U. Proske, Plenum Press, New York, 1995

from the activation of joint nociceptors by the inflammatory process (Schaible and Schmidt 1983; Guilbaud, Iggo and Tegner 1985) which, as part of the flexor reflex afferent (FRA) system (Eccles and Lundberg 1959), tend to enhance flexor tone. This chapter will review evidence to examine whether proprioception is altered by joint abnormalities, whether group II joint afferents reflexly influence the excitability of α and γ motoneurones and whether articular nociceptors can influence these reflex effects.

Proprioception and Joint Abnormalities

Evidence to support the hypothesis of impaired proprioceptive feedback in joint disease comes from studies of kinaesthesis at the knee joint where it was observed that patients with osteoarthritis of the knee had significantly impaired proprioceptive acuity compared to a control group (Barrack, Skinner, Cook and Haddad 1983; Barret, Cobb and Bentley 1991). More recently it has been shown that position sense at the proximal interphalangeal (PIP) joint is substantially impaired in RA patients compared to an age and sex-matched control group (Fig 1).

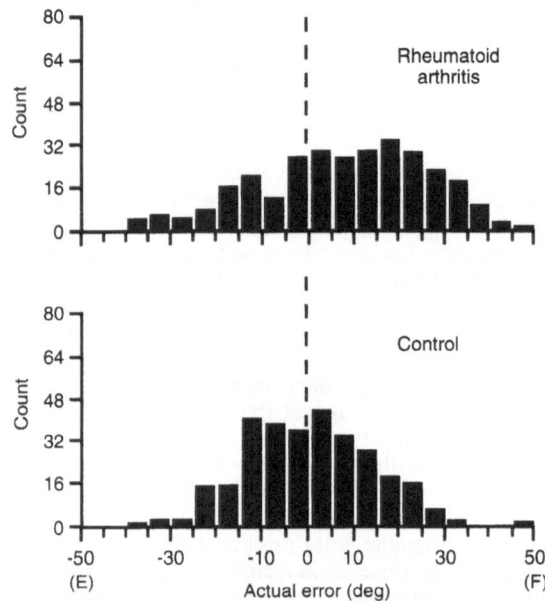

Figure 1. Histograms illustrating the distribution of matching errors in ten patients with rheumatoid arthritis affecting the proximal interphalangeal (PIP) joint compared to ten age- and sex-matched control subjects. Actual error is the difference (in degrees) between the kinaesthetically perceived and the objective position of the PIP joint of the index finger. The joint was rotated at 2°/min from the midposition to one of three randomly chosen positions within the normal range of movement. Subjects could not see the index finger and were required to align a facsimile of the finger, whose axis of rotation coincided with that of the PIP of this finger, to match the actual position of the displaced finger. For a more detailed description of the methodology, see Ferrell and Craske (1992). The arthritic group tended to perceive the PIP joint to be more flexed than it really was *i.e.* they exhibited a flexion bias, whereas the control group were more symmetrically balanced between flexion and extension errors. A total of 300 matches were obtained in each group. Reproduced, with permission, from Ferrell, Crighton and Sturrock (1992).

RA patients not only had poorer acuity of position sense but also showed a systematic flexion bias in their judgements of finger position which is of interest in view of the frequency of occurrence of flexion deformities in RA. In the RA group flexion errors were twice as common as extension errors whereas in the control group flexion and extension errors were equally balanced. Mean actual error in the RA group was 7.2±1.1 degrees (mean ±SEM) but in the control group was 0.3±0.8 degrees and these means differed significantly (P<0.00005;

n=300). Ignoring the direction of error and only taking into account the magnitude of matching error (the modular error = √[actual error^2]) showed ~50% increase in magnitude of matching error for the RA group compared to control (16.5±0.6 and 10.9±0.5 respectively), both groups again differing significantly (P<0.00005; n=300). Such distortion of position sense suggests that proprioceptive feedback from joint afferents (most likely group II afferents) may be altered in these patients.

More recent experiments have focussed on patients with the hypermobility syndrome which is characterised by excessive mobility of joints. Subjects with this condition are sometimes referred to as being "double-jointed" and are believed to show increased incidence of osteoarthritis. Position sense at the PIP joint of the index finger was tested using the same paradigm as above. The Beighton score (Beighton, Solomon and Soskolne, 1973) was used to assess the extent of hypermobility. This involved assessment of the extent of hyperextension of the knees, elbows, thumb, little finger and extent of flexion of the back.

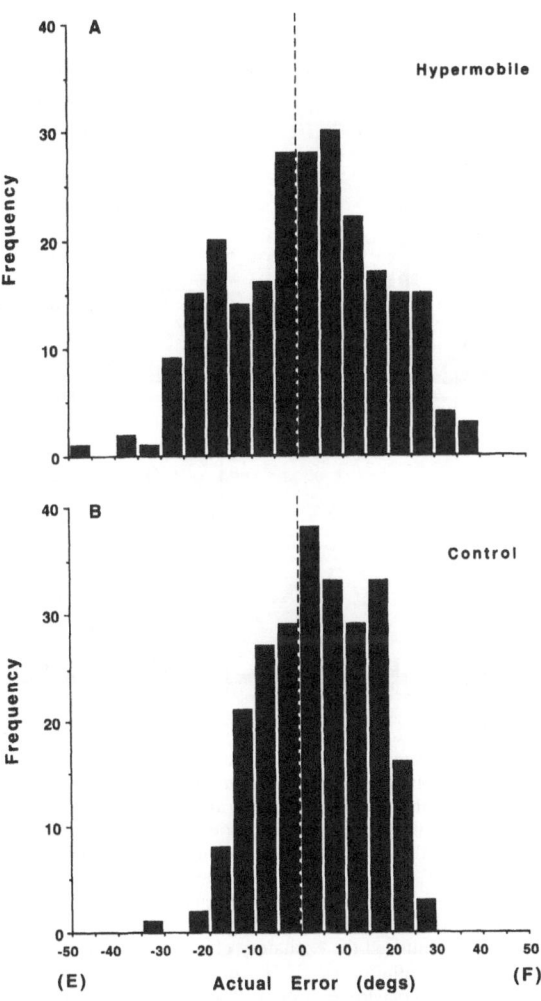

Figure 2: Actual error distribution in eight hypermobile subjects compared to eight age- and sex-matched controls. Same experimental paradigm as described in fig 1. Criteria for selection of patients with the hypermobility syndrome were as described by Beighton, Solomon and Soskolne (1973). This involves assessment of the extent of hyperextension of the knees, elbows, thumb, little finger and degree of flexion of the spine. Hypermobility at each joint scores one point and patients were classified as hypermobile if they scored 4 or more. Although actual errors show little mean difference, the spread of errors in the hypermobile group is clearly greater. Conversion of actual errors to modular errors showed significant differences between the groups (see text).

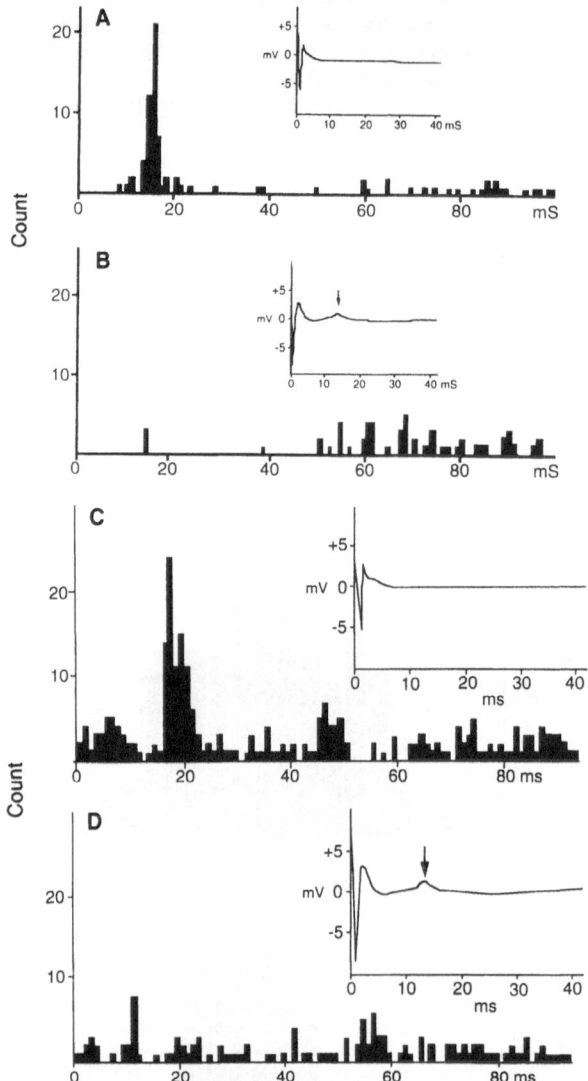

Figure 3. A: Post stimulus time histogram (PSTH) showing the effect of repetitive mechanical indentation of the cat knee joint capsule on the spontaneous discharge of a soleus α-motoneurone. A more detailed description of the methodology is available in Baxendale *et al.*(1987). The mechanical stimulus was preceded by an electrical stimulus, sufficient only to activate group II afferents, applied to the posterior articular nerve (PAN) supplying the knee joint capsule. The compound action potential recorded proximally from PAN in response to this stimulus is shown in the inset. Increased probability of discharge is observed at about 16ms, indicating an oligosynaptic pathway. **B:** Prior to each mechanical stimulus, the intensity of the electrical stimulus to PAN was increased to maximally recruit group IV joint afferents. The inset shows the compound action potential with the C wave arrowed. Such conditioning stimulation, although not suppressing the firing of the motoneurone, nevertheless eliminated the excitatory effects mediated by group II afferents. **C:** PSTH showing the effect of electrical stimulation of PAN, sufficient to activate group II afferents, on the spontaneous discharge of a soleus γ-motoneurone. A more detailed description of the methodology is available from Baxendale *et al.*,(1992). Increased probability of discharge is observed at about 19ms, indicating an oligosynaptic pathway. This test stimulus was itself preceded by another conditioning electrical stimulus, sufficient only to activate group II afferents, applied to PAN. The compound action potential recorded distally from PAN in response to the conditioning stimulus is shown in the inset. **D:** Preceding each test stimulus, a conditioning electrical stimulus, supramaximal for group IV afferents, was applied to PAN. The inset shows the compound action potential with the C wave arrowed. There is no longer any linkage between the test stimulus and the γ-motoneurone discharge, although the motoneurone continues to discharge.

The actual error histograms (fig 2) show that although the distribution of matching errors is roughly symmetrical in both groups, there is clearly a greater spread of errors in the hypermobile group and this was confirmed by calculation of median modular errors (semi-interquartile range) which showed the hypermobile group to have a value of 11.5 (7.5) whilst the control group showed a significantly (P<0.0005; n=240) smaller value of 9.7 (4.6). Thus, although patients with the hypermobility syndrome do not have overt disease affecting the PIP joint, nevertheless their position sense acuity is impaired compared to an age- and sex-matched control group, suggesting that even relatively minor joint abnormalities are associated with degraded proprioceptive feedback.

Thus, kinaesthetic disturbances occur in various forms of joint disease, suggesting that sensory feedback from proprioceptive afferents is distorted by these conditions.

Reflex Effects Arising from Joint Afferents

As it is joint receptors which are most directly affected by articular diseases, it is therefore important to establish whether joint afferents exert significant reflex effects on α and γ motoneurones. Previous work has shown that under appropriate experimental conditions, low threshold joint mechanoreceptors can potently influence the excitability of α motoneurones (Baxendale et al.,1987; Ferrell, Rosenberg, Baxendale, Halliday and Wood 1990) and this is illustrated in fig 3A. In addition, it has been shown that electrical stimulation of group II articular afferents affects γ motoneurones (Johansson, Sojlander and Sojka, 1986; Baxendale et al.,1992). Although some of these effects are inhibitory, the majority are excitatory, as illustrated in fig 3C. In view of these reflex effects, there is the distinct possibility that distorted sensory feedback from group II joint afferents could contribute reflexly to alterations in muscle tone and thereby lead to joint deformity.

However, it is possible in addition to this mechanism, there may also be a contribution from articular nociceptors. It has been known for some time that electrical stimulation of high threshold joint afferents results in flexor excitation and extensor inhibition and it was suggested that these were part of the flexor reflex afferent (FRA) system (Eccles and Lundberg 1959). Group III and IV articular afferents are relatively quiescent in the normal joint as long as noxious stimuli are not applied, but in an acutely inflamed joint become much more active, even within the non-noxious range of movement (Schaible and Schmidt 1983; Guilbaud et al.,1985). Direct evidence for joint inflammation enhancing flexor motoneurone activity comes from experiments where it was shown that flexion reflex excitability in decerebrate, low spinal cats is increased in joints acutely inflamed by intra-articular injection of carrageenan and kaolin (Ferrell, Wood and Baxendale 1988). Activity of flexor α-motoneurones to knee joint rotation is enhanced by the inflammatory process (Ferrell et al., 1988; He, Proske, Schaible and Schmidt 1988) with knee flexor γ-motoneurones showing even more enhancement (He et al.,1988). Sustained activation of joint nociceptive afferents by inflammatory joint disease would therefore tend to promote flexion bias and could directly contribute to flexion deformity of the joint.

However, these nociceptive articular afferents could also exert effects by altering the effectiveness excitatory reflexes mediated by group II joint afferents. To further elucidate the rôle of joint nociceptors in controlling muscle tension, experiments were performed in cats decerebrated under halothane anaesthesia which showed that the excitatory influence of group II joint afferents on both α and γ motoneurones is reduced and often abolished by conditioning stimuli from group III/IV afferents from the same joint (fig 3). Thus, the activation of joint nociceptors, which almost invariably accompanies joint disease, could significantly alter reflex effects mediated via joint proprioceptors, once again resulting in altered muscle tone. Joint deformity could therefore arise by two complimentary mechanisms - altered proprioceptive feedback due to chronic mechanical disturbance of the joint capsule and enhanced nociceptive discharge from the acutely inflamed joint.

CONCLUSIONS

These studies indicate that joint receptors contribute both to proprioceptive sensations and to the control of the musculature and suggest that the sensory innervation of the joint has an important rôle to play in the maintenance of the integrity of the joint. Proprioceptive joint afferents may have a protective function, perhaps preventing hyperextension and hyperflexion of the joint (Baxendale and Ferrell 1981). Joint nociceptors could also be ascribed a protective function as they limit movement of the inflamed and painful joint, but it is also clear that by reflexly altering the tone of muscles acting at the joint they could contribute to joint deformity and thus have detrimental effects.

ACKNOWLEDGEMENTS. The recent work described in this paper was performed in conjunction with the following individuals to whom the author is indebted: R.H. Baxendale, A. Crighton, A.K. Mallik, D.T. Scott and R.D. Sturrock. The work was supported by the MacFeat Bequest of the University of Glasgow, the Wellcome Trust and the Medical Research Council.

REFERENCES

Barrack R.L., Skinner H.B., Cook S.D. and Haddad R.J. (1983). Effect of articular disease and total knee arthroplasty on knee joint-position sense. *Journal of Neurophysiology* **50,** 684-687.

Barret D.S., Cobb A.G. and Bentley G. Joint proprioception in normal, osteoarthritic and replaced knees. *Journal of Bone and Joint Surgery* **73-B,** 53-57.

Baxendale R.H., Davey N.J., Ellaway P.H. and Ferrell W.R. (1992). The interaction between joint and cutaneous afferent input in the regulation of fusimotor neurone discharge. In: *Muscle afferents and spinal control of movement* eds Jami L., Pierrot-Deseilligny E., and Zytnicki D. ch. 1, 95-104. Pergamon Presss, Oxford.

Baxendale R.H. and Ferrell W.R. (1981). The effect of knee joint afferent discharge on transmission in flexion reflex pathways in decerebrate cats. *Journal of Physiology* **315,** 231-242.

Baxendale R.H., Ferrell W.R. and Wood L. (1987). The effect of mechanical stimulation of knee joint afferents on quadriceps motor unit activity in the decerebrate cat. *Brain Research* **415,** 353-356.

Beighton P., Solomon L. and Soskolne C.L. (1973). Articular mobility in an African population. *Annals of the Rheumatic Diseases* **32,** 413-418.

Eccles R.M. and Lundberg A. (1959). Supraspinal control of interneurones mediating spinal reflexes. *Journal of Physiology* **147,** 565-584.

Ferrell W.R. and Craske B. (1992). Contribution of joint and muscle afferents to position sense at the human proximal interphalangeal joint. *Experimental Physiology* **77,** 331-342.

Ferrell W.R., Crighton A. and Sturrock R.D. (1992). Position sense at the proximal interphalangeal joint is distorted in patients with rheumatoid arthritis of finger joints. *Experimental Physiology* **77,** 675-680.

Ferrell W.R., Rosenberg J.R., Baxendale R.H., Halliday D.M. and Wood L. (1990). Fourier analysis of the relation between the discharge of quadriceps motor units and periodic mechanical stimulation of cat knee joint receptors. *Experimental Physiology* **75,** 739-750.

Ferrell W.R., Wood L. & Baxendale R.H. (1988). The effect of acute joint inflammation on flexion reflex excitability in the decerebrate cat. *Quarterly Journal of Experimental Physiology* **73,** 95-102.

Guilbaud G., Iggo A. and Tegner R. (1985). Sensory receptors in ankle joint capsules of normal and arthritic rats. *Experimental Brain Research* **58,** 29-40.

He X., Proske U., Schaible H-G and Schmidt R.F. (1988). Acute inflammation of the knee joint in the cat alters responses of flexor motoneurons to leg movements. *Journal of Neurophysiology* **59,** 326-340.

Johansson H., Sjolander P. and Sojka P. (1986). Actions on γ-motoneurones elicited by electrical stimulation of joint afferent fibres in the hindlimb of the cat. *Journal of Physiology* **375,** 137-152.

Schaible H-G and Schmidt R.F. (1983). Effects of an experimental arthritis on the sensory properties of fine articular afferents. *Journal of Neurophysiology* **49,** 1118-1126.

ASSESSING ACCURACY OF POSITIONING JOINTS AND LIMBS

F. J. CLARK

Department of Physiology
University of Nebraska College of Medicine
Omaha, Nebraska, 68198-4575, USA

SUMMARY

The accuracy of positioning a joint or a limb is usually specified in terms of the mean error between positions of the target and the matches, a measure well suited to reveal any bias or offset in the calibration of position. However, position calibration is labile; it can drift over time and it is readily adjusted by the nervous system. As a result, measures of calibration, which is what a mean-error accuracy metric provides, however useful, do not tell us much about the exactness of position sense. To formulate a metric for assessing the exactness of proprioception, one can treat a position matching task as an information processing exercise, and use information theory to devise a well-defined criterion for exactness. Noise is what limits information transfer in any information processing system, so the metric would necessarily incorporate the effects of noise, which is evident in position matching tasks as matching error variance. In this report, I describe one such metric called "target resolution", which is based on variance and which specifies the maximum number of discrete, equally spaced targets that a subject could discriminate within a span. Measures of target resolutions for various joints of the arm and fingers show surprisingly low resolutions; in some cases subjects could discern only that a joint was in the lower, middle or upper portion of its range.

INTRODUCTION

To set the scene for this presentation, consider hypothetical results from a simple position matching task using one target, as portrayed in Fig. 1, which shows three examples of mean matching errors ± 1 SD. If we compare A and B in Fig. 1 and ask which was the more "accurate", by the usual mean-error criterion of accuracy, A would prevail. Now, suppose Fig. 1 came not from a test of proprioception but were outcomes from a shooting match on a rifle range. By a mean-error criterion, A would still be more accurate than B, though we might agree that B was really the better shooter. B could easily achieve the outcome shown in C simply by adjusting the sight on his rifle. Were we now to compare A and C in Fig. 1 and ask which was the more accurate, again, gauged solely by a criterion of mean error, they would have equivalent accuracies. However, in the shooting match, unless the target was very large, B would hit the bull's eye more often than A and likely win the day.

The shooting analogy applies also to proprioception. The outcome for shooter B reflects an offset or bias in his aim relative to the bull's eye that could be corrected by adjusting the rifle sight. Similarly for proprioception, mean error represents an offset or bias in calibration that can be adjusted by the nervous system. Proprioceptive calibration is labile. Calibration as measured by mean error can drift over time (Wann and Ibrahim, 1992), it can vary according to previous positions of the joint or limb (aftereffects) (Craske and Cranshaw, 1974; Howard and Anstis, 1974), and most important, calibration is subject to learning (Redding and Wallace, 1990). An accuracy measure based upon mean error tells the current status of this calibration and reveals any offset or bias (Poulton, 1979) that might be present, and one may well be vitally interested in this calibration, depending on the question being asked. However, a labile, plastic calibration could hardly represent a stable, defining characteristic of the proprioceptive mechanism.

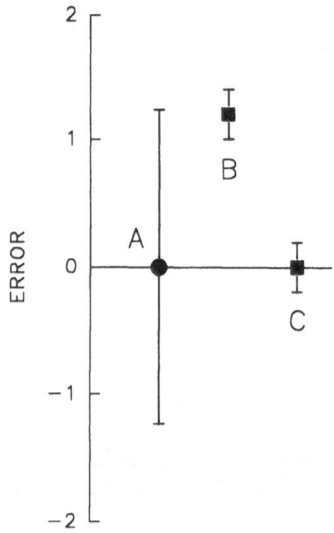

Figure 1. Matching errors from a hypothetical task. Means ± 1 SD.

Error variance can provide a defining measure of the innate capabilities of the proprioceptive mechanism. Error variance reflects the effects of noise that is inherent in every information communication system, including the proprioceptive system, and which can never be entirely eliminated. It is noise that limits information transfer in any communication system (Shannon and Weaver, 1964) and that ultimately limits proprioceptive ability. Matching error variance gives a measure of noise effects, so it seems reasonable to use variance in some form as a basic criterion for proprioceptive performance. Error variance has been incorporated in many studies to establish significance in statistical comparisons of the means, or in some cases to provide an adjunct measure of discriminibility (Gescheider, 1976, p. 34). Nonetheless, accuracy computed from the mean error between indicated positions and true positions of the joint or limb remains the primary criterion for proprioceptive performance.

One can turn to Information Theory (Shannon and Weaver, 1964) and the elegant models of auditory intensity perception developed by Durlach and his colleagues (Durlach and Braida, 1969; Braida and Durlach, 1972) to develop a convenient metric for assessing proprioceptive performance based on variance. I use a metric I call "Target Resolution", defined as the fewest number of discrete, equally spaced targets within a range that will provide the maximum possible information transfer from the target set. Target resolution is computed from matching error variance (see Garner and McGill, 1956). The derivation that follows will help clarify the character of this metric.

METHODS

Consider a simple joint angle matching task with n discrete, equally spaced target angles presented in randomized order using an equal number of presentations per target. Though the targets are discrete, matches vary on a continuum. Fig. 2 shows distributions of matches at each of the n targets. One first needs to compute an average amount of information received by the subject after presentation of the full set of targets. The calculations are made using a stimulus-response matrix (Garner and Hake, 1951), which is a square array with a row for each of the n targets and a column for each of n match categories. Matches, which vary along a continuum, are assigned to categories by "binning" (Shannon and Weaver, 1964), with the bin boundaries set halfway between the mean values of matches at adjacent targets. Thus, the stimulus-response matrix gives an orderly listing of how many times a subject responded to each target by matching in category 1 or in category 2, etc.

From the stimulus-response matrix, one computes an average amount of "uncertainty" (U) in the response set which is expressed in BITS, the standard unit of information transfer (Garner, 1962). The maximum uncertainty possible in the matrix would occur when the count in each cell was the same, as might happen if the subject received no information from presentation of the targets and simply guessed (in a very unbiased manner) at a match for each target. This maximum uncertainty is calculated as: $U_{max} = \log_2(n)$ bits. On the opposite side, if all the matches for target 1 fell into response category 1, and matches for target 2 into response category 2, etc., the average uncertainty in this response set would be zero. In most cases, subjects "miss" a few targets and thus the uncertainty falls between these extremes. If we knew the maximum uncertainty that could exist if the subject received no information from the target set, and also the uncertainty that existed after presentation of the targets, the difference would be the amount of information the subject received from the target set. For example, with 8 targets, the maximum uncertainty (or the maximum amount of information that could be transmitted) would be 3.0 bits. Now, if the average uncertainty remaining in the response set after presentation of the targets computed to 1.3 bits, the information received would have been 3.0 - 1.3 = 1.7 bits.

The target resolution metric specifies the fewest number of targets needed to transfer the maximum amount of information that can be received by the system. Let us examine how information transfer varies when one progressively increases the number of targets, n, keeping range, m, and variance, σ^2, fixed (see Fig. 2). Simulations, verified by actual data, show that as n is increased from a small value (e.g. 2), information transfer increases at first but then reaches a plateau. The value of the plateau (which is related to the variance) indicates the maximum amount of information that can be transferred irrespective of the number of targets used, and the value of n where the curve begins to plateau represents the fewest number of targets that will give this maximum information transfer. This latter value of n would be the target resolution though further analysis is necessary to derive an equation for n.

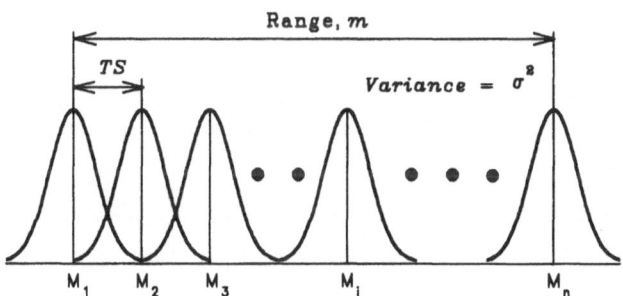

Figure 2. Distributions of matches at each of n discrete targets. Distributions are assumed normal with means M_1 through M_n and variance σ^2, the same for each target. S is the perceived target spacing, and m is the range measured along the response continuum.

The plateau in information transferred as the number of targets is increased and the target resolution idea are akin to the familiar "channel capacity" concept (Garner, 1962). However, I wish to avoid the term "channel capacity" because that concept was developed using the Absolute Judgement paradigm (Garner and Hake, 1951) which itself appears to limit information transfer, giving rise to Miller's "magic number" of 7±2 stimulus categories as an upper limit for category resolution, irrespective of stimulus modality, number of targets, target spacing, target range or training (Miller, 1956). If subjects are allowed to indicate targets by pointing or matching, the number of resolvable targets becomes dependent on target spacing, and no upper limit, like the 7±2 limit, appears, though many positioning tasks do show resolutions much less than 7 (see below, also Durlach, et al., 1989; Sackitt et al., 1983). At issue here is not that many, or even most tasks reveal low resolutions, but whether there exists some global upper limit on resolution that is largely independent of target spacing.

Examination of Fig. 2 indicates that the number of targets one could gainfully fit into a fixed range is limited by the target spacing relative to the variance. As the number of targets increases, target spacing must decrease, producing more "misses" as adjacent matching distributions overlap, causing more matches for any target to fall into the response categories for adjacent targets. Analysis of the relationship between target spacing (S) and variance σ^2 in simulations reveals that for all combinations of range and number of targets, there exists a single optimal relationship between target spacing and variance, which is conveniently expressed as a dimensionless ratio of target spacing to the square root of the variance: S/σ = 4.93. Spacing targets wider than optimal to increase the ratio does not increase information transfer, though spacing targets closer to decrease the ratio decreases information transfer.

RESULTS AND CONCLUSIONS

One can now write an expression for target resolution. Note from Fig. 2, that for any range, m, and target spacing, S, the number of targets, n is given by equation 1. By replacing S with its optimal value of 4.93 x σ, one can obtain a value for the optimal number of targets n_{opt} expressed in terms of the range, m, and square root of the variance, σ (eqn 2). The associated information transfer, I, is given by equation 3.

$$n = 1 + \frac{m}{S} \quad (1) \qquad\qquad n_{opt} = 1 + \frac{m}{\sigma \times 4.93} \quad (2)$$

$$I = \log 2\ (nopt) \qquad = \qquad \log_2 (1 + \frac{m}{\sigma \times 4.93}) \quad (3)$$

Let us now examine example target resolutions for the finger, wrist, elbow and shoulder with two types of targets (Fig. 3): kinaesthetic - where subjects matched the position of a joint on the left side using the corresponding joint on the right side, or visual - where subjects aligned a pointer (the index finger) to visual targets moving only the indicated joint. In all cases the limb itself was blocked from view. A striking feature of these findings is the low resolution one sees, apart from the shoulder. A useful though non-rigorous interpretation of target resolution is the number of categories a subject resolved without error. With the fingers and wrist, subjects could resolve only about 3 target categories; they could distinguish with certainty only that the joint was in the lower, middle or upper third of the range.

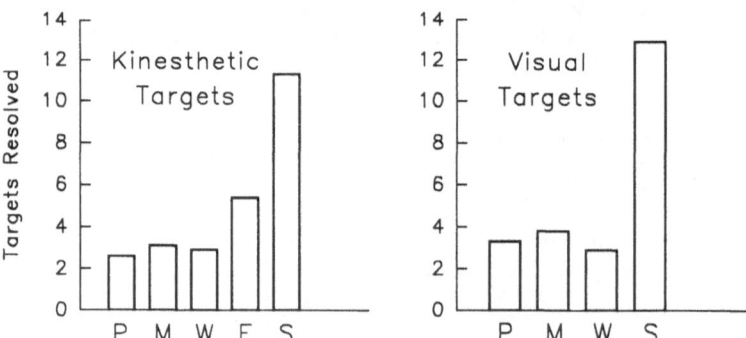

Figure 3. Number of resolvable targets for different joints using kinaesthetic or visual matching. Legend: P = PIP, M = MCP (index finger), W = Wrist, E = Elbow, S = Shoulder.

Resolution with the shoulder appeared much greater than with the fingers, wrist and elbow. This is puzzling considering that subjects had to position the finger, wrist and elbow joints as well as the shoulder to point to a target. How could joints that show low resolution when tested individually do so well in combination (even accounting for different lever arms)? Perhaps when pointing with the shoulder, subjects held the other joints in a position they sensed with unusual precision, like full extension. However, when subjects were asked to point to visual targets where they bent the arm a different way for each trial, they did as well as with the arm straight!

REFERENCES

Braida L.D. and Durlach N.I. (1972). Intensity perception. II. Resolution in one-interval paradigms. *Journal of the Acoustical Society of America* **51,** 372-383.

Craske B. and Cranshaw M. (1974). Differential errors of kinesthesis produced by previous limb positions. *Journal of Motor Behavior* **6,** 273-278.

Durlach N.I. and Braida L.D. (1969). Intensity perception. I. Preliminary theory of intensity resolution. *Journal of the Acoustical Society of America* **46,** 372-383.

Durlach N.I., Delhorne L.A., Wong A., Ko W.Y., Rabinowitz W.M. and Hollerbach J. (1989). Manual discrimination and identification of length by the finger span method. *Perception and Psychophysics* **46,** 29-38.

Garner W.R. (1962). *Uncertainty and Structure as Psychological Concepts.* Wiley:New York.

Garner W.R. and Hake H.W. (1951). The amount of information in absolute judgements. *Psychological Review* **58,** 446-459.

Garner W.R. and McGill W.J. (1956). The relation between information and variance analyses. *Psychometrika* **21,** 219-228.

Gescheider G.A. (1976). *Psychophysics: Method and Theory,* Lawrence Erlbaum Associates: Hillsdale, New Jersey.

Howard I.P. and Anstis T. (1974). Muscular and joint-receptor components in postural persistence. *Journal of Experimental Psychology* **103,** 167-170.

Miller G.A. (1956). The magic number seven, plus or minus two: some limits on our capacity for processing information. *Psychological Review* **63,** 81-97.

Poulton E.C. (1979). Models for biases in judging sensory magnitude. *Psychological Bulletin* **86,** 777-803.

Redding G.M. and Wallace B. (1990). Effects on prism adaptation of duration and timing of visual feedback during pointing. *Journal of Motor Behavior* **22,** 209-224.

Sackitt B., Lestienne F. and Zeffiro A. (1983). the information transmitted at final position in visually triggered forearm movements. *Biological Cybernetics* **46,** 111-118.

Shannon C.E. and Weaver W. (1964). *The Mathematical Theory of Communication.* University of Illinois Press: Urbana, Illinois.

Wann J.P. and Ibrahim S.F. (1992). Does limb proprioception drift? *Experimental Brain Research* **91,** 162-166.

THE BEHAVIOUR OF CUTANEOUS AND JOINT AFFERENTS IN THE HUMAN HAND DURING FINGER MOVEMENTS

V. G. MACEFIELD

Prince of Wales Medical Research Institute
Randwick, New South Wales 2031, Australia

SUMMARY

Movements of the fingers cause strains in the skin and joints, exciting mechanoreceptors within these tissues. Although both cutaneous and joint afferents have the capacity to encode changes in joint angle, microneurographic recordings from single afferents in the median and ulnar nerves of human subjects have uncovered some limitations in their information content. Afferents in the glabrous skin of the fingers and palmar aspects of the interphalangeal joints respond primarily towards the limits of rotation during free movements of the fingers. Conversely, afferents from the hairy skin on the back of the hand, and afferents related to the metacarpophalangeal joints, respond to movements throughout the physiological range, offering these afferents potentially important roles in proprioception and motor control. However, the signalling capacities of cutaneous and joint afferents from the palmar aspect of the hand may increase during manipulation, when strain forces in the skin and joint tissues are higher than during free (unloaded) movements. For instance, when pulling forces are applied to an object that is gripped between finger and thumb, both cutaneous and joint afferents in these digits can respond to the imposed load forces - operating well within the physiological range of joint angles.

INTRODUCTION

As the foregoing chapters have shown, a great deal of research has been directed to working out the mechanisms of action of the fusimotor system and its place in the grand schemes of motor control and proprioception. Muscle spindle afferents, and tendon-organ afferents, are undoubtedly important, but one needs to keep in mind the fact that in the real world they are never activated in isolation: movements produced either by muscular contraction or external forces induce tensile and compressive strains in the skin and joints which can excite mechanoreceptors within these tissues.

Relatively little work has been done on the behaviour of cutaneous and joint afferents during the kinds of movements in which the hand is engaged in normal life, and it is worth pointing out that rarely are movements of the hand passive. What is known about these receptors has been learnt through the technique of microneurography, introduced by Vallbo and Hagbarth (1968) as a means of recording the discharge of single primary afferents, via tungsten microelectrodes inserted percutaneously into a peripheral nerve, in awake human subjects. With microstimulation, an extension of microneurography, one can also selectively stimulate the same sensory axon through the microelectrode, and from this we know that single cutaneous afferents (Torebjörk, Vallbo and Ochoa, 1987) and joint afferents (Macefield, Gandevia and Burke, 1990), but not single muscle afferents (Macefield *et al.*, 1990), can project to perceptual levels.

Neural Control of Movement, Edited by W.R. Ferrell
and U. Proske, Plenum Press, New York, 1995

Cutaneous Afferents from the Palmar Aspect of the Hand

Microelectrode recordings from the median and ulnar nerves have demonstrated the existence of four classes of low-threshold mechanosensitive afferents in the glabrous skin of the human hand: the fast adapting types FA I and FA II and the slowly adapting types SA I and SA II. Quantitative studies by Johansson and Vallbo (1979) showed that each of these classes has a different mean threshold to forces normal to the skin, with the FA II afferents, innervating Pacinian corpuscles, having the lowest thresholds, and the SA II afferents, innervating Ruffini endings, the highest. Each class of cutaneous afferent is also differentially responsive to forces in the plane of the skin, the most sensitive afferents to tensile strain being the SA II afferents: the Ruffini endings are anchored to the longitudinal bundles of collagen within the dermis, providing an optimal mechanical arrangement for transducing tensile strain (Johansson and Vallbo, 1983).

It is the tensile strain produced by joint rotations, rather than forces normal to the skin, that is primarily responsible for the discharge behaviour of cutaneous afferents during finger movements. Quantitative observations on afferent responsiveness to finger movements were first made by Hulliger, Nordh, Thelin and Vallbo (1979), who found that many afferents from the glabrous skin of the hand responded to voluntary isotonic movements. Rapidly-adapting afferents tended to fire in either direction of rotation, that is they were sensitive to movement *per se*, whereas the slowly-adapting afferents often displayed directional sensitivity and hence some capacity to encode angular position. Specifically, it was those receptors with large receptive fields, the Pacinian corpuscles (FA II afferents) and Ruffini endings (SA II afferents), that showed the greatest sensitivity to movements of the fingers, with 81% of Ruffini endings also exhibiting static position sensitivity. However, relatively few of these SA II afferents responded throughout the physiological range of joint rotation, most being recruited towards the limits of rotation.

Similar behaviour was found during passive finger movements (Burke, Gandevia and Macefield, 1988). Fig. 1 shows the response of an SA I afferent to passive flexion and extension movements. The receptor, located over the proximal interphalangeal joint of the index finger, responded only towards the limits of rotation about this joint. The same study also demonstrated that approximately one quarter of the cutaneous afferents responded to passive movements about more than one axis of joint rotation (e.g. abduction-adduction as well as flexion-extension of the metacarpophalangeal joint), a feature that could add ambiguity to the signalling capacities of these afferents.

Cutaneous Afferents from the Dorsal Aspect of the Hand

It should be noted that the skin on the palmar surface of the hand is mechanically different to that on the back of the hand. Unlike glabrous skin, the hairy skin is only loosely connected to the subcutaneous tissues, thereby allowing greater stretch and presumably greater activation of stretch-sensitive cutaneous afferents. Indeed, Edin and Abbs (1991) and Edin (1992) have recently assessed the responsiveness of cutaneous afferents on the dorsum of the hand, recording from these through a microelectrode in the superficial branch of the radial nerve. There do appear to be important differences in the behaviour of afferents in the hairy skin and in the glabrous skin. For instance, 92% of the afferents on the back of the hand responded to finger movements (Edin and Abbs, 1991), compared to only 68-77% of afferents on the front of the hand (Hulliger *et al.*, 1979; Burke *et al.*, 1988).

Unlike afferents from the glabrous skin of the hand, cutaneous receptors on the dorsum mostly responded *throughout* the physiological range of joint rotation, increasing their firing rate in flexion as the skin was stretched. Whereas the rapidly-adapting afferents with small receptive fields, the FA I afferents, responded exclusively to movements of the joint over which they were located, both classes of slowly-adapting afferent, SA I as well as SA II, were very sensitive to the skin stretch associated with finger movements - whether this was produced by rotation of the nearest joint or by movements of remote digital joints. The static sensitivity to stretch was quite high for both classes of afferent (Edin, 1992), whereas in the glabrous skin of the hand very few SA I afferents displayed this property (Hulliger *et al.*, 1979). Measured at an equivalent joint angle, the static sensitivity of the SA I and SA II afferents on the dorsum of the hand was comparable to that of muscle spindle endings in the long extensors of the fingers, 0.2-0.5 Hz per degree of rotation of the metacarpophalangeal joint (Edin and Vallbo, 1990; Edin, 1992).

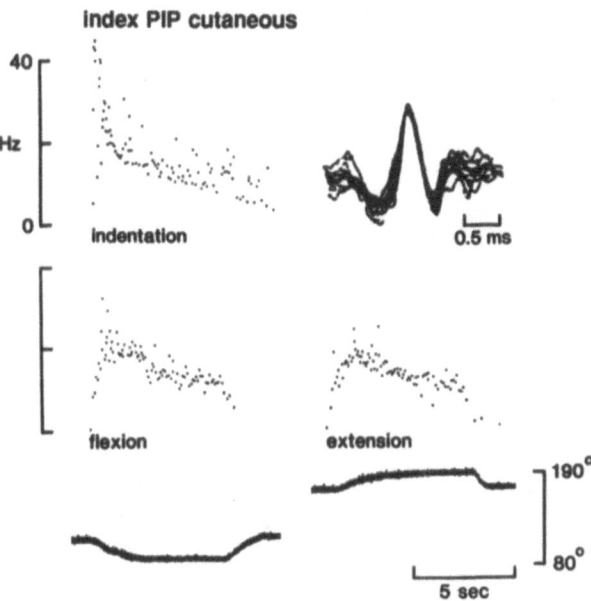

Index PIP cutaneous

Figure 1. Responses of a slowly-adapting type SA I cutaneous afferent to indentation over its receptive field on the proximal interphalangeal joint of the index finger and to passive flexion and extension of the joint. The goniometer records in the bottom panels indicate that the afferent responded to the passive movements only at the limits of the physiological range of joint rotation. Superimposed action potentials in the top right panel indicate the unitary integrity of the recording.

Joint Afferents from the Palmar Aspect of the Hand

Two microneurography studies have detailed the properties of a total of 23 deep receptors associated with the interphalangeal and metacarpophalangeal joints of the human hand, 21 of which were recorded from cutaneous fascicles and two from motor fascicles of the median and ulnar nerves (Burke *et al.*, 1988; Macefield *et al.*, 1990). Joint afferents were defined as such if they had discrete receptive fields over a digital joint and failed to respond to stroking or lifting the skin over the joint and to deep probing of non-articular bone or of the intrinsic muscles (Burke *et al.*, 1988; Macefield *et al.*, 1990). The Ruffini ending is the major class of mechanoreceptor in these joints (Stilwell, 1957; Sathian and Devanandan, 1983), which largely accounts for their overall discharge properties.

Interphalangeal and metacarpophalangeal joint afferents have very high mechanical thresholds to pressure over their receptive fields on the front and sides of the digits, often requiring a rigid probe (such as a pencil) to activate them in a discrete area. Fig. 2 shows the responses of an interphalangeal joint afferent to deep pressure and passive flexion and extension movements. The afferent was silent in the position of rest (but about a third of the sample were tonically active) and during much of the physiological range of joint rotation: it was recruited only towards the limits of flexion and extension, but its greatest response was to flexion, and microstimulation of the afferent with brief trains of constant frequency evoked a frequency-dependent perception of joint flexion. This pattern of responding only towards the limits of the physiological range was characteristic of the majority of joint afferents: only four of the 23 afferents sampled (17%) responded *throughout* the range, and of these only two increased their firing rate in one direction of rotation and could thereby have provided a signal related to absolute joint angle (McCloskey, Macefield, Gandevia and Burke, 1987; Burke *et al.*, 1988). In addition to this overall poor sensitivity to joint movements, the majority of joint afferents responded in both directions of angular excursion and in two or three axes of joint rotation, contributing ambiguity to any proprioceptive signals these afferents may provide. For sake of comparison, spindle afferents from the intrinsic muscles of the hand respond unidirectionally throughout the physiological range, but the majority also possess multi-axial movement sensitivity (Burke *et al.*, 1988; Macefield *et al.*, 1990). With respect to the responsiveness of joint afferents during *active* finger movements, the only observations come from Hulliger *et al.*,(1979), who found two deep receptors associated with the interphalangeal joints that were recruited only at the limits of flexion or extension.

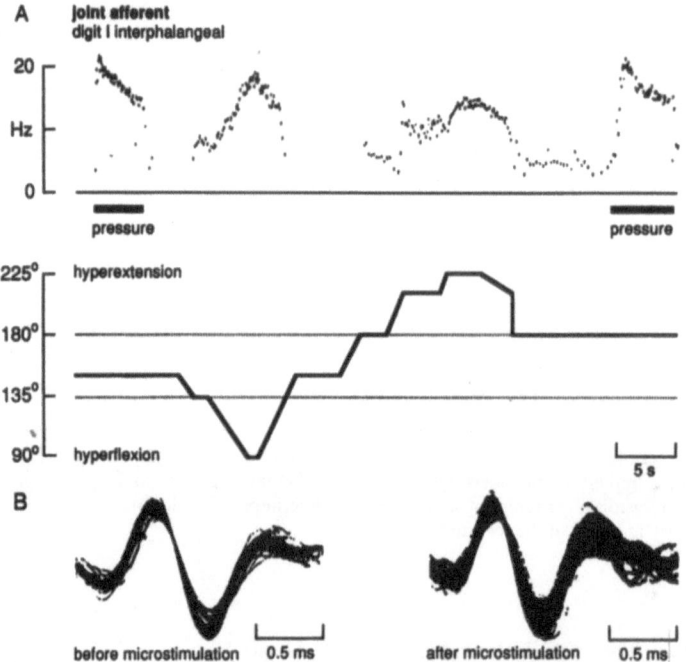

Figure 2. **A:** responses of a joint afferent associated with the interphalangeal joint of the thumb to pressure over its receptive field and to passive flexion and extension of the joint. The broken horizontal lines associated with the schematic representation of the changes in joint angle indicate the range over which the afferent did not respond to movement. **B:** superimposed action potentials indicate that the recording site remained stable after the afferent was electrically stimulated through the microelectrode. (Figure reproduced with permission from the Journal of Physiology).

Are the above findings representative of afferents in other human joints? This will be considered further below, but I recently happened upon a joint afferent from a toe whilst recording in the peroneal nerve (Fig. 3). This afferent, found within the motor fascicle supplying extensor digitorum brevis, was spontaneously active (with a very regular discharge characteristic of Ruffini endings) and responded to firm pressure over the metatarsophalangeal joint of the fourth toe. It responded to passive movements in a manner similar to that of joint receptors in the fingers, that is towards the limits of rotation. In Fig. 3 the afferent is shown responding in both directions to passive movements about the flexion-extension axis and about the longitudinal axis of the toe.

Joint Afferents from the Dorsal Aspect of the Hand

Microelectrode recordings from the superficial (cutaneous) branch of the radial nerve revealed a population of afferents that had small receptive fields on the dorsal surface of the metacarpophalangeal joints (Edin, 1990). These receptors (n=8) had much lower mechanical thresholds than the joint afferents related to the palmar aspects of the joints, and there are reasons to believe that joint afferents in the radial nerve and those in the median and ulnar nerves belong to two different populations of receptors. For instance, the joint-related receptors on the dorsal aspects of the joints all responded *throughout* the physiological range of joint rotation, rather than just towards the extremes. Moreover, their response to passive movements was unidirectional: their static discharge increased towards flexion and decreased towards extension, so these afferents could provide signals related to absolute joint position. None of the afferents responded to extorsion-intorsion of the joints, whereas many afferents on the palmar aspects of the joints did so. During *active* movements of the fingers, most of the joint-related afferents within the radial nerve responded either exclusively to *extension*, or to both extension *and* flexion, partly due to the close association of some of these receptors to the extensor tendons (Edin, 1990).

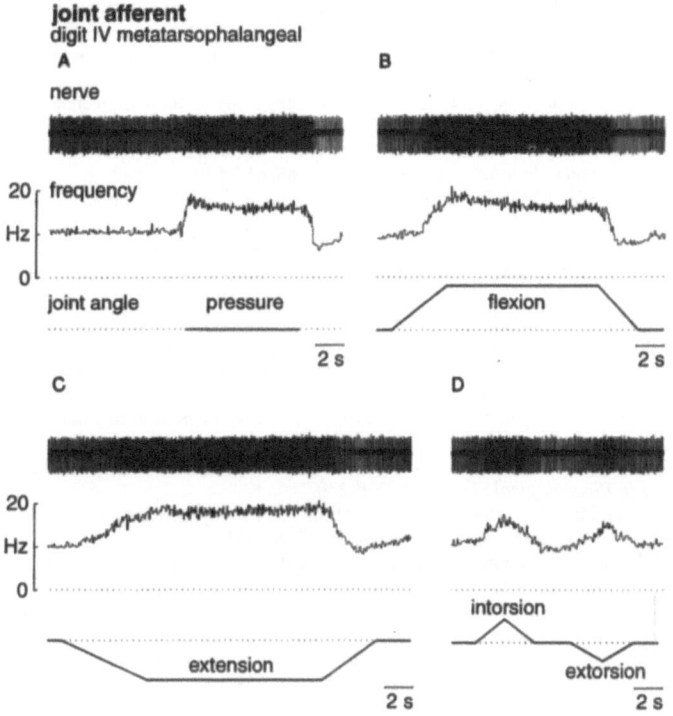

Figure 3. Responses of a joint afferent associated with the metatarsophalangeal joint of the fourth toe to pressure over its receptive field (**A**) and to passive flexion (**B**), extension (**C**), and longitudinal rotation (**D**) of the joint. Changes in joint angle are represented schematically. As for joint afferents recorded from the interphalangeal and metacarpophalangeal joints of the hand, this afferent responded only towards the limits of joint rotation.

CONCLUSIONS

Cutaneous afferents in the glabrous skin of the hand, and digital joint afferents associated with the palmar aspect of the hand, respond mostly towards the limits of joint rotation during passive or active movements of the fingers, being silent throughout much of the physiological range. Conversely, cutaneous and joint-related afferents on the back of the hand respond to movements throughout the physiological range: in particular, both classes of slowly-adapting cutaneous afferent are exquisitely sensitive to skin stretch, offering these afferents potentially important roles in proprioception and motor control, but the pattern of activation of the joint-related afferents depends largely on whether the finger movements are generated passively or actively.

Although receptors on the dorsum of the hand appear better suited for monitoring finger movements, recent observations on cutaneous and joint afferents from the palmar aspect of the hand indicate that these too can respond to changes in tensile strain *within* the physiological range of joint rotation *when the subject is involved in restraining a gripped object* (Macefield, Johansson and Häger, unpublished observations). Presumably, in this situation the local tensile strains in the skin and articular tissues are much higher than those associated with free 'unloaded' movements of the fingers, such as employed in gesticulative actions of the hand. The implications are that the contributions of cutaneous and joint afferents from the palmar aspect of the hand to proprioception and motor control are greatest during the handling of tools or manipulation of objects - the most common actions in the motor repertoire of the human hand.

REFERENCES

Burke, D., Gandevia, S.C. and Macefield, G. (1988). Responses to passive movement of receptors in joint, skin and muscle of the human hand. *Journal of Physiology* **402,** 347-361.

Edin, B.B. (1990). Finger joint movement sensitivity of non-cutaneous mechanoreceptor afferents in the human radial nerve. *Experimental Brain Research* **82,** 417-422.

Edin, B.B. (1992). A quantitative analysis of static strain sensitivity in human mechanoreceptors from hairy skin. *Journal of Neurophysiology* **67,** 1105-1113.

Edin, B.B. and Abbs, J.H. (1991). Finger movement responses of cutaneous mechanoreceptors in the dorsal skin of the human hand. *Journal of Neurophysiology* **65,** 657-670.

Edin, B.B. and Vallbo, Å.B. (1990). Dynamic response of human muscle spindle afferents to stretch. *Journal of Neurophysiology* **63,** 1297-1306.

Hulliger, M., Nordh, E., Thelin, A.-E. and Vallbo, Å.B. (1979). The responses of afferent fibres from the glabrous skin of the hand during voluntary finger movements in man. *Journal of Physiology* **291,** 233-249.

Johansson, R.S. and Vallbo, Å.B. (1979). Detection of tactile stimuli. Thresholds of afferent units related to psychophysical thresholds in the human hand. *Journal of Physiology* **297,** 405-422.

Johansson, R.S and Vallbo, Å.B. (1983). Tactile sensory coding in the glabrous skin of the human hand. *Trends in Neuroscience* **6,** 27-32.

Macefield, G., Gandevia, S.C. and Burke, D. (1990). Perceptual responses to microstimulation of single afferents innervating joints, muscles and skin of the human hand. *Journal of Physiology* **429,** 113-129

McCloskey, D.I., Macefield, G., Gandevia, S.C. and Burke, D. (1987). Sensing position and movements of the fingers. *News in Physiological Sciences* **2,** 226-230.

Sathian, K. and Devanandan, M.S. (1983). Receptors of the metacarpophalangeal joints: a histological study in the bonnet monkey and man. *Journal of Anatomy* **137,** 601-613.

Stilwell, D.L. (1957). The innervation of deep structures of the hand. *American Journal of Anatomy* **101,** 75-99.

Torebjörk, H.E., Vallbo, Å.B. and Ochoa, J.L. (1987). Intraneural microstimulation in man: its relation to specificity of tactile sensations. *Brain* **110,** 1509-1529.

Vallbo, Å.B. and Hagbarth, K.-E. (1968). Activity from skin mechanoreceptors recorded percutaneously in awake human subjects. *Experimental Neurology* **21,** 270-289.

LIMITATIONS IN THE NEURAL CONTROL OF HUMAN THUMB AND FINGER FLEXORS

S.C. GANDEVIA AND S.L. KILBREATH*

Prince of Wales Medical Research Institute
University of New South Wales
*Faculty of Health Sciences
The University of Sydney
Sydney
Australia

SUMMARY

Of all primates, humans have the greatest capacity to make independent movements. Factors contributing to this capacity include morphological changes in humans, e.g. the presence of flexor pollicis longus in humans. However, there are limitations to independent movement due to kinaesthetic and neural factors as well as to biomechanical linkages. Psychophysical studies have been used to study kinaesthesia in human subjects. The subjects' ability to grade forces produced by intrinsic and extrinsic muscles of the hand was limited although forces produced by the thumb were more accurately perceived than those produced by the index finger. Also, there was limited kinaesthetic independence between forces generated by neighbouring muscles. Using EMG recordings from fine-wire intramuscular electrodes, neural limitation to voluntary recruitment of the long flexor muscles can be demonstrated: co-contraction of adjacent muscles occurred when weak contractions were made by a long finger flexor. In summary, although humans have greater capacity than other primates to perform independent movements with their digits, this ability is limited.

INTRODUCTION

The terms 'independent' and 'fractionated' movements of the fingers and thumb have frequently been used by neurologists and neurophysiologists to refer to the ability of primates to move one digit selectively. The neural substrate for this ability is said to be laid down in the evolution of cortical control of motoneurones, via corticospinal connections. In the words of the comparative anatomist Wood Jones (1949), "it is not the hand that is perfect, but the whole nervous mechanism by which movements of the hand are evolved, co-ordinated, and controlled." This widely held view ignores the obvious biomechanical changes that have occurred with evolution of the human hand.

Evolution within the primate series has lead to anatomical modifications of the structure of the hand, particularly of the thumb. These can be seen in the increased width and length of the human thumb so that its tip is relatively closer to that of the index finger. In addition, it can more readily oppose the medially located fingers than other primates (Napier, 1972; Kapandji, 1981). There is some evidence that the capacity for fine manipulation of objects by the hand correlates with certain aspects of the evolution of the pyramidal tract across species (Heffner and Masterton, 1975, 1983; Nudo and Masterton, 1988, 1990). However, such evidence does not explain the difference in dexterity within the primate species (Heffner and Masterton, 1975, 1983; Nudo and Masterton, 1988, 1990) and it suffers from the limitation that correlation does not equal causation: co-evolution of any aspect of the motor system across primates might be of equal relevance.

One important morphological change that has affected the biomechanics of the human thumb is the evolution of its long flexor muscle originating in the forearm and this muscle is anatomically distinct from the adjacent multi-tendoned flexor digitorum profundus. Within the flexor digitorum profundus muscle, the component which gives rise to the tendon for the index finger is anatomically separated from its medial neighbour. In contrast, in the macaque monkey, a species much used for experimental recording of motor cortical discharges, the thumb does not possess a separate long flexor. It receives an appropriate tendon from the distal mid-portion of flexor digitorum profundus; this emerges from the muscle belly distal to the wrist (Serlin and Schieber, 1993). Indeed, the separate tendons for each digit do not emerge from the muscle belly until about wrist level.

In subhuman primates, the motor cortical output to motoneurones innervating muscles of the forearm and hand contains a direct corticomotoneuronal component (Phillips and Porter, 1964; Fetz and Cheney, 1980; Buys, Lemon, Mantel and Muir, 1986). Furthermore, there is evidence that the motor cortical projections to the intrinsic muscles of the thumb in the monkey contain relatively less divergence than to other muscles of the hand (Buys *et al.*, 1986; Lemon, Bennett and Werner, 1992). The importance of these connections in control of the digits appears to demonstrated by sectioning the pyramidal tract in the monkey. This procedure leads to a permanent inability to 'fractionate' movements of the index finger and thumb, regardless of whether the lesion is made in adult monkeys (Towers, 1940; Lawrence and Kuypers, 1968; cf. Schwartzman, 1978; Chapman and Wiesandanger, 1982) or shortly after birth (Lawrence and Hopkins, 1976; Passingham, Perry and Wilkinson, 1983). In human patients, the marked deficit in hand function following a stroke involving corticospinal output is well known. There is some evidence that the muscles which move the human hand also receive a relatively large direct corticospinal innervation (Palmer and Ashby, 1992). It is interesting that even the corticospinal projections to motoneurones of midline muscles in human subjects have evolved a monosynaptic component (e.g. Gandevia and Rothwell, 1987a; Plassman and Gandevia, 1989; Gandevia, 1993). However, it should not be forgotten that the anatomical differences between the human and monkey thumb may make comparison between their cortical control somewhat questionable.

Many factors, anatomical and neural, will contribute to the capacity to make relatively independent movements of the tips of the fingers. Apart from biomechanical linkages between the muscles which control the digits, there is the kinaesthetic limitation to movement: i.e. how independently do we know the positions and forces associated with one digit? In addition, the capacity to deliver a motor command to recruit only motor units which cause movement of one finger is limited (Gandevia and Rothwell, 1987b). We have undertaken both psychophysical and electromyographic studies to examine the ability of subjects to grade forces with intrinsic and extrinsic muscles acting on the thumb and index finger. In addition, a method has been developed to assess the degree of neural co-activation among the long extrinsic muscles which flex the distal joints of the digits. We hypothesised that neural control of the musculature of the thumb may be more developed than that of other digits.

In this chapter we consider first the accuracy of force judgements with a range of muscles which act on the hand, and secondly, the degree of kinaesthetic independence between the judgements for the different fingers is described. Finally, the limits to the voluntary recruitment of the long flexor muscles which act on the digits will be described.

Judgement of Forces with Intrinsic and Extrinsic Hand Muscles

Much evidence indicates that judgements of force or heaviness are biased by signals related to the centrally-generated motor command or 'effort' (for review see McCloskey, 1981; Gandevia, 1987). This evidence has been derived from many psychophysical studies using weight- and force-matching. The common theme is that judgements of force are overestimated when the muscle is effectively weakened either by factors acting on the muscle (such as fatigue) or on the CNS (such as cutaneous and muscle reflexes). Although there is a unique ability to recruit motoneurones of the intrinsic muscles of the hand (Gandevia and Rothwell, 1987b) and the corticomotoneuronal input to these muscles may be relatively selective (Buys *et al.*, 1986; Lemon *et al.*, 1992), our recent studies do not suggest that judgements of force are necessarily made with greater accuracy for these muscles (Gandevia and Kilbreath, 1990; Kilbreath and Gandevia, 1993a). Subjects lifted weights in four ways (Fig. 1). They were designed to limit torque production to first dorsal interosseous (FDI) and adductor pollicis (AP) which are intrinsic muscles acting on the index finger and thumb respectively. Torque was also limited to two extrinsic hand muscles, the index portion of flexor digitorum profundus (FDP) and flexor pollicis longus (FPL) which acts on the thumb. A reference weight lifted on the right side was matched to a variable weight lifted simultaneously and in the same way on the left. The reference weights matched represented a wide range of force, from 2.5% to 50% of maximal voluntary force (MVC). The coefficient of variation from repeated observations provided the index of reproducibility or accuracy.

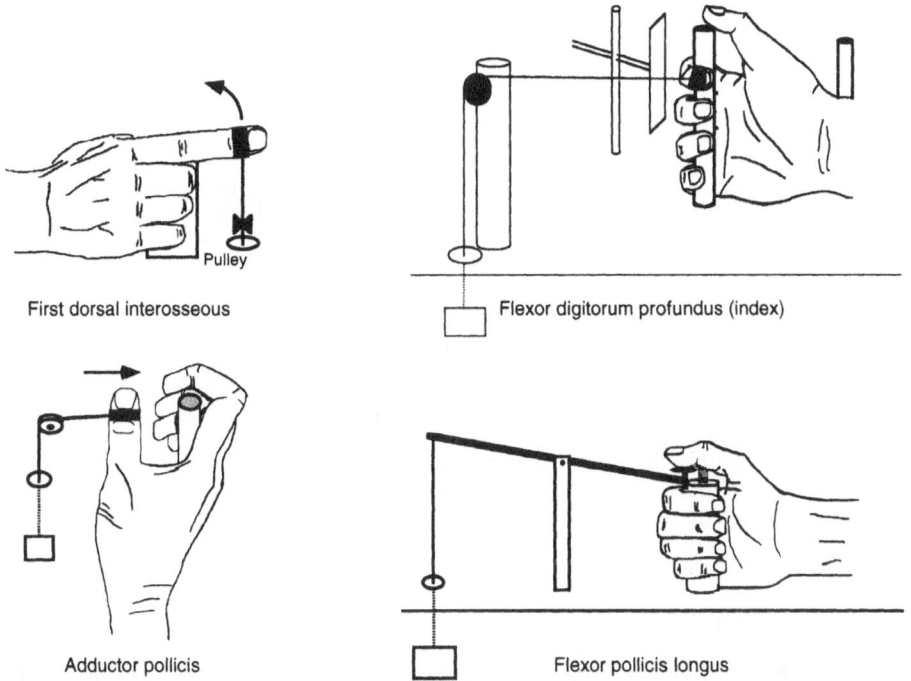

First dorsal interosseous

Flexor digitorum profundus (index)

Adductor pollicis

Flexor pollicis longus

Figure 1. Diagrammatic representation of hand postures used to limit lifting to the muscle being tested. The hand postures used for lifting with the two intrinsic hand muscles, FDI and AP are shown in the left upper and lower panel respectively, and for lifting with the two extrinsic hand muscles, the index portion of FDP and FPL, in the right upper and lower panel respectively. The view for FDI and AP is from above and for the index portion of FDP and FPL from the side. Note that FDI and AP lift in the horizontal plane, i.e. with gravity 'eliminated'. Kilbreath and Gandevia (1993a), reproduced with permission

Data from a group of naive subjects are summarized in Fig. 2. Accuracy was greatest with FPL, least with FDP, and intermediate for the two intrinsic muscles (for details see Kilbreath and Gandevia, 1993a). When considered together, accuracy was greater for muscles acting on the thumb than on the index. For all muscles except FPL, the accuracy of judgements decreased significantly as the size of the reference weight decreased. For FPL, accuracy remained high for weights equivalent to 1% - 50% MVC. We have confirmed this general observation in similar studies with different subjects. The decrement in accuracy at low weights has been previously noted for isometric forces (e.g. Newell and Carlton, 1984). The uniformly high accuracy shown when weights are lifted by thumb flexion does not appear to depend on the cutaneous input from the large thumb pad as it is still present following anaesthesia of the digital nerves of the thumb (Wirianski, Kilbreath and Gandevia, unpublished observations).

Many mechanisms could theoretically subserve the special performance of FPL. Muscle afferents can provide signals of intramuscular tension and they can indicate the time that a force is generated or an object lifted (e.g. McCloskey, Ebeling and Goodwin, 1974;

Gandevia and McCloskey, 1978). Thus, an intriguing possibility is that muscle afferent signals of intramuscular tension are particularly important for FPL. There are no recordings of tendon organ afferents from FPL to refute or to support the suggestion that their discharge provides a more accurate index of intramuscular force than for other muscles, or that their discharge is especially sensitive at very low force levels. Thus, it is presently reasonable to suggest that the performance may reflect an enhanced capacity of the CNS to access signals related to corollaries of the motor command for this unusual human muscle.

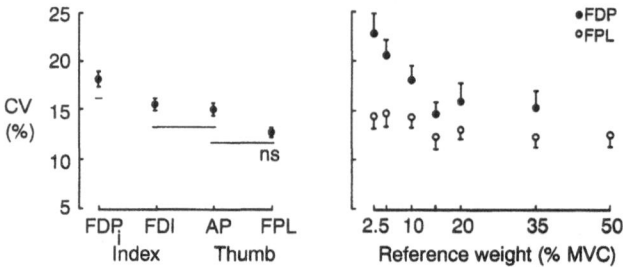

Figure 2. A.: Accuracy in judgements of weights expressed in terms of muscles used to lift them. The CV (mean ± SEM), i.e. the estimate of accuracy, are shown. **B:** CV (mean ± SEM) for each reference weight lifted by the index portion of FDP (filled circles) and FPL (open circles). Accuracy was poor when light weights were lifted with the index portion of FDP, indicated by the high CVs but accuracy remained high when light weights were lifted by FPL. Horizontal lines indicate data which were not statistically different (ns, P>0.05, Duncan's multiple-range test). Adapted from Kilbreath and Gandevia (1993a), with permission.

Whatever the neural mechanism, it is interesting to speculate why this capacity to judge forces accurately from 1% to 50% MVC occurs for the main thumb flexor. One possibility relates to the biomechanical requirement for the thumb flexor to contract 'against' the other digits. This means that the thumb flexor would be naturally exposed to a wide range of forces exerted against it by from one to four digits, depending on the specific grasp employed (Kilbreath and Gandevia, 1993a).

Non-Independent Judgement of Forces Exerted by the Long Flexor Muscles

To establish whether the force generated by one digital portion of the deep digital flexors can be judged independently from that exerted by its neighbours, subjects lifted a reference weight with flexion of one digit on the reference side and simultaneously lifted a 'concurrent' weight with another flexor. The subject was asked to match only the reference weight with a variable weight lifted by thumb flexion on the matching side. For these studies the hand on the reference side was positioned so that the finger extensors were unable to exert torque and the radial nerve was anaesthetized to paralyse the extensors of the thumb (for details see Kilbreath and Gandevia, 1991, 1992). An illusory increase in perceived heaviness occurred whenever the reference and concurrent weights were lifted by two digital flexors. This has been measured under a number of conditions including when the reference and concurrent weights were lifted by the thumb and index finger and by the index and middle fingers, and by the index and fourth fingers. An example of the raw data for one combination in a typical subject is given in Fig. 3. The illusion was not quantitatively altered when any unintended extensor torques were eliminated (see above), nor when the digital nerves of the relevant digits on the reference side were anaesthetized. In addition, the increase in perceived heaviness was accentuated when the muscle lifting the concurrent weight was fatigued (Kilbreath and Gandevia, 1991). The one condition tested in which the illusion did not occur was when the concurrent weight was lifted by a remote muscle in the lower limb. The illusion did not adversely affect the accuracy of heaviness judgements (measured as the coefficient of variation) in any of the digital combinations.

These data suggest that the capacity to assess the force generated by one digital portion of the long finger flexors independently is limited. This is unlikely to be imposed by reflexes arising from the lifting muscles (which would be expected to be mutually facilitatory) or by mechanical interactions between adjacent tendons of the deep flexors. Such mechanical interactions would 'assist' the lifting and thus make the reference weight seem lighter. Furthermore, the illusion was not influenced by anatomical boundaries as it was present for the anatomically distinct FPL and the index portion of FDP, as well as occurring amongst digital portions of FDP.

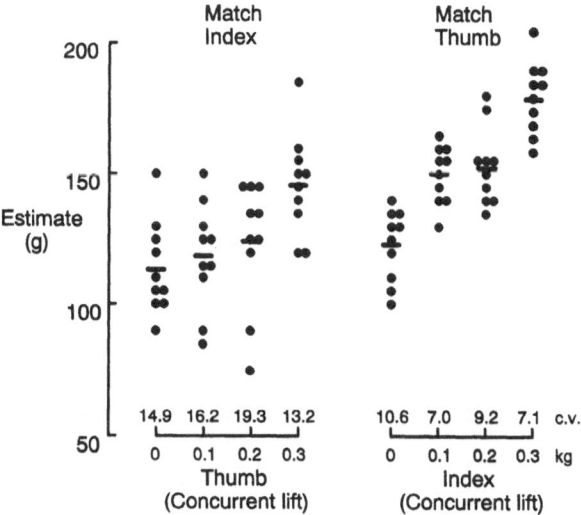

Figure 3. Data from a single subject for two studies in which FDP and FPL lifted the reference and concurrent weights. The subject's radial nerve was anaesthetised for each study. On the left, the reference weight (200g) was lifted by the index and the concurrent weight was lifted by the thumb, and on the right, the converse. The coefficient of variation for each condition is shown on the horizontal axis. The mean estimates, indicated by the horizontal bars, increase when a weight is concurrently lifted. Kilbreath and Gandevia (1991), reproduced with permission.

The simplest explanation for the illusory increase in heaviness is that while it is possible to determine the overall motor command to the digital flexors, there is an inability to specify fully the intended destination of the output. Thus, for thumb and index combinations, the subject misinterprets some of the command destined for the concurrently lifting index flexor as being associated with the output to the thumb flexor lifting the reference weight. Given the functional usage of the long flexors in co-operative tasks such as grasping, this phenomenon may not seem so surprising. Further studies are needed to define the limitations to this lack of kinaesthetic awareness for the digits of the hand. There is already some evidence that the cutaneous input from one finger may facilitate the detection of movements applied to the adjacent one (Gandevia and McCloskey, 1976).

Capacity to Move the Individual Digits Separately

While it is frequently stated that humans have the capacity for selective movement of the digits, this ability has not, to our knowledge, been quantified. At one extreme, human subjects can learn, in the absence of peripheral feedback, to deliver liminal commands (subthreshold for motoneuronal recruitment) to one of a pair of intrinsic muscles of the hand (Gandevia and Rothwell, 1987b). By contrast, this cannot be learned for pairs of muscles in the forearm.

Connections between the tendons of the long flexor muscles of the digits occur in the forearm and palm (Rank, Wakefield and Hueston, 1968; Linburg and Comstock, 1979; Brand, Beach and Thompson, 1981; Gandevia and Kilbreath, unpublished observations). Such inter-tendinous connections are better known for the components of extensor digitorum communis (e.g. Wood Jones, 1949; Edin and Vallbo, 1987). These connections must restrict the ability to move one digit via a long tendon without affecting its neighbours. Hence, in recent studies we have concentrated on the recruitment of low-threshold motor units associated with one digital component of the long flexor muscles when the other flexors act in a graded fashion (Kilbreath and Gandevia, 1993b). This allows an estimate of how well the CNS can command the relevant motoneurones without the complexities imposed by their mechanical actions: it allows measurement of a threshold for co-activation of related motoneurone pools.

Intramuscular wire electrodes were used to record selectively the digital components of the muscles which flex the distal joints of the digits (i.e. FDP and FPL). Recordings were made from low-threshold motor units associated with flexion of one distal joint against gravity (termed the 'test' digit), and their behaviour was monitored when the subjects lifted weights with the other four digits (termed 'lifting' digits). Perhaps, not surprisingly, recruitment of the 'test' motoneurones occurred when weak contractions were made by adjacent digits. For muscles more remote from the test digit this recruitment occurred with progressively stronger contractions. This sort of co-activation was observed not only when the hand was restrained to prevent the necessity for muscular contraction for stabilization, but also when it was resting freely.

Inability to prevent this co-contraction, in the presence of intact cutaneomuscular reflexes, indicates that there is a fundamental limitation to the capacity to recruit motoneurones whose action is associated principally with one digit. Co-activation of this type occurred for the index-thumb combination (with anatomically distinct muscles) just as readily as for combinations involving different digital portions of flexor digitorum profundus. However, it was not possible with this method to assess the recruitment 'gain' once co-contraction had occurred. It is likely, although not formally measured, this is different for the various combinations acting on the fingers and thumb. This co-contraction will limit the functional capacities of the hand (along with the biomechanical interactions between adjacent finger flexors). It must reflect a limitation of the human motor system to direct sufficiently focal commands without 'spillover' to other muscles. This limit may depend on the known propensity of corticospinal (and presumably other) paths to diverge and facilitate motoneurones in more than one pool (e.g. Fetz and Cheney, 1980; Buys, Lemon, Mantel and Muir, 1986).

In summary, to echo the suggestions of Wood Jones (1949) quoted earlier, our data support the view that a particular degree of neural specialization has evolved for the human thumb. This permits it to work with accuracy over a vast range of forces. However, in other respects the behaviour of all the long flexor muscles (of fingers or thumb) was similar: forces produced by one digit cannot be judged independently of the actions of its neighbour, and the capacity to recruit the motor units which flex only one digit is also limited.

REFERENCES

Brand, P.W, Beach, R.B. and Thompson, D.E. (1981). Relative tension and potential excursion of muscles in the forearm and hand. *Journal of Hand Surgery* 6, 209-219

Buys, E.J., Lemon, R.N., Mantel, G.W.H. and Muir, R.B. (1986). Selective facilitation of different hand muscles by single corticospinal neurones in the conscious monkey. *Journal of Physiology* 381, 529-549

Chapman, C.E. and Wiesendanger, M. (1982). Recovery of functions following unilateral lesions of the bulbar pyramid in the monkey. *Electroencephalography and Clinical Neurophysiology* 53, 374-387

Edin, B.B. and Vallbo, Å.B. (1987). Twitch contraction for identification of human muscle afferents. *Acta Physiologica Scandinavica* 131, 129-138

Fetz, E.E. and Cheney, P.D. (1980). Postspike facilitation of forelimb muscle activity by primate corticomotoneuron cells. *Journal of Neurophysiology* 44, 751-772

Gandevia, S.C. (1987). Roles for perceived voluntary motor commands in motor control. *Trends in Neurosciences* 10, 81-85

Gandevia, S.C. (1993). Assessment of corticofugal output: testing and transcranial stimulation of the motor cortex. In: *Science and Practice in Clinical Neurology,* ed. Gandevia, S.C., Burke, D. and Anthony, M., pp. 75-88. Cambridge University Press, London

Gandevia, S.C. and McCloskey, D.I. (1976). Joint sense, muscle sense, and their combination as position sense, measured at the distal interphalangeal joint of the middle finger. *Journal of Physiology* 260, 387-407

Gandevia, S.C. and McCloskey, D.I. (1978). Interpretation of perceived motor commands by reference to afferent signals. *Journal of Physiology* 283, 493-499

Gandevia, S.C. and Rothwell, J.C. (1987a). Activation of the human diaphragm from the motor cortex. *Journal of Physiology* 384, 109-118

Gandevia, S.C. and Rothwell, J.C. (1987b). Knowledge of motor commands and the recruitment of human motoneurons. *Brain* 110, 1117-1130

Gandevia, S.C. and Kilbreath, S.L. (1990). Accuracy of weight estimation for weights lifted by proximal and distal muscles of the human upper limb. *Journal of Physiology* 397, 113-226

Heffner, R. and Masterton, B. (1975). Variations in form of the pyramidal tract and its relationship to digital dexterity. *Brain, Behavior and Evolution* **12**, 161-200

Heffner, R. and Masterton, B. (1983). The role of the corticospinal tract in the evolution of human digital dexterity. *Brain, Behavior and Evolution* **23**, 165-181

Hepp-Reymond, M.-C. and Wiesendanger, M. (1982). Unilateral pyramidotomy in monkeys: effect on force and speed of a conditioned precision grip. *Brain Research* **36,** 117-131

Kapandji, I.A. (1981). Biomechanics of the thumb. In: *The Hand,* vol 1, ed. Tubiana, R., pp. 404-422. W.B. Saunders Company, Philadelphia

Kilbreath, S.L. and Gandevia, S.C. (1991). Independent digit control: failure to partition perceived heaviness of weights lifted by digits of the human hand. *Journal of Physiology* **442**, 585-599

Kilbreath, S.L. and Gandevia, S.C. (1992). Independent control of the digits: changes in perceived heaviness over a wide range of forces. *Experimental Brain Research* **91**, 539-542

Kilbreath, S.L. and Gandevia, S.C. (1993a). Neural and biomechanical specializations in the human thumb. *Journal of Physiology* **472**, 537-556

Kilbreath, S.L. and Gandevia, S.C. (1993b). The lack of independence between flexor pollicis longus and digital portions of flexor digitorum profundus in human subjects. Proceedings of the XXXII International Union of Physiological Sciences, 248.9

Lawrence, D.G. and Hopkins, D.A. (1976). The development of motor control in the rhesus monkey: evidence concerning the role of corticomotoneuronal connections. *Brain* **99**, 235-254

Lawrence, D.G. and Kuypers, H.G.J.M. (1968). The functional organization of the motor system in the monkey. Part 1. The effects of bilateral pyramidal lesions. *Brain* **91**, 1-14

Lemon, R., Bennett, K.M. and Werner, W. (1991). The cortico-motor substrate for skilled movements of the primate hand. In: *Tutorials in Motor Neuroscience,* ed. Requin, J. and Stelmach, G.E., pp. 477-495. Kluwer Academic Publishers, Netherlands

Linburg, R.M. and Comstock, B.E. (1979). Anomalous tendon slips from the flexor pollicis longus to the flexor digitorum profundus. *Journal of Hand Surgery* **4**, 79-83

McCloskey, D.I. (1981). Corollary discharges: motor commands and perception. In: *The Nervous System.* Handbook of Physiology, Section 1, Volume 2, Part 2, ed. Brooks, V.B., pp.1415-1447. Bethseda, MD: American Physiological Society

McCloskey, D.I., Ebeling, P. and Goodwin, G.M. (1974). Estimation of weights and tensions and apparent involvement of a 'sense of effort'. *Experimental Neurology* **42**, 220-232

Napier, J.R. (1972). Primates and Their Adaptation. Oxford University Press, London Newell, K.M., Carlton, L.G. and Hancock, P.A. (1984). Kinetic analysis of response variability. *Psychological Bulletin* **96**, 133-151

Nudo, R.J. and Masterton, R.B. (1990). Descending pathways to the spinal cord IV: Some factors related to the amount of cortex devoted to the corticospinal tract. *Journal of Comparative Neurology* **296**, 584-597

Palmer, E. and Ashby, P. (1992). Corticospinal projections to upper limb motoneurones in humans. *Journal of Physiology* **418**, 397-412

Passingham, R..E., Perry, V.H. and Wilkinson, F. (1983). The long-term effects of removal of sensorimotor cortex in infant and adult rhesus monkeys. *Brain* **106**, 675-705

Phillips, C.G. and Porter, R. (1964). The pyramidal projection to motoneurones of some muscle groups of the baboon's forelimb. In: *Physiology of Spinal Neurones.* Progress in Brain Research 12, ed. Eccles, J.C. and Schadé, J.P., pp. 222-245 Elsevier, Amsterdam

Plassman, B.L. and Gandevia, S.C. (1989). Comparison of human motor cortical projections to abdominal muscles and intrinsic muscles of the hand. *Experimental Brain Research* **78**, 301-308

Rank, B.K., Wakefield, A.R. and Hueston, J.T. (1968). *Surgery of Repair as Applied to Hand Injuries.* Third Edition. Edinburgh: E. and S. Livingstone, pp 35-36

Schwartzman, R.J. (1978). A behavioral analysis of complete unilateral section of the pyramidal tract at the medullary level in Macaca mulatta. *Annals of Neurology* **4**, 234-244

Serlin, D.M. and Schieber, M.H. (1993). Morphologic regions of the multi-tendoned extrinsic finger muscles in the forearm. *Acta Anatomica* **146**, 255-266

Tower, S.S. (1940). Pyramidal lesions in the monkey. *Brain* **63**, 36-90

REFLEXES

FUSIMOTOR REFLEXES FROM JOINT AND CUTANEOUS AFFERENTS

P. H. ELLAWAY

Dept. of Physiology
Charing Cross & Westminster Medical School
London W6 8RF, UK

SUMMARY

This article reviews our understanding of the role played by cutaneous afferents and by non-muscular proprioceptors in the regulation of fusimotor drive to muscle spindles in the cat. Also, new observations are presented concerning the actions on gamma motoneurones to the gastrocnemius and soleus muscles of: (1) Pacinian corpuscles in the interosseus membrane, (2) mechanoreceptors in the knee joint, and (3) mechanoreceptors innervated by the sural nerve supplying the skin on the lateral aspect of the heel. Interactions between the inputs from these proprioceptors and cutaneous afferents have also been studied. Pacinian corpuscles appear to exert neither a direct influence on gamma motoneurones nor any modulation of the cutaneous excitatory pathway. The discharge of knee joint afferents excites some gamma motoneurones and inhibits others. The excitatory action alone may be suppressed by concomitant cutaneous input. Experiments in which recordings were made from muscle spindle afferents confirmed other reports that joint and cutaneous afferent input could influence both dynamic and static spindle sensitivity. However, the extent and sign of that action was dependent upon the integrity of supraspinal pathways. It is concluded that the relevance of the regulation exerted by particular sensory modalities may not be evident until studied during specific tasks when the different afferents will be stimulated in a suitably co-ordinated manner and their inputs will be integrated with the central commands being issued from supraspinal structures.

INTRODUCTION

The uncertainties concerning the role of muscle spindles in the control of movement in mammals stem in no small part from our ignorance of the central nervous control of the fusimotor innervation. In contrast to the extensive literature on the structure and internal functioning of the neuromuscular spindle, and the central destination and effects of spindle afferents, we have only a rudimentary knowledge of the supraspinal and segmental regulation of fusimotor neurones. This lack of knowledge has been due mainly to the difficulties encountered in making intracellular recordings from gamma motoneurones (Eccles et al., 1960; Grillner et al., 1969; Appelberg et al., 1977), which are smaller than alpha motoneurones (Cullheim and Ulfhake, 1979; Westbury, 1982), and the difficulty of identifying mixed (Beta) skeletofusimotor neurones. Most investigations into the regulation of the fusimotor innervation have therefore been made by observing the alteration in probability of firing of gamma motoneurones exhibiting a background discharge, or have been made indirectly by monitoring the changes in spindle afferent discharge. Reviews of this literature are not recent (Matthews, 1972; Murthy 1978; Hulliger 1984).

Neural Control of Movement, Edited by W.R. Ferrell
and U. Proske, Plenum Press, New York, 1995

The execution of movement and the maintenance of posture and equilibrium involve peripheral input from a wide range of proprioceptive and cutaneous afferents. At the level of the spinal cord these afferent systems are directed at both alpha and gamma motoneurones, mostly via interneurones. A clear difference in the afferent connections to the two categories of motoneurone is the lack of a monosynaptic, or even a substantial oligosynaptic, input to gamma motoneurones from primary spindle afferents of the homonymous or synergist muscles (Grillner *et al.*, 1969; Trott, 1976; Fromm and Noth 1976; Ellaway and Trott, 1978; Appelberg *et al.*, 1983a). Other muscle afferents including group II spindle (Noth and Thilmann, 1980; Appelberg *et al.*, 1983b), Golgi tendon organ afferents (Ellaway *et al.*, 1979, 1980; Appelberg *et al.*, 1983a; Noth, 1983) and group III mechanoreceptors (Ellaway *et al.*, 1982; Appelberg *et al.*, 1983c) have extensive segmental actions on gamma motoneurones. The patterns to static and dynamic neurones differ and neither conform well with projections to alpha motoneurones of the same muscles. A more extensive review of the influence of muscle afferents on gamma motoneurones has been made by Hulliger (1984). This article assesses the current position with regard to our understanding of the regulation of gamma motoneurone discharge by non-muscular proprioceptors and cutaneous afferents.

It has been a common experience in this laboratory, in a number of animal preparations (decerebrated, spinal, and anaesthetized, cats and rabbits), that muscle stretch and sub-maximal contraction rarely cause any marked change in the discharge of gamma motoneurones. In contrast, innocuous stimulation of the skin can cause a substantial change in the discharge frequency of gamma efferents. Such an evident influence by cutaneous afferents has not been the experience of those recording from spindle afferents in intact, behaving cats (Prochazka *et al.*, 1979; Loeb *et al.*, 1985) and has rarely been observed in man (Aniss *et al.*, 1990). This may simply reflect the fact that segmental actions by skin afferents on gamma efferents are subject to powerful supraspinal control (Grillner *et al.*, 1967; Davey and Ellaway, 1988) and will be task dependent (cf., Aniss *et al.*, 1988, 1990). The relevance of cutaneous afferents in the regulation of fusimotor neurones may well reflect their role in detecting contact with external objects. However, some cutaneous afferents, such as the slowly-adapting type-1 and type-2 mechanoreceptors, appear to respond to movements that stretch skin, particularly around joints (Johansson, 1978; Hulliger *et al.*, 1979; Macefield *et al.*,1990) and may be considered to act more like proprioceptors than cutaneous mechanoreceptors. One of these receptors, the type-1 slowly-adapting mechanoreceptor of the hairy skin of the cat, has a potent influence over gamma motoneurones innervating a muscle that operates at a joint close to the source of those cutaneous afferents (Davey and Ellaway, 1989). Such an association makes it relevant to examine the inter-relation between conventional proprioceptors and proprioceptive-like cutaneous afferents in the regulation of gamma motoneurone discharge.

Pacinian Corpuscles

One proprioceptor for which a role in the automatic regulation of posture or movement remains unclear is the Pacinian corpuscle. The interosseus membrane lying between the fibula and tibia in the hind limb of the cat is richly provided with Pacinian corpuscles (Hunt 1961; Silfvenius, 1970). The afferent information from Pacinian corpuscles ascends to the somatosensory cortex in the dorsal columns (McIntyre 1962; McIntyre *et al.*, 1967) but Pacinian afferents have limited action at a segmental level in the spinal cord (McIntyre and Proske, 1968). Although stimulation of Pacinian corpuscle afferents fails to modify monosynaptic reflex transmission to flexor and extensor hindlimb motoneurones it does evoke dorsal root potentials and depolarization of afferent terminals of fast conducting fibres in the cutaneous sural and the posterior articular nerves (Yeo, 1978). Since cutaneous afferents have potent inputs to gamma efferents it is relevant to assess any indirect action that Pacinian corpuscles might have on gamma motoneurones through the modification of cutaneous reflex inputs, as well as any direct reflex action.

McIntyre *et al.*, (1978) found that electrical stimulation of the interosseous nerve at twice threshold failed to influence the discharge of spindle afferents and concluded that Pacinian mechanoreceptors did not influence fusimotor neurones. Ellaway and Murphy (1978) used vibration applied to the bones of the shank to elicit the selective discharge of Pacinian corpuscles from the interosseus membrane in the otherwise denervated limb of decerebrated and of spinal cats and rabbits. The vibration did not influence the background discharge of gamma motoneurones of the triceps surae (Fig. 1A). Conditioning vibration at 170Hz and 30μm amplitude applied to the shank for 100ms preceding sural nerve stimulation did not alter the firing indices of gastrocnemius/soleus (GS) gamma motoneurones to sural nerve stimulation over a range of stimulus intensities up to four times sural threshold (Fig. 1B).

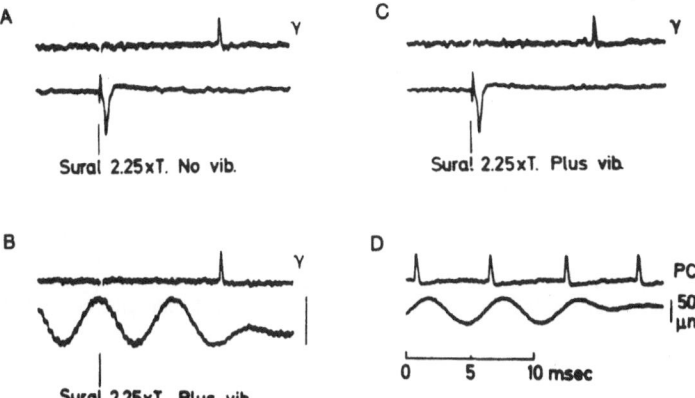

Figure 1. Response of a GS gamma motoneurone (axonal conduction velocity 32m/s) in the decerebrated spinal rabbit to electrical stimulation of the ipsilateral sural nerve at 2.25 times threshold. **A,C:** recording of the gamma efferent (top) and sural nerve volley (bottom) with (C) and without (A) conditioning vibration applied to the tibia. **B:** as in C but with the mechanical displacement of the electromagnetic vibrator probe signalled (bottom). The vibration, applied for 100 ms, terminates approximately 10ms after the sural nerve stimulus. **D:** recording of the discharge of a Pacinian corpuscle from the interosseus membrane (top) and mechanical displacement of the electromagnetic vibrator (below).

Vibration had no effect on the excitation of gamma motoneurones by sural nerve stimulation irrespective of whether the gamma efferent exhibited a background discharge (Fig 2). Direct recordings from dorsal root filaments showed that the threshold of Pacinian corpuscles in the interosseus membrane to vibration was 8μm with one-to one driving occurring at 15μm.

Rudomin and co-workers have evidence showing that presynaptic inhibition can be quite specific in terms of target primary afferents (Rudomin et al., 1981; see Rudomin, 1990). Thus the presumed presynaptic inhibitory influence of Pacinian corpuscles on cutaneous sural afferents (Yeo, 1978) appears to be directed at pathways other than the excitatory connection to GS gamma motoneurones. We conclude that impulses in Pacinian corpuscles do not modulate the discharge of gamma motoneurones or regulate the reflex actions of skin afferents.

If our findings, based on the Pacinian corpuscles of the interosseous membrane, can be extrapolated to Paciniform afferents in general there are implications for the interpretation of the effects of articular afferents on the fusimotor system. The two principal types of joint afferent are Ruffini endings and Paciniform corpuscles (Boyd, 1953; Boyd & Roberts 1954), together with a smaller number of Golgi type endings. Our results may indicate therefore that any regulation of gamma efferent discharge evident through the discharge of articular afferents comes preferentially from Ruffini endings rather than Paciniform endings.

Articular Afferents

The continuing debate over the role of joint afferents in motor control and kinaesthesia (Proske et al., 1988; Gandevia and Burke, 1992) has barely addressed the extent to which proprioceptors in joints and ligaments regulate fusimotor neurones. Results from experiments using electrical stimulation of specific articular nerves have not lead to a clear

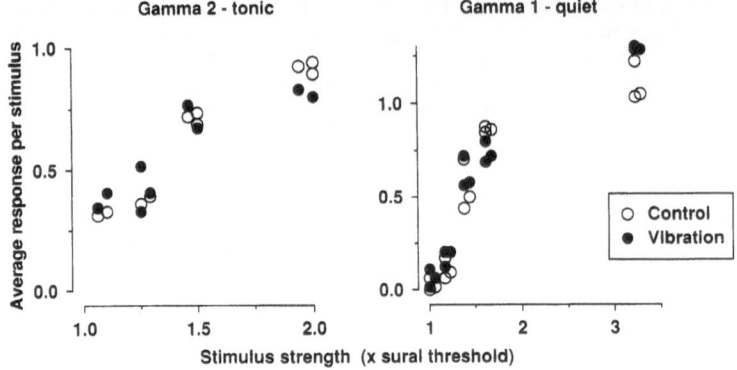

Figure 2. Magnitude of the excitatory response of GS gamma efferents to electrical stimulation of the ipsilateral sural nerve in the decerebrated spinal rabbit. Left: Response of a tonically firing efferent (23 m/s). Right: response of a gamma efferent (22m/s) with no background discharge. The average response per stimulus of a gamma efferent is expressed as the ratio of the number of responses to the number of stimuli in 64 trials. Open symbols: no vibration. Filled symbols: conditioning vibration applied to the tibia close to the interosseus membrane. Vibration was applied for 90ms before and 10ms after nerve stimuli.

picture of any organization (Grillner *et al.*, 1969; Johansson *et al.*, 1986). Electrical stimulation of low threshold articular afferents evokes excitation and inhibition of both static and dynamic gamma motoneurones with a predominance of excitation in flexor but not extensor muscles (Johansson *et al.*, 1986). Natural stimulation, such as stretch of cruciate ligaments has also revealed an influence of articular afferents on both static and dynamic gamma motoneurones (Sojka *et al.*, 1989; Johansson *et al.*, 1990) again with a mixture of effects on static and dynamic sensitivity of muscle spindles. This variety of action and the difficulties faced in attributing any action to specific receptor types preclude us from making a precise proposal as to the role of articular afferents in the regulation of muscle spindles.

An additional consideration is that cutaneous and proprioceptive inputs interact in their regulation of fusimotor drive. We have shown (Baxendale *et al.*, 1992) that excitation of GS gamma motoneurone discharge by electrical or natural stimulation of articular afferents from the knee joint is suppressed by natural stimulation of skin afferents on the lateral aspect of the heel. Inhibitory responses to electrical stimulation of the posterior articular nerve (PAN) innervating the knee joint were, however, unaffected by cutaneous stimulation. In an attempt to decide whether these excitatory and inhibitory inputs from articular afferents were directed at static or dynamic gamma motoneurones we have now examined the responses of GS spindle afferents to stimulation of the knee joint capsule during stretch of the GS muscle. Fig. 3 shows the averaged response of primary spindle afferents to sinusoidal stretch of the muscle both in the presence and absence of mechanical stimulation of the knee joint. Typically, in the cat with a complete spinal cord transection, continuous light pressure applied to the posterior aspect of the knee joint capsule caused an increase in the depth of modulation of the spindle afferent, although this response was invariably weak in comparison with the response to cutaneous stimulation, and was often absent.

The action by joint afferents was not evident in the decerebrated cat with intact spinal cord. The interpretation of such changes in spindle afferent responses is difficult. Using the patterns of spindle discharge caused by stimulation of individual gamma motoneurones as a guide (Hulliger *et al.*, 1977a,b), one interpretation would be that joint stimulation had excited or facilitated the discharge of dynamic fusimotor neurones. However, this dynamic action may have been accompanied by a reduction in static fusimotor drive. The absence of an effect in the cat with intact spinal cord presumably indicates that the input from joint afferents to gamma motoneurones is subject to supraspinal regulation. However, it is possible that some increases in dynamic fusimotor activity may have gone undetected in

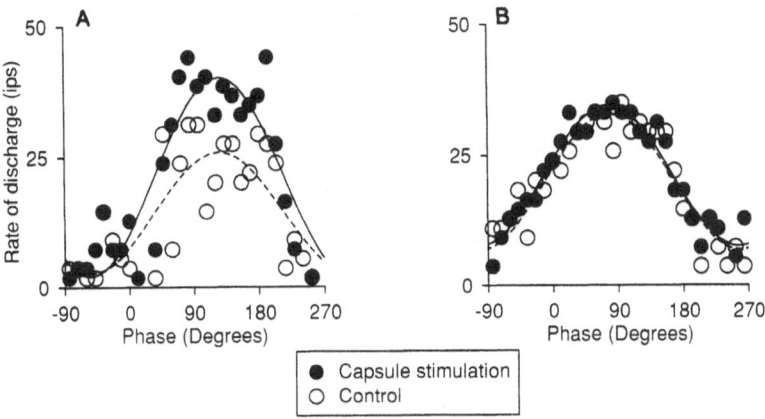

Figure 3. Cycle histograms showing the discharge frequency of gastrocnemius/soleus primary spindle afferents plotted against the phase of a sinusoidal stretch of the muscle at 1Hz over ± 0.95mm in a decerebrate-spinal (A) and a decerebrate (B) cat. Each histogram is the average of 20 trials. Sine curves have been fitted to the histogram data using a least squares algorithm. Dashed lines/open symbols: control responses. Solid lines/filled symbols: responses during light mechanical indentation of the posterior aspect of the ipsilateral knee joint capsule. Note the lack of effect of joint stimulation in the decerebrate animal. In the spinal animal joint stimulation caused an increase in modulation and small increase in mean rate of discharge of the primary afferent. Axonal conduction velocities: 114m/s, A; 83m/s, B.

both the spinal and non-spinal cat due to limitations of the sinusoidal stimulation technique in revealing fusimotor actions (Rodgers *et al.*, 1993).

Cutaneous Afferents

Regulation of fusimotor drive by skin afferents has received repeated attention due to the potent effects exerted by electrical stimulation of peripheral cutaneous nerves (Hunt and Paintal, 1958; Grillner, 1969; Appelberg *et al.*, 1977; Catley and Pascoe, 1978; Johansson and Sojka, 1985; Johansson *et al.*, 1989). The organization of cutaneous inputs to gamma motoneurones revealed by these studies is complex. Static and dynamic neurones to both flexor and extensor muscles, receive inputs that cannot readily be categorized. This has lead to the proposal that cutaneous inputs may be organized differently for individual gamma motoneurones to a particular muscle (Johansson & Sojka, 1985). However, the confusing plethora of observations may have arisen from the use of electrical stimulation of peripheral nerves which excite indiscriminately, cutaneous axons of different modality. In a study of individual sensory modalities, type-1 slowly-adapting mechanoreceptors, but not type-2 mechanoreceptors or hair follicle afferents, were found to exert an excitatory action on gamma motoneurones. The action was revealed by spike triggered averaging of the discharges of single gamma efferents triggered by discharges of single type-1 cutaneous afferents (Davey & Ellaway, 1989). We have now carried out experiments to find out whether natural stimulation that would excite these cutaneous afferents is directed at static or dynamic fusimotor neurones. Fig. 4 shows the responses of a primary and secondary spindle afferent in the same spinal cat to sinusoidal stretch of the muscle. The protocol was the same as that used to investigate joint afferent input (see fig. 3). Non-noxious mechanical stimulation of the lateral aspect of the ipsilateral heel, within the sural nerve field of innervation, caused an increase in the mean rate of discharge of both spindle endings. There was also a small increase in depth of modulation of discharge of the primary ending and a reduction for the secondary ending. These changes can best be interpreted as an increase in both dynamic and static fusimotor drive. The dynamic effect is almost certainly stronger than would appear to be the case from the small increase in modulation of the primary ending. This results from the concomitant static drive, evident from the action on the secondary ending, which would have occluded the dynamic action (Hulliger *et al.*, 1977b)

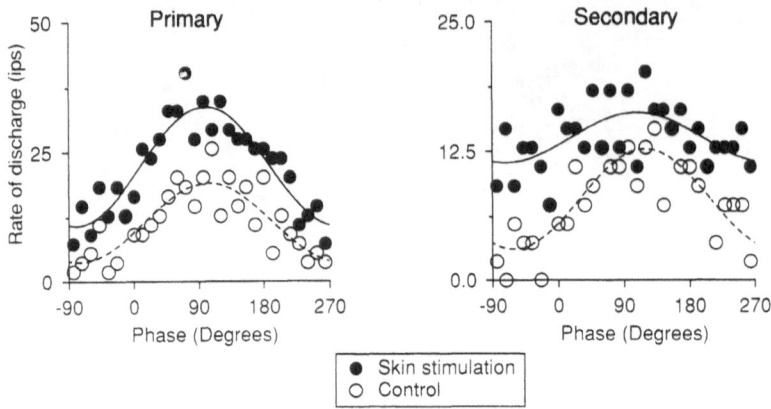

Figure 4. Cycle histograms showing the discharge of primary (left) and secondary (right) spindle afferents (gastrocnemius medialis) plotted against the phase of a sinusoidal stretch of the muscle at 1Hz over ± 0.2mm in a spinal cat. Histograms and fitted curves constructed as in fig. 3. Dashed lines/open symbols: control responses. Solid lines/filled symbols: responses during light mechanical stimulation of the skin over the lateral aspect of the calcaneum of the ipsilateral heel. Skin stimulation causes a increase in both modulation and mean rate of the discharge of the primary afferent but a decrease in modulation and increase in mean rate of discharge of the secondary afferent.

CONCLUSIONS

The work reviewed here indicates that a number of sensory modalities influence the discharges of static and dynamic fusimotor neurones. However, certain sensory systems such as muscle spindle primaries, Pacinian corpuscles and hair follicle afferents appear to have little if any part to play in the regulation of gamma efferents. Of those that do have a role it is becoming clear that their influences interact at least at the level of the spinal cord. The modulation of joint afferent input by cutaneous stimulation (Baxendale *et al.*, 1992) is an example. Such interaction suggests that the relevance of the action of particular sensory modalities should become apparent when studied during specific motor tasks. Only then will the appropriate mix of sensory inputs be achieved and the contribution to the control of movement through the integrated action on fusimotor neurones be evident.

ACKNOWLEDGEMENTS. Previously unpublished observations came from experiments in collaboration with P. R. Murthy, N. J. Davey and M. Lubjsavlijevic. Our work was supported by the MRC, the Wellcome Trust and the European Science Foundation. I gratefully acknowledge the expert technical assistance of Maria Catley, Steven Rawlinson and Helen Thomas.

REFERENCES

Aniss, A.M., Diener, H-C., Hore, J., Burke, D. and Gandevia, S.C. (1990). Reflex activation of muscle spindles in human pretibial muscles during standing. *Journal of Neurophysiology* **64,** 671-679

Aniss, A.M., Gandevia, S.C. and Burke, D. (1988). Reflex changes in muscle spindle discharge during a voluntary contraction. *Journal of Neurophysiology* **59,** 908-921

Appelberg, B., Johansson, H. and Kalistratov, G. (1977). The influence of group II muscle afferents and low threshold skin afferents on dynamic gamma motoneurones to the triceps surae of the cat. *Brain Research* **132,** 153-158

Appelberg, B., Hulliger, M., Johansson, H. and Sojka, P. (1983a). Actions on gamma motoneurones elicited by electrical stimulation of group I afferent fibres in the hindlimb of the cat. *Journal of Physiology* **335,** 237-253

Appelberg, B., Hulliger, M., Johansson, H. and Sojka, P. (1983b). Actions on gamma motoneurones elicited by electrical stimulation of group II afferent fibres in the hindlimb of the cat. *Journal of Physiology* **335,** 255-273

Appelberg, B., Hulliger, M., Johansson, H. and Sojka, P. (1983c). Actions on gamma motoneurones elicited by electrical stimulation of group III afferent fibres in the hindlimb of the cat. *Journal of Physiology* **335**, 275-292

Baxendale, R.H., Davey, N.J., Ellaway, P.H. and Ferrell, W.R. (1992). Interaction between joint and cutaneous afferent input in the regulation of fusimotor neurone discharge. In: *"Muscle afferents and spinal control of movement".* eds. L. Jami, E. Pierrot-Deseilligny & D. Zytnicki, Pergamon Press. pp 95-104

Boyd, I.A. (1953). The histological structure of the receptors in the knee joint of the cat correlated with their physiological response. *Journal of Physiology* **124**, 476-488.

Boyd, I.A. and Roberts, T.D.M. (1954). Proprioceptive discharge from stretch receptors in the knee joint of the cat. *Journal of Physiology* **122**, 38-58

Catley, D.M. and Pascoe, J.E. (1978). The reflex effects of sural nerve stimulation upon gastrocnemius fusimotor neurones of the rabbit. *Journal of Physiology* **276**, 32P

Cullheim, S. and Ulfhake, B. (1979). Observations on the morphology of intracellularly stained gamma motoneurones in relation to their axon conduction velocity. *Neuroscience Letters* **13**, 47-50

Davey, N.J. and Ellaway, P.H. (1988). Control from the brain stem of synchrony of discharge between gamma motoneurones in the cat. *Experimental Brain Research* **72**, 249-263

Davey, N.J. and Ellaway, P.H. (1989). Facilitation of individual gamma motoneurones by the discharge of single slowly-adapting type-1 mechanoreceptors in cats. *Journal of Physiology* **411**, 97-114

Ellaway, P.H. and Murphy, P.R. (1978). Lack of reflex segmental action from Pacinian corpuscle afferents on gamma motoneurones in the cat and rabbit. *Journal of Physiology* **285**, 60P

Ellaway, P.H. and Murphy, P.R. (1980). Autogenetic effects of muscle contraction on extensor gamma motoneurones in the cat. *Experimental Brain Research* **38**, 305-312

Ellaway, P.H., Murphy, P.R. and Tripathi, A. (1982). Closely coupled excitation of gamma motoneurones by group III muscle afferents with low mechanical threshold in the cat. *Journal of Physiology* **331**, 481-498

Ellaway, P.H., Murphy, P.R. and Trott, J.R. (1979). Inhibition of gamma motoneurone discharge by contraction of the homonymous muscle in the decerebrated cat. *Journal of Physiology* **291**, 425-441

Ellaway, P.H. and Trott, J.R. (1978). Autogenetic reflex action onto gamma motoneurones by stretch of triceps surae in the decerebrate cat. *Journal of Physiology* **276**, 49-66

Eccles, J.C., Eccles, R.M., Iggo, A. and Lundberg, A. (1960). Electrophysiological studies on gamma motoneurones. *Acta Physiologica Scandinavica* **50**, 32-40

Fromm, C. and Noth, J. (1976). Reflex responses of gamma motoneurones to vibration of the muscle they innervate. *Journal of Physiology* **256**, 117-136

Gandevia, S.C. and Burke, D. (1992). Does the nervous system depend on kinesthetic information to control natural limb movements? *Behavioural & Brain Sciences* **15**, 614-632

Grillner, S. (1969). The influence of DOPA on the static and dynamic fusimotor activity to the triceps surae of the spinal cat. *Acta Physiologica Scandinavica* **77**, 490-509

Grillner, S., Hongo, T. and Lund S. (1969). Descending monosynaptic and reflex control of gamma motoneurones. *Acta Physiologica Scandinavica* **75**, 592-618

Grillner, S., Hongo, T. and Lundberg, A. (1967). The effect of DOPA on the spinal cord. 7. Reflex activation of static gamma motoneurones from the flexor reflex afferents. *Acta Physiologica Scandinavica* **70**, 403-411

Hulliger, M. (1984). The mammalian muscle spindle and its central control. *Reviews of Physiology, Biochemistry & Pharmacology* **101**, 1-110

Hulliger, M., Matthews, P.B.C. and Noth, J. (1977a). Static and dynamic fusimotor action on the response of Ia fibres to low frequency sinusoidal stretching of widely ranging amplitude. *Journal of Physiology* **267**, 811-838

Hulliger, M., Matthews, P.B.C. and Noth, J. (1977b). Effects of combining static and dynamic fusimotor stimulation on the response of the muscle spindle primary endings to sinusoidal stretching. *Journal of Physiology* **267**, 839-856

Hulliger, M., Nordh, E., Thelin, A-E. and Vallbo, A.B. (1979). The responses of afferent fibres from the glabrous skin of the hand during voluntary finger movements in man. *Journal of Physiology* **291**, 233-249

Hunt, C.C. (1961). On the nature of vibration corpuscles in the hindlimb of the cat. *Journal of Physiology* **155**, 175-186

Hunt, C.C. and Paintal, A.S. (1958). Spinal reflex regulation of fusimotor neurones. *Journal of Physiology* **143**, 195-212

Johansson, R.S. (1978). Tactile sensibility in the human hand: receptive field characteristics of mechanoreceptive units in the glabrous skin area. *Journal of Physiology* **281**, 101-123

Johansson, H. and Sojka, P. (1985). Actions on gamma motoneurones elicited by electrical stimulation of cutaneous afferent fibres in the hind limb of the cat. *Journal of Physiology* **366**, 343-363

Johansson, H., Sjolander, P. and Sojka P. (1986). Actions on gamma motoneurones elicited by electrical stimulation of joint afferent fibres in the hind limb of the cat. *Journal of Physiology* **375**, 137-152

Johansson, H., Sjolander, P., Sojka, P. and Wadell, I. (1990). Activity in receptor afferents from the anterior cruciate ligament evokes reflex effects on fusimotor neurones. *Neuroscience Research* **8**, 54-59

Johansson, H., Sjolander, P., Sojka, P. & Wadell, I. (1989). Effects of electrical and natural stimulation of skin afferents on the gamma-spindle system of the triceps surae muscle. *Neuroscience Research* **6**, 537-555

Loeb, G.E., Hoffer, J.A. and Marks, W.B. (1985). Activity of spindle afferents from cat anterior thigh muscles. III. Effects of external stimuli. *Journal of Neurophysiology* **54**, 578-591.

Matthews, P.B.C. (1972). *Mammalian muscle receptors and their central actions*. Monograph of the Physiological Society. Number 23. Edward Arnold: London

Macefield, G., Gandevia, S.C. and Burke, D. (1990). Perceptual responses to microstimulation of single afferents innervating joints, muscles and skin of the human hand. *Journal of Physiology* **429**, 113-129

McIntyre, A.K. (1962). Cortical projection of impulses in the interosseous nerve of the cat's hindlimb. *Journal of Physiology* **163**, 49-60

McIntyre, A.K., Holman, M.E. and Veale, J.L. (1967). Cortical responses to impulses from single Pacinian corpuscles in the cat's hind limb. *Experimental Brain Research* **4**, 243-255.

McIntyre, A.K. and Proske, U. (1968). Reflex potency of cutaneous afferent fibres. *Australian Journal of Biological & Medical Science* **46**, 19

McIntyre, A.K., Proske, U. & Tracey, D.J. (1978). Fusimotor responses to volleys in joint and interosseous afferents in the cat's hindlimb. *Neuroscience Letters* **10**, 287-292.

Murthy, K.S.K. (1978). Vertebrate fusimotor neurones and their influences on motor behaviour. *Progress in Neurobiology* **11**, 249-307

Noth, J. (1983). Autogenetic inhibition of extensor gamma motoneurones revealed by stimulation of group I fibres. *Journal of Physiology* **342**, 51-65

Noth, J. and Thilmann, A. (1980). Autogenetic excitation of extensor gamma motoneurones by group II muscle afferents in the cat. *Neuroscience Letters* **17**, 23-26

Prochazka, A., Stephens, J.A. & Wand, P. (1979). Muscle spindle discharge in normal and obstructed movements. *Journal of Physiology* **287**, 57-66

Proske, U., Schaible, H-G. and Schmidt, R.F. (1988). Joint receptors and kinesthesia. *Experimental Brain Research* **72**, 219-224

Rodgers, J.F., Fowle, A.J., Durbaba, R. and Taylor, A. (1993). Sine versus ramp stretches for characterizing fusimotor actions on muscle spindles in the anaesthetized cat. *Journal of Physiology*. 467, 298P

Rudomin, P. (1990). Presynaptic control of synaptic effectiveness of muscle spindle and tendon organ afferents in the mammalian spinal cord. In: *"The segmental motor system"* ed. M. Binder. Oxford University Press Oxford. pp 349-380

Rudomin, P., Engberg, I. and Jimenez, I. (1981). Mechanisms involved in presynaptic depolarization of group I and rubrospinal fibres in cat spinal cord. *Journal of Neurophysiology* **46**, 532-548

Silfvenius, H. (1970). Characteristics of receptors and afferent fibres of the forelimb interosseous nerve of the cat. *Acta Physiologica Scandinavica* **79**, 6-23

Sojka, P., Johansson, H., Sjolander, P., Lorentzon, R. and Djupsjobacka, M. (1989). Fusimotor neurones can be reflexly influenced by activity in receptor afferents from the posterior cruciate ligament. *Brain Research* **483**, 177-183

Trott, J.R. (1976). The effect of low-amplitude muscle vibration on the discharge of fusimotor neurones in the decerebrate cat. *Journal of Physiology* **255**, 635-649

Westbury, D. (1982). A comparison of the structures of alpha and gamma spinal motoneurones of the cat. *Journal of Physiology* **325**, 79-91

Yeo, P.T. (1978). Central actions of impulses in Pacinian corpuscles. In: *"Studies in Neurophysiology"*. ed. R. Porter. Cambridge University Press. pp 143-153

REFLEX PERFORMANCE OF THE CHRONICALLY ISOLATED HUMAN SPINAL CORD

R. H. BAXENDALE

Institute of Physiology
University of Glasgow
G12 8QQ, Scotland, U.K.

SUMMARY

Almost all knowledge about the functioning of the spinal cord comes from experimentation in cats and rats and most commonly from acute experimental procedures. It is clear that in several respects the acute rat and cat cord provides a poor experimental model for the normal functioning of the human cord. In addition, animal models fail to reproduce many clinical features of disordered function of the human cord. Most obviously, there is no good animal model for spasticity.

The material presented here has been gathered over the last 10 years of attempts to restore aspects of normal motor behaviour like standing or walking in adult humans who have suffered partial or complete spinal cord lesions. The reflex behaviour of the human cord has afforded many surprises. These have often been inconvenient from the point of view of rehabilitation of the injured person but have yield insights in to the functional organisation of the lower spinal cord segments in man.

INTRODUCTION

Over the last 30 years improvements in the care of persons who have suffered neurological injury has resulted in substantial populations of adult humans with damaged spinal cords. The European Community is estimated to have 250,000 persons with partial or complete lesions. The single most common cause of these injuries is road traffic accidents, but industrial and sporting injuries are also common as are developmental problems of the nervous system such as spina bifida and problems in later life such as multiple sclerosis. Given this variety of causes of damage there is considerable variation in the reflex performance between individuals but certain groups can be identified fairly easily.
This population of paraplegics can be investigated by appropriate experimental techniques to examine the reflex performance of the intact spinal cord segments.

The main thrust behind work of this sort lies in the hope that a more complete understanding of the reflex capabilities of isolated cord segments might lead the incorporation of residual segmental reflex properties in to a pattern of treatment or rehabilitation. For example, several research groups in Europe, the USA and Japan have succeeded in restoring a primitive gait in paraplegics by activating muscles via microcomputer controlled stimulators (see Popovic 1990, Pedotti and Ferrarin 1992 for reviews). The principal problem in this task lies in coping with the number of channels of stimulation needed to generate a reasonably smooth movement. Considerable economy in controlling the swing phase of stepping can be achieved if a single channel of stimulation is used to initiate a flexion reflex rather than using concurrent activation of hip, knee and ankle dorsiflexors. In addition, tasks requiring co-ordinated activity in several muscles like emptying of the bladder, could also be devolved to segmental reflex control, if the characteristics of the segmental mechanisms could be understood. The potential personal, clinical and economic benefits of learning to understand the residual behaviour to be found in isolated human spinal cord segments is enormous. This paper will concentrate on results obtained whilst studying the flexion reflex.

METHODS

The results described here come from human volunteers were patients at the Spinal Injuries Unit at Philipshill Hospital or who are now patients at the Queen Elizabeth National Spinal Injuries Unit at the Southern General Hospital, Glasgow. All experimental procedures have been approved by the local ethics committee and the projects have received financial support from The Medical Research Council and Action Research. The co-operation of patients and clinical staff is gratefully acknowledged.

Volunteers were supported in an upright position throughout each experiment. This was achieved in a variety of ways; by being strapped to a harness on a rigid metal support frame, by supporting themselves with a long leg calliper on one limb and their arms using parallel bars, by parallel bars alone or by using electrical stimulation of extensor muscles combined with minimal mechanical bracing of the limb with a lightweight ankle/foot orthosis. Reflex activity was recorded as the mechanical movement of the hip or knee joint in the unrestrained limb as detected by self-centring electrogoniometers (Penny & Giles Ltd, U.K.). These mechanograms were preferred because principal aim of the project is to incorporate reflexes into functional limb control. Thus any reflex must be strong enough to move or brace the limb for load bearing if it is to be useful. These mechanograms could be combined with multichannel EMG recording when necessary. The EMG analyses were most commonly used to search for co-contraction of antagonist muscles.

Attempts to elicit stretch reflexes in knee extensor muscles were made during routine clinical investigations of spasticity. The stretches could be slow, as in the standard Ashworth test, or rapid in the Fast Passive Movement test or the Wartenberg pendulum test. Of these, the pendulum test is the easiest to quantify and angular velocities exceeding 300° per second were regularly employed.

Flexion reflexes were elicited by transcutaneous stimulation of the common peroneal nerve, saphenous or sural nerves. Pulses lasting between 100 microseconds and 5 milliseconds were delivered via self-adhesive electrodes 5cm in diameter on the skin over the nerve. The stimulation was delivered as trains of pulses at 20, 50 or 100Hz, lasting for 0.5sec and the intensity was increased until flexor responses were obtained, up to a maximum of 120 milliamperes.

RESULTS

Stretch Reflexes

Whilst almost all paraplegics showed stronger patellar tendon jerks than normal most of the time, their stretch reflexes were more difficult to characterise. No consistent pattern of stretch reflexes seen with Ashworth tests, fast passive movement tests or pendulum tests in paraplegic volunteers. Some individuals exhibited almost no stretch reflex behaviour with an almost silent quadriceps EMG even during fast stretches. In other individuals fast stretching caused powerful long-lasting contractions which would support the lower limb rigidly in a horizontal position.

Clinical or biomechanical assessment of stretch reflexes proved to be impossible since there was almost no correlation between the results of the different tests, even when they were all performed by the same experimenter in the most consistent circumstance which could be arranged. The most obvious conclusion is that the tests measure different features of spinal cord behaviour, ie that tendon jerk amplitude is not related to the sensitivity or intensity of the stretch reflex and that tonic and phasic stretch reflexes as elicited by fast or slow stretching of quadriceps are similarly independent. In addition, it is worthwhile considering that the increased connective tissue which develops in paraplegic muscle might be a complicating factor here since it will provide similar mechanical properties to tonic stretch reflexes. Some authorities consider that about half of the 'stiffness' opposing rotations of the ankle joint during voluntary contractions comes from connective tissues with the balance being provided by segmental stretch reflexes (Sinkjaer, personal communication).

Repeated testing of the same volunteers by the same experimenters on successive occasions yielded very different results. Whilst this can be partly explained by day to day variations in the clinical judgement of the investigator, it also suggests that stretch reflexes in as much as they are related to quadriceps spasticity are not stable even in paraplegics several years after the spinal lesion.

In practice, the residual spinal stretch reflexes in quadriceps are very rarely strong enough to stabilise the knee position sufficiently to support the weight of a standing paraplegic adult. This is not due to any muscular weakness since the force generated under continuous direct electrical stimulation can support an upright posture for upwards of 10 minutes. Such continuous stimulation is a very inefficient way of activating muscle since it quickly fatigues muscle. Indeed, during quiet standing in normal people there is remarkably little quadriceps activity, mostly occurring in short bursts.

We have constructed orthoses incorporating an 'artificial quadriceps stretch reflex' so that the muscle is stimulated only when the knee begins to flex (Andrews and Baxendale 1986). This rapidly returns the knee to full extension where the body weight is transmitted through the bones and so the muscle activity can be shut off. Typically, the stimulator is on for less than 10% of the standing time and so muscle fatigue ceases to the limiting factor. Thus incorporating of the 'artificial stretch reflex' serves as a mechanism for reducing muscle activity and so delaying the onset of fatigue at least as much as a mechanism for stabilising joint position.

Flexion Reflexes

Undergraduate textbooks frequently discuss the flexor reflex in the lower limb of neurologically normal humans, often reproducing an illustration from Descartes original work. The experience in this laboratory is that flexor reflexes are extremely difficult to elicit in the lower limbs of normal adults. Electrical stimulation applied to the peroneal, saphenous or tibial nerves can cause very painful sensations but rarely elicits frank flexor responses. When these do occur they are generally recorded as short bursts of EMG activity lasting 20-30msecs with a latency of 50-60msec. They almost never result in movement of the leg. This poor reflex performance is not connected to posture since similar results are seen with the subjects standing or seated. Naive subjects show no stronger a reflex than more experienced volunteers do.

Similar experiments repeated in paraplegics with intact lumbar cord segments reveals a rather different picture. In some cases it is impossible to elicit flexor reflexes even with the most intense stimulation, even though afferent and efferent pathways are intact. However, in the great majority of investigations the flexor response can be elicited. The response is usually a co-ordinated flexion of hip, knee and ankle. Since the principal interest here is gait restoration the most significant experimental task is recording hip movement. Whilst the limb flexion allows the leg to swing forward with the toes clearing the ground, hip flexion is necessary to reposition the foot ahead of the body ready for the next stance phase (Andrews et al.,1988, Granat et al.,1993).

Unlike stretch reflexes, flexion reflexes are of great practical use in gait restoration. They can be elicited in almost all paraplegics with intact lumbar cord segments by a burst of 5-10 shocks at 20Hz at about 5 times motor threshold delivered to the common peroneal nerve. However, crossed extensor reflexes are either absent or very weak and are never of any functional value.

The practical difficulty arises on repeated stimulation. After 10 to 20 bursts of stimulation habituation of the response often sets in and magnitude of the reflex declines. If the hip flexion is less than 10-15^0 then the limb will not swing through for the next step and so gait stops. Several strategies have been tried in an attempt to avoid habituation. Applying the stimulation to other nerve trunks eg the saphenous and sural or rotating the stimulation amongst these sites or stimulating them in combination does not significantly delay habituation. This suggests that the fall off in reflex magnitude is not due to a loss of excitability of flexor reflex afferent fibres. Neither can the decline be a consequence of muscle fatigue since repeated stimulation of the peroneal nerve causes little decrement in the direct motor response in the pretibial dorsiflexors. The most likely interpretation is that habituation is a product of loss of responsiveness in a common pool of interneurones shared by flexor reflex afferents from a wide territory in the lower limb.

Increasing the intensity or the frequency of stimulation is also of little use in delaying the onset of habituation. However, it is possible reverse the habituation in some circumstances by single, very high intensity shocks (up to 5 msec duration and 120 milliamperes) to the peroneal nerve (Granat *et al.*, 1991). These shocks do not elicit flexor responses directly, but they can reverse the habituation process and restore the original magnitude of the flexion. This appears very similar to a potentiation of flexor responses ascribed to C fibre stimulation in spinal rats first described by Wall & Woolf (1986). No attempts have been made to record a C fibre volley following the high intensity shocks delivered to the paraplegics but their use is often accompanied by a cutaneous flare reaction suggesting that they may have been excited. The principal difference in humans is that the flexor potentiation lasts for only a minute or two whereas in the rat the potentiation may extend for a hour or more.

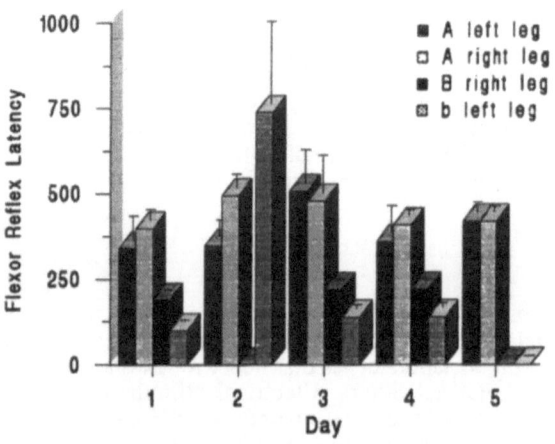

Figure 1. This figure shows the mean latency to the onset of movement at the hip when flexion reflexes were elicited in left and right legs of 2 subjects over 5 successive experimental days. In all cases the latency is variable day to day. On some days no reflex could be leicited in one leg even though short latency responses were seen in the other.

One additional problem encountered in setting up an artificial gait incorporating residual flexor reflexes is the considerable day to day variation in the latency from stimulation to hip flexion. Figure 1 shows mean and standard deviations of flexor latencies in both legs of 2 paraplegics measured on 5 separate days. The relatively long delays reflect the recording of mechanograms rather than EMGs, but the variation is still there even in the EMG. Electrical recordings are difficult to make from moving human subjects during intermittent strong electrical stimulation.

Since afferent, efferent and muscle contraction times will be subject to very little day to day change, most of the variation in latency must be due to different central delays. In one subject the flexion reflex is absent on 3 experimental days though it is present on 5 other days. It is present in one leg whilst unobtainable in the other. These observations are hard to understand if flexor reflexes are simple segmental processes.

Intriguingly, after 20-30 repetitions of stimulation, many paraplegics show a pattern of responses of alternating flexions which show little or no decrement with complete absences of flexion. This is shown in figure 2. When the rate of application of bursts of stimulation was slowed for one burst every 2 seconds to one burst every 4 seconds then each burst elicited a flexion. The most probable explanation must be that these movements are not due to simple segmental circuitry but are more the product of a pattern generator which allows flexion if the afferent volley arrives at the appropriate time and also prevents flexion if the timing the volley is inappropriate. The observations suggest that the cycle time of such a generator is longer than 2 seconds but shorter than 4 seconds. In unhurried walking, normal humans typically take a step every 2 seconds.

One additional observation supporting the belief that these observations are of a human stepping generator in action is that synchronous stimulation of left and right peroneal nerves leads to almost no reflex movement. Simple segmental pathways would allow simultaneous flexion of both limbs.

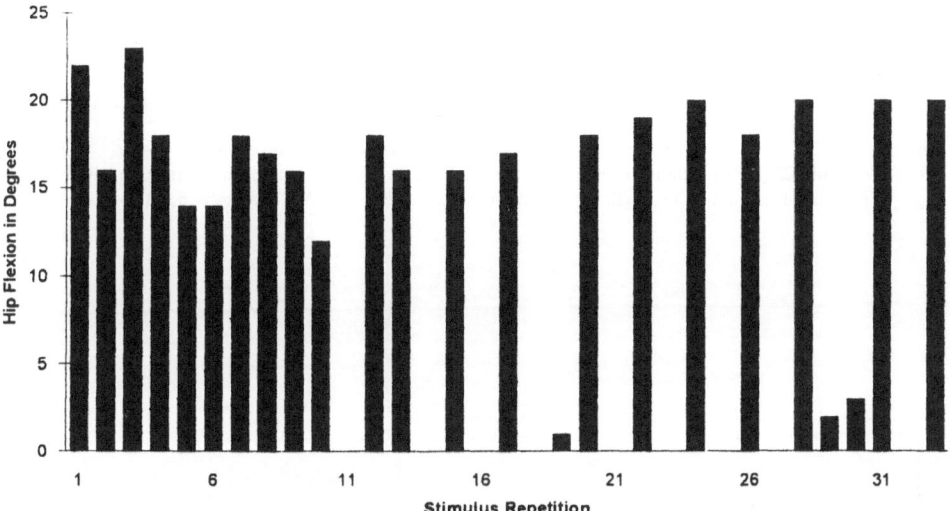

Figure 2. This figure shows the magnitude of hip flexion in degrees elicited by successive burst of stimulation applied to the peroneal nerve of a paraplegic volunteer. At the 11th burst the reflex is absent but reappears on alternating presentations of the stimulus.

DISCUSSION

The observations described here show that segmental stretch reflexes in the lower limbs of human adults never become sufficiently strong to be of use in stabilising posture, even when they appear to be exaggerated in cases of strong spasticity. The residual stretch reflexes can be sufficiently strong to make movements generated by muscles still under voluntary control or by muscles under direct electrical stimulation inconveniently slow or difficult. The conclusion must be that in normal individuals segmental stretch reflexes are under considerable descending control of an inhibitory nature to allow muscles to lengthen during movement. When stretch reflexes are desired then the segmental components may provide early resistance but this will provide too little force to be of much use. The greatest part of the functional stretch reflex must be due to supra-spinal pathways interrupted by the spinal lesion in paraplegics.

Flexor responses are common in the paraplegic population making the segmental flexion reflex response of functional significance. However, many aspects of flexor reflex behaviour suggest that not all flexor behaviour can be attributed to simple spinal circuits. The alternating responses seen during extended periods of repeated stimulation resemble the behaviour of 'stepping generators'. Such a spinal generator mechanism would explain the variable latencies recorded, the absences of responses on certain days and the near extinction of flexor responses when stimulation is applied simultaneously to both limbs. The upright posture with the legs bearing part of the body weight is probably important in allowing this behaviour to appear since seated volunteers show flexor responses with habituation but not the generator-like behaviour.

In conclusion, incorporation of residual spinal reflexes in strategies for motor restoration in paraplegics appears to have some practical applications. These are likely to be ones where low power is needed and where accurate timing of movements is not critical.

REFERENCES

Andrews, B.J. & Baxendale R.H. (1986) A hybrid orthosis incorporating artificial reflexes for spinal cord damaged patients. *Journal of Physiology* **367,** 86P.

Andrews, B.J., Baxendale R.H., Barnett R., Phillips G.F., Yamazaki T., Paul J.P. & Freeman P.A. (1988) A hybrid orthosis incorporating closed loop control and sensory feedback. *Journal of Biomedical Engineering* **10,** 189-195.

Granat M.H., Nicol D.J., Baxendale R.H. & Andrews B.J. (1991) Dishabituation of the flexion reflex in spinal cord injured man and its application in the restoration of gait. *Brain Research* **559,** 344-346

Granat MH, Heller BW, Nicol DJ, Baxendale & Andrews BJ (1993) Improving limb flexion in FES gait using the flexion withdrawal response for the spinal cord injured person. *Journal of Biomedical Engineering.* **15,** 51-56.

Pedotti A. and Ferrarin M. (1992) *Restoration of Walking for Paraplegics.* IOS Press Amsterdam ISBN 90 5199 094 4

Popovic D. (1990) *Advances in the Control of Human Extremities.* Nauka, Belgrade ISBN 86-7621-001-2

Wall PD & Woolf CJ (1986) Relative effectiveness of C fibre primary afferents of different origins in evoking a prolonged facilitation of the flexion reflex in the rat. *Journal of Neuroscience* **6,** 1433-1442

ENKEPHALINERGIC AND MONOAMINERGIC CONTROL OF SEGMENTAL PATHWAYS FROM FLEXOR REFLEX AFFERENTS (FRA)

E.D. SCHOMBURG

Institute of Physiology
University of Göttingen
D-37073 Göttingen, Germany

SUMMARY

The influence of DOPA, μ- and δ-agonistic enkephalins and naloxone on short- and long latency FRA pathways and spinal rhythm generation was comparatively analyzed in high spinal cats. The results showed that DOPA and the enkephalins are synergistic with respect to a depression of short latency FRA pathways but antagonistic with respect to long latency FRA pathways and spinal rhythm generation which are facilitated by DOPA but depressed by the enkephalins. DOPA and naloxone, on the other hand, are synergistic with respect to a facilitation of long latency FRA pathways and rhythm generation, but antagonistic with respect to the short latency FRA pathways, which are depressed by DOPA but not by naloxone. Without prior DOPA application naloxone may induce high frequency (up to 5.9 cycles/sec) rhythmic activity. The results support the hypothesis that the long latency FRA pathways and their release by DOPA form the basis for a rhythmic motor activity and that spinal motor functions of the enkephalinergic systems go beyond a plain nocifensive engagement. They may subserve a mechanism which blocks the influence of the afferent FRA feed back and its spinal motor actions during the performance of specific movements.

INTRODUCTION

In the sixties Anders Lundberg and co-workers demonstrated that there are two distinctly different pathways from cutaneous, joint and high threshold muscle afferents, being subsumed as flexor reflex afferents (FRA), to α-motoneurones: a short latency oligosynaptic one and a long latency one via which long-lasting responses were induced. These pathways were shown to be under a strong descending monoaminergic control. In the spinal animal the transmission in the short latency pathway was heavily depressed by DOPA which is assumed to increase the turnover and release of transmitter from the terminals of descending noradrenergic pathways, while DOPA released the long latency pathways from a tonic inhibition which is characteristic for these animals (Andén, Jukes, Lundberg and Vyklicky, 1966). The long latency pathways revealed a reciprocal "half-centre" organization which was assumed to form the basis for rhythmic spinal motor activity up to a locomotor pattern occurring after injection of DOPA (Jankowska, Jukes, Lund and Lundberg, 1967; for further details and references see reviews: Baldissera, Hultborn and Illert 1981; Schomburg, 1990). The enkephalinergic influence on spinal motor control has been less extensively investigated. However, regarding the interactions between the monoaminergic and the enkephalinergic systems at the different levels of the central nervous system (Duggan, 1985; Fields and Basbaum, 1978; Yaksh, 1981; Clarke, Ford and

Neural Control of Movement, Edited by W.R. Ferrell
and U. Proske, Plenum Press, New York, 1995

Taylor, 1988; Chesselet, Cheramy, Reisine and Glowinski, 1981) it is of interest to compare the influence of both systems on the segmental FRA pathways and spinal rhythm generation.

TECHNICAL COMMENTS

In order to avoid interactions with supraspinal motor centres, movement related afferent activity and anaesthetic influences all experiments were performed in anaemically decapitated high spinal paralyzed cats (Kniffki, Schomburg and Steffens, 1981). Recording techniques included intracellular motoneuronal recording, monosynaptic reflex testing and neurogram recording. Nociceptive cutaneous afferents were activated by radiant heat, other cutaneous and muscle afferents by graded electrical stimulation. The enkephalins (D-Ala^2, N-Me-Phe^4, Gly^5-ol)-enkephalin (DAGO, μ-receptor agonist) and (D-Ser^2)-Leuenkephalin-(Thr^6) (DSLET, δ-receptor agonist) were either injected i.v. (0.5 - 3.6mg/kg) or suffused over the lumbar spinal cord (10^{-3} - 10^{-5}M in Ringer's solution).

RESULTS

Short Latency FRA Pathways

Since the investigations of Lundberg et al., (see above) the sensitivity of a spinal pathway to DOPA could almost be taken as a sign for the attribution of that pathway to the FRA system. Therefore the fact should be stressed that the transmission in segmental reflex pathways from group II muscle afferents is also blocked by i.v. injection of DOPA or clonidine (Schomburg and Steffens 1988) or local intraspinal application of DOPA (Bras, Cavallari and Jankowska, 1988). As shown in Fig. 1 the blockade by DOPA concerned the short latency, probably disynaptic group II pathway as well as the polysynaptic ones. In fact, the depression of the transmission in reflex pathways from group II muscle afferents was generally more complete than that in cutaneous reflex pathways (Schomburg and Steffens 1988). These results support the attribution of group II muscle afferents to the FRA.

However, it turned out that the fairly specific monoaminergic action on the short latency FRA pathways is not unique. The enkephalins DSLET and DAGO with a δ- and μ-opioid receptor specificity exert quite similar specific effects on these FRA pathways. Fig. 2 summarises the results of 14 experiments, six experiments with i.v. injection of DSLET (0.5-3.6mg/kg) and eight experiments with superfusion of DSLET (10^{-4}-10^{-5} M solution) over the lumbar spinal cord. It shows that the transmission in the excitatory FRA pathways from nociceptive afferents, low threshold cutaneous afferents and group II muscle afferents to the flexor PBSt (A) is heavily depressed by DSLET. Thereby it is noteworthy that the depression of the transmission in the pathway from group II muscle afferents is almost as high as that in the pathway from nociceptive afferents. Pathways from joint afferents (not shown in the figure) were affected in the same range.

Inhibitory FRA pathways to extensors (B) were generally less affected than the excitatory FRA pathways, but on average they nevertheless underwent a significant depression. On the other hand the "autogenetic" inhibitory pathway from group Ib muscle afferents to extensors (the pathway to flexors is not open in the preparation used) was almost never affected by DSLET (D).

In order to underline the fact that the opioid action is really selective with respect to FRA pathways and not primarily to nociceptive pathways and just additionally to FRA pathways, it was necessary to show that nociceptive pathways not belonging to the FRA system were not influenced by the opioids or at least to a distinctly less extent than FRA pathways. Indeed this could be demonstrated. There is an excitatory nociceptive pathway from the central pad of the foot to plantaris and to intrinsic foot extensors which does not belong to the FRA system. As shown in Fig. 2C the transmission in this pathway is virtually not depressed by DSLET. In some experiments it was even slightly increased, probably due to an overlapping parallel FRA inhibition which was depressed by the opioid (Schmidt, Schomburg, Steffens, Strohmeyer and Wada, 1987).

The depressive effect of the opioids on the short latency FRA pathway was fully antagonized by naloxone (opioid antagonist with preferential μ-receptor specificity, 0.1-0.5

PBSt Motoneurone

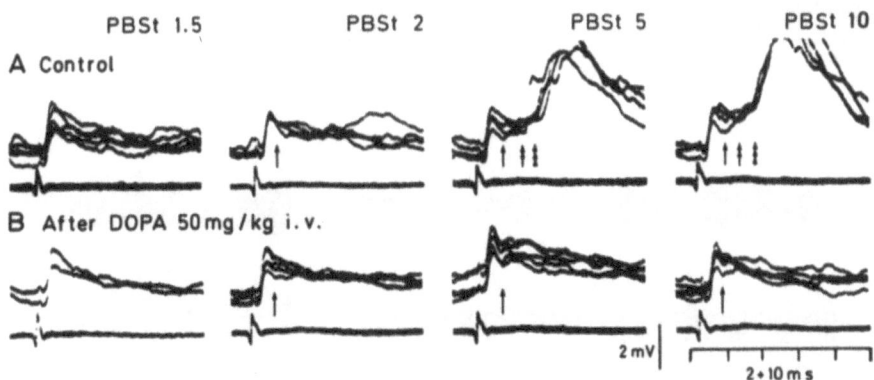

Figure 1. Inhibition of reflex transmission in short latency probably disynaptic (second, double arrow) and longer latency oligosynaptic (third, triple arrow) pathways from group II muscle afferents by DOPA, while monosynaptic group I EPSPs and early probably disynaptic group I and/or monosynaptic group II EPSPs (first, simple arrow) were not affected. High spinal cat; intracellular recording of a posterior biceps semitendinosus (PBSt) motoneurone; stimulation of the PBSt nerve with indicated strength (modified from Schomburg and Steffens, 1988).

mg/kg i.v.). Naloxone given without prior opioid application generally had a facilitatory action on the transmission in these pathways, particularly in the excitatory ones to the flexor (Schmidt *et al.*, 1991).

Long Latency FRA Pathways

While DOPA and the enkephalins had a synergistic action with respect to the short latency FRA pathways and naloxone had an antagonistic influence, the pattern of effects

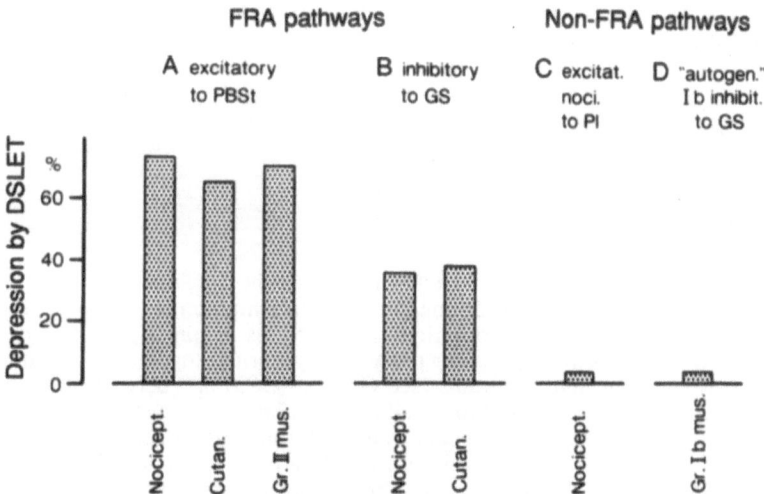

Figure 2. Inhibitory influence of the δ-opioid receptor agonist DSLET on the transmission in segmental FRA pathways (**A** and **B**) versus non-FRA pathways, an excitatory nociceptive pathway from the central pad of the foot to plantaris and intrinsic foot extensors (**C**) and the "autogenetic" Ib inhibitory pathway (**D**). The height of the columns represents the percentage of depression of the facilitation (**A** and **C**) or inhibition (**B** and **D**), respectively, of the amplitude of monosynaptic reflexes of the indicated muscles by conditioning stimulation of the corresponding afferents after application of DSLET (data taken from Schmidt, Schomburg and Steffens, 1991). Activation of nociceptive cutaneous afferents by radiant heat, activation of low threshold cutaneous afferents (stimulus strength < 1.2 times threshold) and muscle group Ib and II afferents by graded electrical nerve stimulation.

was different with respect to the long latency FRA pathways. These pathways, which were activated by stimulation of FRA with trains and higher stimulus strength, typically became manifest after DOPA (Andén et al., 1966). By application of the enkephalins the transmission in these pathways was depressed. An action which was antagonized by naloxone.

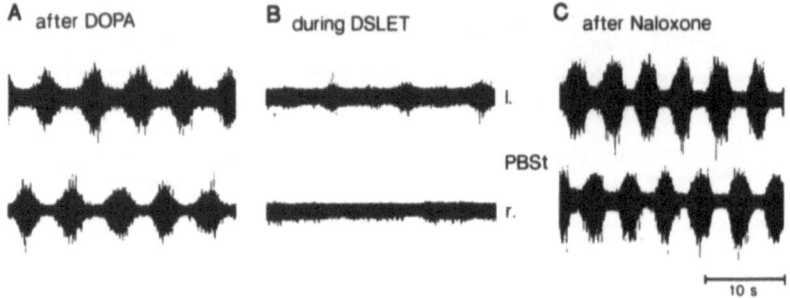

Figure 3. Influence of DOPA and the δ-opioid receptor agonist DSLET on spinal rhythm generation. Neurograms of left (l.) and right (r.) posterior biceps semitendinosus (PBSt) in a high spinal paralyzed cat. **A:** Stable alternating rhythm after injection of nialamide (50mg/kg i.v.) and DOPA (30mg/kg i.v.), **B:** depression of that rhythm by superfusion of the lumbar spinal cord with DSLET (10^{-3}M in Ringer's solution), **C:** more than plain antagonization of the DSLET effect by naloxone (0.5mg/kg i.v., cord being still covered with DSLET solution).

Naloxone on its own, without prior opioid application, facilitated the transmission in the long latency FRA pathways. Interestingly it facilitated a rhythmic modulation of the long lasting response evoked via these pathways.

Table 1 Influence of DOPA, enkephalins and naloxone on segmental FRA pathways and spinal motor rhythm generation (+ = facilitation; - = depression)

	Short latency FRA pathways	Long latency FRA pathways	Spinal motor rhythm generation
DOPA	-	+	+
Enkephalins	-	-	-
Naloxone	+	+	++

Spinal Rhythm Generation

As mentioned in the introduction DOPA facilitates the generation of rhythmic spinal motor activity. Particularly after premedication with nialamide, a monoaminoxidase blocker, DOPA can induce long lasting rhythmic activities in spinal paralyzed cats. These activities largely resemble the efferent activity during normal locomotion including a co-ordination between fore- and hindlimbs (see Grillner 1985) and between limb and trunk muscles (Köhler, Schomburg and Steffens, 1984). Fig. 3A shows such as stable alternating rhythm between left and right PBSt being induced by injection of nialamide and DOPA. Superfusion of the lumbar spinal cord with a solution of the δ-opioid receptor agonist DSLET (B) severely depressed this rhythm. Intravenous injection of that substance or of the μ-opioid receptor agonist DAGO had the same or often even a stronger depressing effect on DOPA induced rhythmic activity. Naloxone (0.5mg/kg i.v.) injected while the cord was still covered with the DSLET solution (C) did not only antagonize the DSLET effect, but induced a rhythmic activity which was more pronounced than that observed before application of DSLET indicating some further naloxone action. The conclusion that naloxone has a rhythm facilitating potency beyond the antagonization of exogenously supplied enkephalins was confirmed by injections of naloxone in non-paralyzed and paralyzed cats before application of nialamide and DOPA. In both cases naloxone induced rhythmic motor activity with a distinctly higher frequency (up to 5.8 cycles/s in the paralyzed spinal cat) than that observed after DOPA (around 0.2 - 0.3 cycles/s, Schomburg and Steffens 1992; *cf.* also Pearson, Jiang and Ramirez, 1992).

DISCUSSION and CONCLUSIONS

A comparison of the influence of DOPA, eg enkephalins and naloxone on short- and long latency FRA pathways and spinal rhythm generation (see Table) shows, that DOPA and the enkephalins are synergistic with respect to the depression of short latency FRA pathways but antagonistic with respect to long latency FRA pathways and spinal rhythm generation which are facilitated by DOPA but depressed by the enkephalins. DOPA and naloxone, on the other hand, are synergistic with respect to the facilitation of long latency FRA pathways and rhythm generation, but antagonistic with respect to the short latency FRA pathways, which are depressed by DOPA but not by naloxone.

If one regards the results concerning motor rhythm generation they support the hypothesis that the long latency FRA pathways and their release by DOPA form the basis for a rhythmic motor activity (Jankowska *et al.,* 1967). A substance like naloxone which facilitates the long latency FRA pathways also facilitated rhythm generation, while substances depressing the long latency pathways like the enkephalins also depressed spinal rhythm generation. If the generation of high frequency rhythms after naloxone in the absence of DOPA is due to the additional facilitation of short latency FRA pathways by this substance cannot yet be decided.

The results support the assumption that spinal motor functions of the enkephalinergic system go beyond a plain nocifensive engagement, but that they subserve a mechanism which may block the influence of the quite unspecific input from FRA during a movement in cases in which a centrally induced movement has to be performed without a disturbance by the afferent FRA feed back and its spinal motor actions (Schmidt *et al.,* 1991).

ACKNOWLEDGEMENT. Supported by the Deutsche Forschungsgemeinschaft (Scho 37/3-3).

REFERENCES

Andén, N.-E., Jukes, M.G.M., Lundberg, A. and Vyklicky, L. 1966). The effect of DOPA on the spinal cord. 1. Influence on transmission from primary afferents. *Acta Physiologica Scandinavica* **67**, 373-386

Baldissera, F., Hultborn, H. and Illert, M. (1981). Integration in spinal neuronal systems. In V.B. Brooks (Ed.). Handbook of Physiology, Vol. 2, Sect. I, Nervous System, Motor Control, Part 1, American Physiological Society, Bethesda, MD, pp. 509-595

Bras, H., Cavallari, P. and Jankowska, E. (1988). An investigation of local actions of ionophoretically applied DOPA in the spinal cord. *Experimental Brain Research* **71**, 447-449

Chesselet, M.F., Cheramy, A., Reisine, T.D. and Glowinski, J.(1981). Morphine and delta-opiate agonists locally stimulate in vivo dopamine release in cat caudate nucleus. *Nature* **291**, 320-322

Clarke, R.W., Ford, T.W. and Taylor, J.S. (1988). Adrenergic and opioidergic modulation of a spinal reflex in the decerebrated rabbit. *Journal of Physiology* **404**, 407-414

Duggan, A.W. (1985). Pharmacology of descending control systems. *Philosophical Transactions of the Royal Society of London B* **308**, 375-391

Fields, H.L. and Basbaum A.I. (1978). Brain stem control of spinal pain transmission neurons. *Annual Review of Physiology* **40**, 193-221

Grillner, S. (1985). Neural control of vertebrate locomotion - central mechanisms and reflex interaction with special reference to the cat. In W.J.P. Barnes and M.H. Gladden (Eds). Feedback and Motor Control in Invertebrates and Vertebrates, Croom Hall, London, pp. 35-56

Jankowska, E., Jukes, M.G.M., Lund, S. and Lundberg, A. (1967).The effect of DOPA on the spinal cord. 6. Half-centre organization of interneurones transmitting effects from the flexor reflex afferents. *Acta Physiologica Scandinavica* **70**, 389-402

Kniffki, K.-D., Schomburg, E.D. and Steffens, H. (1981).Effects from fine muscle and cutaneous afferents on spinal locomotion in cats. *Journal of Physiology* **319**, 543-554

Koehler, W.J., Schomburg, E.D. and Steffens, H. (1984).Phasic modulation of trunk muscle efferents during fictive spinal locomotion in cats. *Journal of Physiology* **353**, 187-197

Pearson, K.G., Jiang, W. and Ramirez, R.M. (1992). The use of naloxone to facilitate the generation of the locomotor rhythm in spinal cats. *Journal of Neuroscience Methods* **42**, 75-81

Schmidt, R..F., Schomburg, E.D. and Steffens, H. (1991). Limitedly selective action of a δ-agonistic leu-enkephalin on the transmission in spinal motor reflex pathways in cats. *Journal of Physiology* **442**, 103-126

Schmidt, R..F., Schomburg, E.D., Steffens, H., Strohmeyer, A. and Wada, N. (1987). A nociceptive non-FRA pathway to plantaris in the cat. *Journal of Physiology* **390**, 49P

Schomburg, E.D. (1990). Spinal sensorimotor systems and their supraspinal control. *Neuroscience Research* **7**, 265-340

Schomburg, E.D. and Steffens, H. (1988). The effect of DOPA and clonidine on reflex pathways from group II muscle afferents to alpha-motoneurones in the cat. *Experimental Brain Research* **71**, 442-446

Schomburg, E.D. and Steffens, H. (1992). Facilitation of rhythmic motor activity by naloxone in cats. *Acta Physiologica Scandinavica* **146**, Suppl. 608, C 7.3

Yaksh, T.L. (1981). Spinal opiate analgesia: Characteristics and principles of action. *Pain* **11**, 293-346

CAN THE SYMPATHETIC NERVOUS SYSTEM ACTIVATION CONTRIBUTE TO "CONTEXT-RELATED" MODULATIONS OF THE STRETCH REFLEX ?

M. PASSATORE, C. GRASSI[*] and F. DERIU

Dipartimento di Anatomia e Fisiologia Umana
Università di Torino, Corso Raffaello 30
10125 Torino, Italy

[*]Instituto di Fisiologia Umana
Università Cattolica del Sacro Cuore
Largo F. Vito 1, 00168 Roma, Italy

SUMMARY

Sympathetic stimulation at frequencies within the physiological range, markedly depresses both the jaw jerk and the tonic vibration reflex in the jaw closing muscles of precollicular decerebrate rabbits. In particular, bilateral stimulation of the cervical sympathetic trunk reduces by 60-90% the force reflexly produced by the jaw closing muscles and strongly decreases or suppresses EMG activity on both sides. These effects are mainly mediated by α-adrenergic receptors, a modest contribution of the noradrenaline co-transmitter neuropeptide Y to the late component of the response not being excluded. In addition, the sympathetically-induced depression of the stretch reflex cannot be attributed to an action, direct or secondary to vasomotor changes, exerted by the adrenergic mediator either on jaw muscles contraction or on the central nervous structures mediating the studied reflexes.

Data reported in the present paper suggest that the sympathetic nervous system can influence the stretch reflex in the jaw closing muscles through an action exerted at the peripheral level, probably on spindle afferent information. The possible functional role of such sympathetic effect is discussed within the context of the numerous modulatory actions exerted on the stretch reflex gain by central command and sensory information during intentional movements.

INTRODUCTION

There is an increasing body of evidence which shows that the gain of the stretch reflex exhibits a large variability during voluntary movements, depending on the type and difficulty of the task and on the context in which it is to be executed, e.g. prior instruction to subjects, external disturbances, attention and psychological factors (Dufresne *et al.*,1980; Lund and Olson, 1983; Taylor and Gottlieb, 1985; Meunier and Pierrot-Deseilligny, 1989; references in Prochazka, 1989 and Davidoff, 1992).

The modulation of the gain of the myotatic feedback is operated by central commands and sensory information. Such modulation is exerted both at peripheral level, mainly by changing the operational condition of the neuromuscular spindles, and at central level, by changing either the transmission of proprioceptive information to motoneurones or the level of excitability of the motoneurone itself. Our recent studies on the stretch reflex changes induced

Neural Control of Movement, Edited by W.R. Ferrell
and U. Proske, Plenum Press, New York, 1995

109

by the sympathetic nervous system activation contribute in providing experimental evidence of one of the modulatory actions exerted at the peripheral level. In particular, we studied the effect of cervical sympathetic trunk (CST) stimulation on the jaw jerk reflex (JJR) and on the tonic vibration reflex (TVR) in jaw elevator muscles.

METHODS AND RESULTS

In precollicular decerebrate rabbits the JJR and the TVR were evoked by applying respectively to the lower jaw ramp-and-hold stretches (downward mandibular displacements of 1-2mm at a speed of 0.1-0.2mm/ms, maintained for 0.5-1s and repeated every 4-6s) and vibratory stimulations (170Hz, 15-50μm amplitude, lasting 5s and repeated every 10-30s). These mechanical stimuli were delivered by a puller rigidly connected to the mandibular symphysis, servocontrolled by the length signal of the jaw position. Electromyographic activity (EMG) from the masseter muscle was bilaterally recorded and force developed by the reflex contraction of the jaw muscles was measured through an isometric transducer placed in series with the puller (for further methodological details, see Grassi et al., 1993a,b).

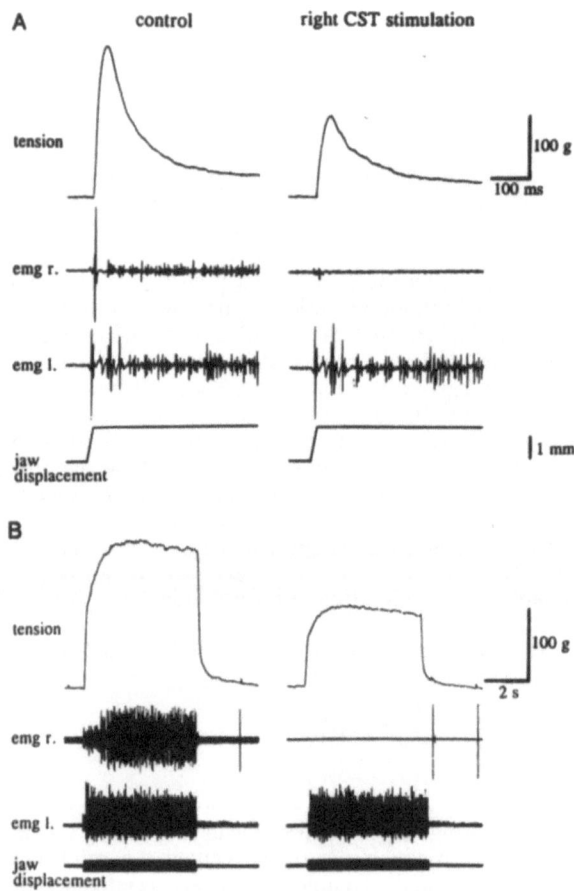

Figure 1. Effect of unilateral sympathetic stimulation at 10s^{-1} on the jaw jerk reflex (A) and on the tonic vibration reflex (B) in the jaw elevator muscles. These reflexes are induced by applying respectively to the mandible a downward movement (1.5mm at a speed of 0.1mm ms^{-1}, maintained for 1s) and vibrations (170Hz, 40μm peak-to-peak amplitude). The electromyograms are recorded from the right (emg r.) and left (emg l.) masseter muscles. Tension records in A were obtained by subtracting the passive tension from the total recorded force (for further details, see Grassi et al., 1993a).

Stimulation of the peripheral stump of the CST at 3-10s^{-1}, lasting 30s - 3min, consistently elicited a frequency-dependent depression of both TVR and JJR. In particular, unilateral stimulation at 10s^{-1} produced a marked reduction or suppression of the EMG activity in the ipsilateral masseter muscle, associated with a 30-40% decrease of the reflexly developed force (Fig. 1), whilst the EMG from the contralateral side was not significantly affected. Bilateral CST stimulation induced a marked decrease or the disappearance of the EMG activity on both sides and reduced the developed force by 60-90%. These effects lasted throughout the stimulation period, then a rapid partial recovery occurred in the first minute, followed by a slower return to the control values. This required 1 to 10 minutes, depending on the duration of the stimulation. The above described TVR and JJR depression was often preceded by a transient enhancement of variable magnitude (see Grassi et al. 1993b) which is not taken into consideration in the present paper. However, depending on whether such activatory effect was present and on how long it lasted, the latency of the reduction of the reflex ranged from 7 to 40s.

In order to define the site of action of the sympathetically-induced reduction in the amplitude of the myotatic reflexes we followed an exclusion criterion, i.e. we performed different trials which allowed us to exclude a number of possibilities.

First of all the observed effect cannot be attributed to an action exerted by noradrenaline on muscle contraction since sympathetic stimulation elicited a small increase in the contractile force of the directly stimulated jaw muscles (Passatore and Grassi, 1991; Grassi et al. 1993a). The possibility that the TVR depression could be secondary to vasomotor effects causing muscle hypoxia or changes in muscle mechanics should also be discarded on the basis of both the latency of the response and the inefficacy of bilateral occlusion of the common carotid artery in reducing the TVR. In decerebrate animals this manoeuvre guarantees, in fact, the interruption of blood supply to jaw muscles.

Moreover, a possible influence exerted on the reflex by sympathetically-affected sensory inputs other than spindles should be ruled out. In fact the elimination of afferent information from temporomandibular joints (by local anaesthesia) and from structures localized in the inferior dental arch (by bilaterally sectioning the inferior alveolar nerve) did not significantly modify the effects of sympathetic stimulation on the TVR (Grassi et al. 1993b).

We also checked whether CST stimulation could affect the stretch reflex by producing a decrease in blood flow at the level of the soma of the afferent neurones from jaw muscle spindles (mesencephalic trigeminal nucleus) or of the motor nucleus of the V nerve. For this purpose the sympathetically-induced changes of blood flow were analysed in this area and compared with those produced in other central nervous system structures whose flow is known to be affected by sympathetic fibre activation (Saeki et al. 1990; references in Sato and Sato, 1992). Blood flow in the microvascular bed was studied by using a laser-Doppler flowmeter. Bilateral stimulation of the CST at 10s^{-1}, which occasionally produced very modest reduction in blood flow in the tested areas of the cerebellum, did not elicit significant changes in the brainstem area where the mesencephalic and the motor trigeminal nuclei are located.

We then studied the adrenergic receptors responsible for the stretch reflex decrease and checked whether neuropeptide Y, co-transmitter of noradrenaline in postganglionic sympathetic fibres, contributes to such an effect. The sympathetically-induced depression in both JJR and TVR was almost completely abolished by α-adrenoceptor blockade obtained with phentolamine (2.0-2.5mg kg^{-1}, i.v.) or prazosin (1mg kg^{-1}, i.v.), the latter drug being a selective antagonist of α_1-adrenoceptors. The blockade of ß-receptors (propranolol, 1-2 mg kg^{-1}, i.v.) however did not significantly affect the reflex reduction. The role of α_1-adrenoceptors is also confirmed by trials in which injection of phenylephrine into the thyroid artery produced a dose-dependent decrease in the TVR (Fig. 2). Neuropeptide Y administration (150-500pM kg^{-1}, i.a.) often induced a long-latency and long-lasting small decrease in the developed force which might be associated with a reduction of the discharge in the recorded motor units.

DISCUSSION

These data show that both phasic and tonic stretch reflexes (JJR and TVR) are consistently and markedly reduced by sympathetic nervous system activation. This effect is largely mediated by α_1-adrenoceptors, a modest contribution of neuropeptide Y to the late component of the response not being excluded. Experimental findings showed however that the reflex changes are not secondary to the α-adrenoceptor mediated vasoconstriction, and consequent hypoxia induced by CST stimulation at the level of either skeletal muscle fibres and muscle spindles (Grassi et al., 1993b) or the central nervous system structures involved in the reflex pathways. In particular, bilateral stimulation of the CST, with the same parameters used in JJR and TVR trials, did not induce significant blood flow reductions in the brainstem areas in which the mesencephalic and the motor trigeminal nuclei are located (unpublished data), in analogy with what has been previously reported for other ponto-mesencephalic structures (references in Sato and Sato, 1992). If we were to consider the possibility that catecholamines released by the stimulated CST may reach and directly influence the central areas containing the mesencephalic and the motor trigeminal nuclei, available experimental evidence suggests that in this case the reflex responses should be enhanced rather than suppressed. In fact microinfusion of noradrenaline directly into the motor nucleus increases the JJR (Stafford and Jacobs, 1990), as does stimulation of the region of the lateral lemniscus which is reported to be the main source of the central noradrenergic innervation to the motor nucleus of the V nerve (Vornov and Sutin, 1983).

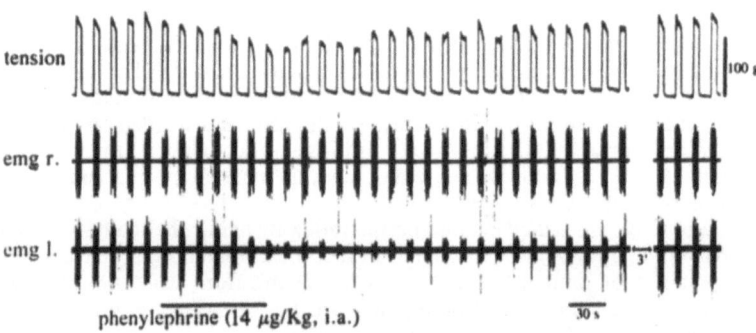

Figure 2. Effect of α_1-adrenergic agonist administration on tonic vibration reflexes, in jaw elevator muscles. TVRs are evoked by 5s lasting vibrations repeated every 15s (170Hz, 30μm peak-to-peak amplitude). Abbreviations as in Fig. 1.

The sympathetically-induced decrease in the stretch reflex should be then ascribed to a direct action exerted by the adrenergic mediators at the peripheral level. A change in the muscle spindle afferent information has been suggested as the mechanism most likely to be responsible for this effect since the contribution of other sensory inputs, possibly modulated by the sympathetic outflow, has been demonstrated not to be relevant (Grassi et al. 1993b). On the other hand, both morphological and functional data are available, pointing to a direct action of catecholamines on muscle spindles (Hunt, 1960; Barker and Saito, 1981; Hunt et al., 1982; Passatore et al.,1985a,b; references in Staderini and Ambrogi Lorenzini, 1969), even though large variability is reported in the density of innervation and in the size of the effects in the different territories. Preliminary data showing that sympathetic stimulation elicits a decrease in jaw muscle spindle sensitivity to sinusoidal stretches (Grassi et al.,1989) seem to be quite in keeping with the above mentioned interpretation.

The sympathetic influence on the stretch reflex may be operating under physiological conditions since effects of increasing magnitude can be detected by using CST stimulation rates ranging from 3 to $10s^{-1}$, i.e. frequencies which match the activity reported under physiological conditions in various sympathetic sections (Passatore, 1976; Polosa, 1979; Kocsis et al.,1990; references in Janig, 1985).

The functional role of the sympathetically-induced marked depression of the stretch reflex should be evaluated in the light of the various and still debated interpretations regarding the functional meaning of this reflex in the control of movements. In several muscles the gain of this reflex exhibits a large modulation during the course of voluntary movements and various mechanisms acting at both central and peripheral level have been proposed, which may produce such changes; the sympathetic nervous system may participate in such control. In considering the possible influences of the noradrenergic pathways on motor control, we should also mention the data indicating that several somatic and visceral afferent inputs are affected by changes in the sympathetic outflow through an action exerted at the receptor level (references in Akoev, 1981). With regard to the orofacial territory, sympathetic fibre stimulation modulates afferent information from neuromuscular spindles, some type of cutaneous mechanoreceptors and receptors located in tooth and periodontal structures (Matthews, 1976; Cash and Linden, 1982; Passatore and Filippi, 1983; Passatore et al., 1985a,b). Moreover, recent data suggest that transmission of sensory information conveyed by certain populations of C fibres is depressed by the activation of the sympathetic system (Shyu et al.,1989). Then the sympathetic system might be viewed as a system able to operate a selection among various afferent inputs, through an action exerted at the peripheral level. Consequently a change in the relative weight of the different afferent inputs, in a movement controlled by multiple feedback systems, could privilege or suppress a particular set of controls. In addition, monoaminergic projections from the brainstem have been found able to control spinal neurones as well as the trigeminal neurones involved in the myotatic reflexes, even though a univocal interpretation of these effects is not yet available (e.g. Clarke, 1988; Bras et al.,1989; Fung et al.,1991; Stafford and Jacobs, 1990; references in Grillner, 1981).

Then, if a given task is performed under conditions of stress or excitement, the ensuing movement should be the result of the combination of the original motor command, assuming it remains the same under these conditions, and the changes produced by the increase in the sympathetic outflow. In other words, within a given task-related modulation of reflexes, the noradrenergic pathways may add a modulation which adjusts the reflex to the context in which that motor task is performed.

The above described modulation of the stretch reflex, which we suggested was due to a sympathetically-induced change in the muscle spindle information, might have different functional relevance in the different muscular districts, also in relation to their specific functions. Many attempts have been made in fact to demonstrate sympathetic effects on spindle activity or on reflex responses in hindlimb muscles (references in Bowman, 1981; Grassi et al., 1993b). Overall, mainly modest effects have been demonstrated and the mechanisms are still being discussed. The effects of sympathetic stimulation we have shown on jaw muscle spindles and on the jaw elevator reflexes (Passatore et al.,1985a,b ; Passatore and Grassi 1989; Grassi et al.,1993 a,b) are noteworthy because they are much larger than any that have been observed for limb muscles. The differences should not be taken to mean that one or the other of the experimental tests are flawed in some way, but rather to indicate the variety of ways that spindle receptors might be modulated by nervous system activity.

The muscle receptors of the hind limb muscles have long been taken as representative of muscle spindles throughout the body. However, as other muscle systems are investigated, exceptions and modifications of this functional and morphological model have been observed (e.g. neck muscles, Richmond and Abrahams, 1979; eye muscles, Bach-y-Rita, 1971; respiratory muscles, Sears, 1964; jaw muscles, Goodwin and Luschei, 1975; Taylor and Gottlieb, 1985). Such differences probably reflect various adaptations to the different functional demands of the various muscle systems. Limb muscles and their reflex control are adapted to functions of locomotion and posture. The role of the spindles and the reflexes in which they participate are likely to be specifically adapted for these functions. The jaw muscles, which provide functions different from the limb muscles are likely to be controlled differently as well. Sympathetic function might also be assumed to have a different relationship to the jaw, the organ of feeding and combat for most mammals, then to the limbs, the organ of flight.

REFERENCES

Akoev, G. N. (1981). Catecholamines, acetylcholine and excitability of mechanoreceptors. *Progress in Neurobiology* **15**, 269-294.

Bach-Y-Rita, P. (1971). Neurophysiology of eye movement. In *The Control of Eye Movements*, eds. Bach-Y-Rita, P., Collins, C. C. and Hyde, J. E., pp. 7-45, Academic Press, New York, London.

Barker, D. and Saito, M. (1981). Autonomic innervation of receptors and muscle fibres in cat skeletal muscle. *Proceedings of the Royal Society*, London B **212**, 317-332.

Bowman, W. C. (1981). Effects of adrenergic activators and inhibitors on the skeletal muscles. In *Handbook of Experimental Pharmacology, Adrenergic activators and inhibitor*, 54/2, ed. Szekeres, L., pp. 47-128. Springer, Berlin Heidelberg New York.

Bras, H., Jankowska, E. and Noga B. R. (1989). Depression of transmission from group II muscle afferents to spinal interneurones by descending monoaminergic pathways. *European Journal of Neuroscience* suppl. **2**, 54.2.

Cash, R. M. and Linden R. W. A. (1982). Effects of sympathetic nerve stimulation on intra-oral mechanoreceptor activity in the cat. *Journal of Physiology* **329**, 451-463.

Clarke, R. W., Ford, T. W. and Taylor, J.S. (1988). Adrenergic and opioidergic modulation of a spinal reflex in the decerebrated rabbit. *Journal of Physiology* **404**, 407-417.

Davidoff, R. A. (1992). Skeletal muscle tone and the misunderstood stretch reflex. *Neurology* **42**, 951-963.

Dufresne, J. R., Soechting, J. F. and Terzuolo, C. A. (1980). Modulation of the myotatic reflex gain in man during ongoing intentional movements. *Brain Research* **193**, 67-84.

Fung, S. J., Manzoni, D., Chan, J. Y. H., Pompeiano, O. and Barnes, C. D. (1991). Locus coeruleus control of spinal motor output. *Progress in Brain Research* **88**, 395-409.

Goodwin, G. M. and Luschei, E.S. (1975). Discharge of spindle afferents from jaw closing muscles during chewing in alert monkeys. *Journal of Neurophysiology* **38**, 560-571.

Grassi, C., Conserva, E. and Passatore, M. (1989). Action of the sympathetic system on gain and phase of the spindle afferent response to sinusoidal changes of length. *Pflügers Archiv* **415**(3), s14,48.

Grassi, C., Deriu, F. , Artusio, E. and Passatore, M. (1993a). Modulation of the jaw jerk reflex by the sympathetic nervous system. *Archives Italiennes de Biologie* **131**, 213-226.

Grassi C., Deriu, F. and Passatore, M. (1993b). Effect of sympathetic nervous system activation on the tonic vibration reflex in rabbit jaw closing muscles. *Journal of Physiology* **469**, 601-613.

Grillner, S. (1981). Control of locomotion in bipeds, tetrapods, and fish. In *Handbook of Physiology,* sect. 1, *The Nervous System,* vol. II, *Motor Control,* ed. Brooks, V. B., pp. 1179-1236, American Physiological Society, Bethesda.

Hunt, C. C. (1960). The effect of sympathetic stimulation on mammalian muscle spindles. *Journal of Physiology* **151**, 332-341.

Hunt, C. C., Jami, L. and Laporte, Y. (1982). Effects of stimulating the lumbar sympathetic trunk on cat hindlimb muscle spindles. *Archives Italiennes de Biologie* **120**, 371-384.

Janig, W. (1985). Organization of the lumbar sympathetic outflow to skeletal muscle and skin of the cat hindlimb and tail. *Reviews of Physiology, Biochemistry and Pharmacology* **102**, 119-213.

Kocsis, B., Gebber, G. L., Barman, S. M. and Kenney, M. J. (1990). Relationships between activity of sympathetic nerve pairs - phase and coherence. *American Journal of Physiology* **259**, R549-R560.

Lund, J. P., and Olson, K.A. (1983). The importance of reflexes and their control during jaw movement. *Trends in Neuroscience* **6**, 458-463.

Matthews, B. (1976). Effect of sympathetic stimulation on the response of intradental nerves to chemical stimulation of dentine. In *Advances in Pain Research and Therapy*, eds. Bonica, J. J., Liebeskind, J. C. and Albe-Fessard, D. G., pp. 195-203. Raven Press, New York.

Meunier, S. and Pierrot-Descilligny, E. (1989). Gating of the afferent volley of the monosynaptic stretch reflex during movement in man. *Journal of Physiology* **419**, 753-763.

Passatore, M. (1976). Physiological characterization of efferent cervical sympathetic fibres influenced by changes of illumination. *Experimental Neurology* **53**, 71-81.

Passatore, M. and Filippi, G. M. (1983). Sympathetic modulation of periodontal mechanoreceptors. *Archives Italiennes de Biologie* **121**, 55-65.

Passatore, M., Filippi, G. M. and Grassi, C. (1985a). Cervical sympathetic nerve stimulation can induce an intrafusal muscle fibre contraction in the rabbit. In *The Muscle Spindle*, eds. Boyd, I. A. and Gladden, M. H., pp. 221-226. Macmillan Press, London.

Passatore, M., and Grassi, C. (1991). Somato-vegetative interaction at the peripheral level: possible effects on motor performance. In *Cardiorespiratory and Motor Coordination* eds. Koepchen, H. P. and Huopaniemi T., pp. 181-187, Springer-Verlag, Berlin, Heidelberg, New York.

Passatore, M., Grassi, C. and Filippi, G. M. (1985b). Sympathetically-induced development of tension in jaw muscles: the possible contraction of intrafusal muscle fibres. *Pflügers Archiv* **405**, 297-304.

Polosa C. (1979). Tonic activity on the autonomic nervous system: functions, properties, origins. In *Integrative functions of the autonomic nervous system*, eds. Brooks, McC, Koizumi, K., and Sato, A., pp. 342-354, University of Tokyo Press and Elsevier/North-Holland Biomedical Press.

Prochazka, A. (1989). Sensorimotor gain control: a basic strategy of motor systems? *Progress in Neurobiology* **33**, 281-307.

Richmond, F. J. R. and Abrahams, V. C. (1979). Physiological properties of muscle spindles in dorsal neck muscles of the cat. *Journal of Neurophysiology* **42**, 604-617.

Saeki, Y., Sato, A., Sato, Y. and Trzebski, A. (1990). Effects of stimulation of cervical sympathetic trunks with various frequencies on the local cortical cerebral blood flow measured by laser Doppler flowmetry in the rat. *Japanese Journal of Physiology* **40**, 15-32.

Sato, A. and Sato, Y. (1992). Regulation of regional cerebral blood flow by cholinergic fibers originating in the basal forebrain. *Neuroscience Research* **14**, 242-274.

Sears, T. A. (1964). Efferent discharges in alpha- and fusimotor fibres of intercostal nerves of the cat. *Journal of Physiology*, **174**, 295-315.

Shyu, B. C., Olausson, B., Huang, K. H., Widerström, E. and Andersson, S. A. (1989). Effects of sympathetic stimulation on C-fibre responses in rabbit. *Acta Physiologica Scandinavica* **137**, 73-84.

Staderini, G. and Ambrogi Lorenzini, C. (1969). Sympathetic nervous system and reflex muscular activity. *Archivio di Fisiologia* **67**, 70-200.

Stafford, I. L. and Jacobs, B. L. (1990). Noradrenergic modulation of the masseteric reflex in behaving cats. I. Pharmacological studies. *Journal of Neuroscience* **10**, 91-98.

Taylor, A. and Gottlieb, S. (1985). Convergence of several sensory modalities in motor control. In *Feedback and Motor Control in Invertebrates and Vertebrates*, eds. Barnes, W. J. P. and Gladden, M. H., pp. 77-92, Croom Helm, London.

Vornov, J. J. and Sutin, J. (1983). Brainstem projection to the normal and noradrenergically hyperinnervated trigeminal motor nucleus. *Journal of Comparative Neurology* **214**, 198-208.

CLASSICAL CONDITIONING OF EYEBLINK IN DECEREBRATE CATS AND FERRETS

G. HESSLOW

Department of Physiology and Biophysics
University of Lund
Sölvegatan 19
S-223 62 Lund, Sweden

SUMMARY

Several lines of evidence suggest that the cerebellum is involved in classical conditioning of the eyeblink response. For instance, lesions to the interpositus nucleus abolish both learning and retention of conditioned responses. There is strong disagreement about the precise nature of the cerebellar involvement, however. Although many now believe that it is the site of learning, is has also been argued that the cerebellum is merely necessary for the normal performance of conditioned responses. In order to clarify the role of the cerebellum in conditioning, electrophysiological techniques were applied to decerebrate cats and ferrets, which can acquire normal conditioned responses. Four small discrete areas of the cerebellar cortex have been identified which seem to control the orbicularis oculi muscle. Electrical stimulation of these areas, which inhibits neurones in the interpositus nucleus, completely suppresses a conditioned response but has only a weak effect on the unconditioned response. Recordings from Purkinje cells in one of these areas show firing patterns which are consistent with their being involved in the learning of the conditioned response. In combination with results from other groups, these findings provide strong support for the cerebellar learning hypothesis.

INTRODUCTION

If a stimulus which elicits a reflex eyeblink response, such as an air-puff to the cornea, is repeatedly preceded by a neutral stimulus, such as a tone, the neutral stimulus will gradually acquire the ability to elicit an eyeblink. This is an example of classical or Pavlovian conditioning, a form of associative learning first described by Pavlov. He originally trained dogs to respond with salivation to sound, but the same basic technique has later been applied to a wide variety of autonomic as well a skeletal muscle responses (Gormezano and Moore, 1976). In all cases, an unconditioned stimulus (US) initially causes an unconditioned response (UR). A neutral conditioned stimulus (CS) initially has no effect, but after training elicits a conditioned response (CR).

Although Pavlov suggested that the neural plasticity mediating this learning was located in the cerebral cortex, it has been known for some time that conditioning, at least of somatic responses, can occur in decorticate, hemispherectomized or hippocampectomized rabbits and cats, leaving the brain stem and cerebellum as the only possible sites of learning (Oakley and Russell, 1972; Norman, Villablanca, Brown, Schwafel and Buchwald, 1974; Schmaltz and Theios, 1972). The cerebellum, with its massive convergence of sensory information via the mossy and parallel fibres to the Purkinje cells, satisfies the anatomical requirements of a structure suitable for associative learning and has also been implicated in other forms of motor learning (Ito 1984). It is therefore a plausible site of classical conditioning.

There is now strong evidence for this hypothesis. Eyeblink conditioning is absent or at least highly abnormal in animals with lesions of the anterior interpositus nucleus (NIA), the hemispheral part of lobule VI of the cerebellar cortex or to the inferior olive (McCormick and Thompson 1984; Yeo, Hardiman and Glickstein 1985a,b, 1986). Pharmacological blockade of the NIA and overlying cortex during conditioning prevents learning, although such blockade of the red nucleus, an output relay from the cerebellum, does not have this effect (Krupa, Thompson and Thompson, 1993).

Several writers have proposed that the CS would reach the Purkinje cells in the cerebellar cortex via the mossy and parallel fibres. When this input is paired with climbing fibre input elicited by the US, changes would occur in the parallel fibre/Purkinje cell synapses.

Both the results and the interpretation of these studies have been challenged by other investigators. For instance, Welsh and Harvey (1989) reported that a substantial proportion of animals with NIA lesions were able to relearn after prolonged training. CRs in these animals occurred with lower frequencies, were smaller and had longer latencies but they were not abolished. But, since there is a bilateral output from the cerebellum, at least in the ferret, it cannot be excluded that these responses were due to the contralateral cerebellar hemisphere (Ivarsson and Hesslow, 1993). Kelly, Zuo and Bloedel (1990) observed that learned responses in animals which had been trained after decerebration survived complete removal of the ipsilateral cerebellum. But, this finding could not be replicated by Hesslow, Hardiman and Yeo, (1990). Furthermore, other evidence suggests that the responses observed by Kelly *et al.*, may not have been true CRs (Nordholm, Lavond, Thompson, 1991).

The precise role the cerebellum in conditioning is still controversial, however, and can only be clarified by physiological investigation of the relevant neuronal elements. A few studies employing physiological techniques have been published previously, for instance recordings from Purkinje cells (Berthier and Moore, 1986), but it was not known if these units were actually related to the control of eyeblink.

METHODS

Applying physiological techniques to intact animals is very difficult, for technical as well as for ethical reasons, and we have therefore used decerebrate cats and ferrets. These preparations permit a higher degree of experimental control and freedom. It has been shown by several investigators that forebrain structures are not necessary for conditioning and that normal conditioning can be obtained in decerebrate cats and rabbits (e.g. Norman *et al.* 1974).

The animals were anaesthetized with halothane and then decerebrated by sectioning the brain stem just rostral to the superior colliculus and the red nucleus. After decerebration, the anaesthesia was terminated. Surface recordings were made with monopolar silver ball electrodes. For single unit recordings, glass-coated tungsten wires were used.

The CS was a 300 ms, 50 Hz train of stimuli applied via needle electrodes inserted through the skin of the medial side of the proximal left forelimb. The US was a 50 Hz, 60 ms train of stimuli (0.2 ms, negative square pulses), to the lower eyelid delivered through a pair of stainless steel needles inserted about 1 mm into the skin. The CS-US interval was 300 ms. It was ensured, by observing extinction after unpaired presentations of the CS and US, that the responses were 'true' CRs and not due to sensitization or pseudoconditioning.

Identification of Eyeblink Areas in the Cerebellar Cortex

Studies of the climbing fibre projection to the cerebellar cortex have revealed the presence of sagitally oriented zones of Purkinje cells, each receiving input from a distinct part of the olive and each projecting to a specific portion of the cerebellar nuclei (Oscarsson, 1980; Ito, 1984). The zones can be further subdivided into microzones on the basis of the topography of the climbing fibre input. Here, we will only consider the c1 and c3 zones of the intermediate part of the cerebellar cortex. Lesion studies have implicated the NIA and the dorsal accessory olive in normal conditioning (Yeo *et al.*, 1985a,b, 1986). The NIA is innervated by Purkinje cells in the c1 and c3 zones, which are precisely the zones which receive their climbing fibre input from the dorsal accessory olive. It is these zones, therefore, that are likely to participate in the control of eyeblink.

It has been suggested that the climbing fibre input to a certain area of the cerebellar cortex is related to its functions, such that, for instance, the areas of c3 receiving forelimb input would control forelimb muscles and further that each microzone in the forelimb area would control a single muscle or muscle group. (Oscarsson 1980; Ito, 1984; Ekerot, Garwicz and Schouenborg, 1991). Thus, one would expect the zones projecting to the NIA to contain microzones which have climbing fibre input from the cornea and the periorbital area and which control the orbicularis oculi muscle. This has now been demonstrated in the cat.

The recording and stimulation arrangement is illustrated in Fig. 1 A. Four areas in the cerebellar cortex of the cat, which control the orbicularis oculi muscle are shown in B. These areas were identified by recording climbing fibre input from the cornea and skin around the eye and by recording 'delayed' EMG responses in the eyelid. The latter can be evoked by train stimulation at certain sites in the cerebellar cortex. They have long latencies, typically 30-50 ms after termination of the stimulation and they can be delayed by prolonging the stimulus train. Delayed responses probably result from activation of Purkinje cell axons which causes hyperpolarization followed by rebound excitation of interpositus neurones. A site from which such responses can be evoked therefore probably projects to those neurones in the NIA which control eyeblink (Hesslow, 1994a).

Inhibition of CRs by Cortical Stimulation

If the cerebellum has a critical role in the performance of CRs, cortical stimulation during the execution of a CR would be expected to inhibit it. Furthermore, cortical stimulation should inhibit CRs more effectively than URs. Conversely, if the cerebellum has no particular relationship with the CR and solely provides background excitation and facilitation of the motoneurones in the facial nucleus, CRs should be inhibited to the same degree as the URs.

Fig. 2 shows rectified and averaged EMG recordings of CRs and URs in a cat. When a single shock was applied to the cerebellar cortex, the CR was completely suppressed. This effect on the CR was topographically specific. When a stimulus was applied to a site in the periphery or outside an eyeblink area, there was little or no inhibition of the CR (Hesslow, 1994b).

When the UR was preceded by a cortical stimulus, some depression of this response was observed, but it was slight compared to the effect on the CR. It could be argued that the difference in effect on CR and UR reflects a difference in excitatory drive. If the motoneurones are excited far above threshold during the UR and only slightly above threshold during the CR, a certain amount of inhibition could have a much stronger effect on the latter. This suggestion may be rejected, however, since the inhibition of the UR was insignificant even when a weak US was used, and the EMG activity in the UR was considerably smaller than in the CR. This suggests that the effect on the CR is not just due to a loss of background facilitation of the motoneurones in the facial nucleus. The CR is more critically dependent upon the cerebellum than is the UR.

It cannot be excluded, of course, that the cerebellum exerts a tonic facilitation of some other brain stem structure, which is involved in generating the CR but not the UR. If such a structure used the red nucleus as an output pathway, cerebellar cortical stimulation would cause a loss of background excitation which could block the CR. In the absence of a plausible candidate structure, apart from the cerebral cortex, which we know is not necessary for conditioning, this suggestion must be regarded as *ad hoc*. The most reasonable explanation for the inhibition of CRs is, therefore, that they are generated in cells in the NIA, which are inhibited by stimulation of the cerebellar cortex.

Single Unit Recordings

Microelectrode recordings have been made from Purkinje cells, Golgi cells and from NIA neurones in decerebrate ferrets. The number of cells in the latter two categories is still too small to permit any conclusions, but there are observations, consistent among the more 150 Purkinje cells now studied in an eyeblink-related area of the c3 zone (Hesslow and Ivarsson 1994). These may, with some simplification, be summarized as follows:

a) Purkinje cells in naive animals respond weakly or not at all to the CS. An example of a cell recorded in a naive animal is shown in Fig. 2 A. Slight increases or decreases in firing were observed in many neurones, but the changes in firing frequency were less than 10%. It is very difficult to judge the importance of such responses and it cannot be excluded that even weak changes in firing are important at the population level.

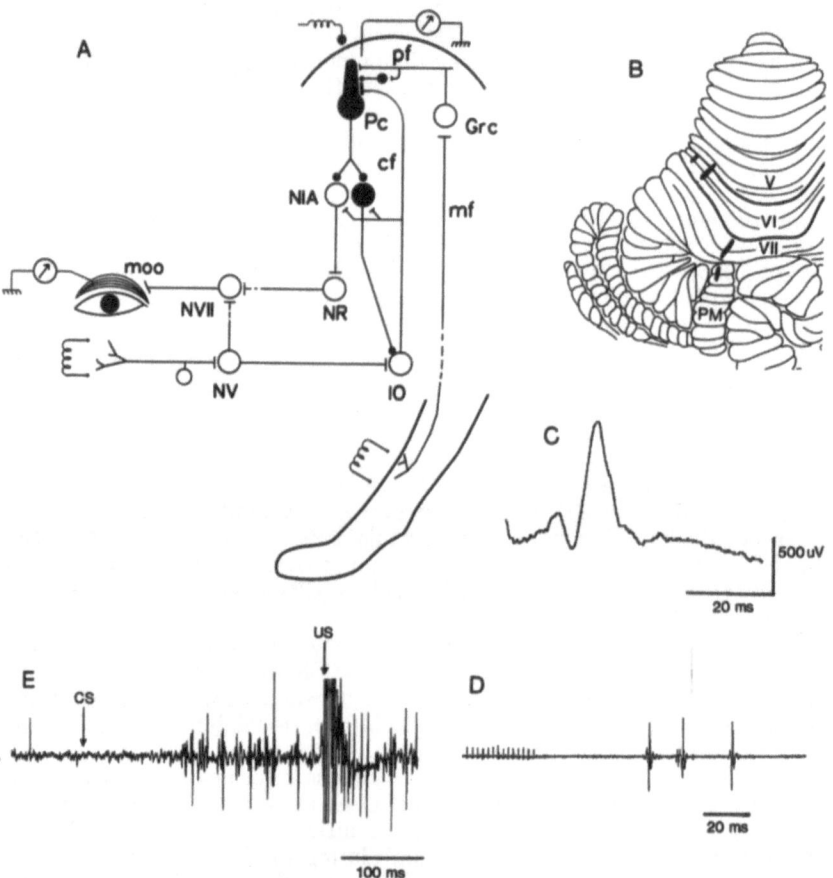

Figure 1. A: Experimental setup and diagram of the circuit controlling eyeblink. Recording electrodes in the orbicularis oculi muscle (moo) and in the cerebellar cortex. Stimulation electrodes are in the cerebellar cortex, the lower eyelid (US) and the forelimb (CS). The US pathway is through the trigeminal nucleus (NV), the inferior olive (IO) and the climbing fibres (cf.) to the Purkinje cells (Pc). A hypothetical CS pathway from the forelimb is via mossy fibres (mf), granule cells (Grc) and parallel fibres (pf). Output from the cerebellar cortex is to the anterior interpositus nucleus (NIA), red nucleus (NR) and facial nucleus (NVII). **B:** Outline of the cerebellar surface of the cat with eyeblink areas indicated in black. **C:** Climbing fibre response recorded from the cerebellar surface in this area. **D:** Delayed EMG response in the eyelid evoked by stimulation of the same area. **E:** Conditioned response.

b) In the trained animals, a certain proportion of Purkinje cells responded with a marked (sometimes complete) suppression of simple spike firing during the later parts of the CS-US interval. An example is shown in Fig. 2 B. The suppression was sometimes preceded by a period of excitation. About 15-25% of the cells (depending on the criteria used) showed this inhibitory response. Because of the variability between cells in the degree, latency and duration of the simple spike suppression, this figure unavoidably involves some simplification.

c) The remaining Purkinje cells in conditioned animals show the same kind of behaviour as cells in naive animals, that is weak or no responses to the CS.

These observations are consistent with the cerebellar hypothesis of conditioning. Suppression of simple spikes would be expected to cause a disinhibition of the target neurones in the NIA which would lead to excitation, via the red nucleus, of motoneurones in the facial nucleus. It cannot be excluded, that the Purkinje cell responses observed were effects of changes in other brain stem structures such as feedback from the motoneurones, but since the latency of the inhibition was often considerably shorter than the CR latency this seems unlikely.

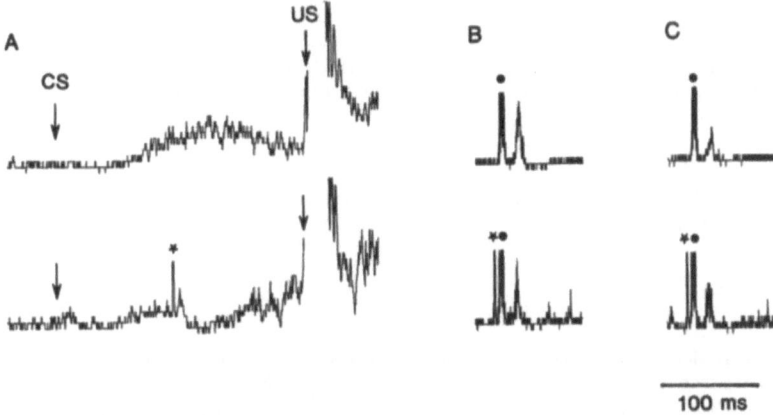

Figure 2. Comparison of the effect of cerebellar cortical stimulation on conditioned and unconditioned responses. **A**: The upper trace is an average of rectified EMG recordings from 20 conditioning trials. The lower trace shows effect of cortical stimulation. Arrows indicate onset of CS, US and cortical stimulus (Ctx). **B**: The upper trace shows the rectified average of 40 EMG responses to a very weak US and the lower trace is the response to the same stimulus preceded by a cortical stimulus.

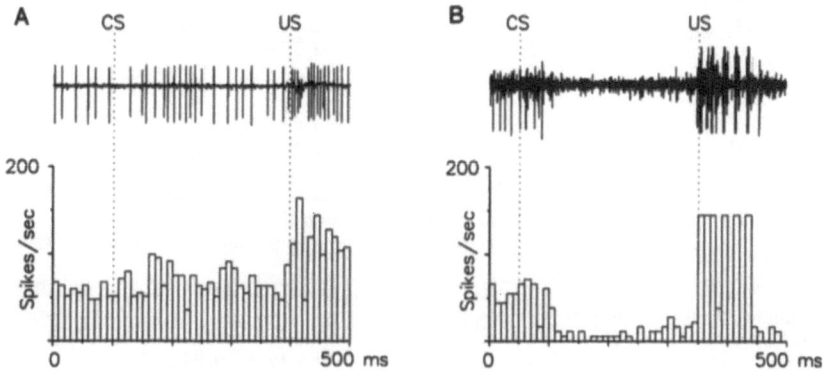

Figure 3. Sample records and time histograms from non-responsive (**A**) and responsive (**B**) Purkinje cells in conditioned ferrets.

These findings may be simply and powerfully interpreted in terms of the long term depression (LTD) of parallel fibre/Purkinje cell synapses, which is induced by simultaneous parallel fibre and climbing fibre input to the Purkinje cells, a mechanism which probably plays a role in the adaptation of the vestibulo-ocular reflex (Ito 1984). As suggested by several authors, and confirmed by data described above, the CS activates the mossy fibre/parallel fibre input while the US activates climbing fibres to the Purkinje cells controlling the orbicularis oculi muscle. Parallel fibres activated by the CS will influence the Purkinje cells directly via the excitatory synapses as well as indirectly via inhibitory stellate and basket cells. It is conceivable that the CS before training activates both these mechanisms so that excitation and inhibition are balanced and the net effect on the Purkinje cell therefore is close to zero. During conditioning, the paired parallel fibre and climbing fibre input would then cause a progressively more pronounced depression of the excitatory parallel fibre - Purkinje cell synapses whereafter the dominating effect of the CS would be interneuronal inhibition of the Purkinje cells.

121

CONCLUSIONS

The hypothesis, that the neural plasticity underlying classical conditioning is located in the cerebellum, has met fierce opposition, and the critics have often pointed out that the empirical evidence is not conclusive. However, since the cerebellum is the only structure remaining in the decerebrate, which is known to have the requisite anatomical connections as well as a mechanism of synaptic plasticity (LTD), it is an *a priori* plausible site of motor learning. Furthermore, although evidence in empirical science is never conclusive in any strict sense, there are now a number of diverse findings in the literature which can be effectively and parsimoniously explained by the cerebellar hypothesis, and no plausible alternative has so far been suggested.

ACKNOWLEDGEMENTS. This study was supported by grants from the Medical Faculty, University of Lund and the Swedish Medical Research Council (project no. 09899)

REFERENCES

Berthier, N.E. AND Moore, J.W. (1986). Cerebellar Purkinje cell activity related to the classically conditioned nictitating membrane response. *Experimental Brain Research* **63**, 341-350.

Ekerot, C.-F., Garwicz, M. AND Schouenborg, J. (1991). Topography and nociceptive receptive fields of climbing fibres projecting to the cerebellar anterior lobe in the cat. *Journal of Physiology* **441**, 257-274.

Gormezano, I. AND Moore, J.W. (1976). Classical conditioning. In *Learning: Processes*, ed. Marx, M.H., Macmillan, New York.

Hesslow, G. (1994a). Correspondence between climbing fibre input and motor output in eyeblink related areas in cat cerebellar cortex. *Journal of Physiology* **476**,229-244.

Hesslow, G. (1994b). Inhibition of classically conditioned eyeblink responses by stimulation of the cerebellar cortex in the cat. *Journal of Physiology* **476**, 245-256.

Hesslow, G., Hardiman, M. AND Yeo, C.H. (1990). Cerebellar lesions abolish eyeblink conditioning in the decerebrate rabbit. *European Journal of Neuroscience* Suppl. 3:301.

Hesslow, G. AND Ivarsson, M. (1994). Suppression of cerebellar Purkinje cells during conditioned responses in ferrets. *Neuroreport* **5**, 649-52.

Ito, M. (1984). *The Cerebellum and Neuronal Control*. Raven Press, New York.

Ivarsson, M. AND Hesslow, G. (1993). Bilateral control of the orbicularis oculi muscle from one cerebellar hemisphere in the ferret. *Neuroreport* **4**, 1127-1130.

Kelly, T.M., Zuo, C.-C. AND Bloedel, J.R. (1990). Classical conditioning of the eyeblink reflex in the decerebrate-decerebellate rabbit. *Behavioural Brain Research* **38**, 7-18.

Krupa D.J., Thompson, J.K., Thompson, R.F. (1993). Localization of a Memory Trace in the Mammalian Brain. *Science* **260**, 989-991.

McCormick, D.A. AND Thompson, R.F. (1984). Cerebellum: essential involvement in the classically conditioned eyelid response. *Science* **223**, 296-9.

Nordholm, A.F., Lavond, D.G., Thompson, R.F. (1991). Are eyeblink responses to tone in the decerebrate, decerebellate rabbit conditioned responses? *Behavioural Brain Research* **44**, 27-34.

Norman, R.J., Villablanca, J.R., Brown, K.A., Schwafel, J.A. AND Buchwald, J.S. (1974). Classical eyeblink conditioning in the bilaterally hemispherectomized cat. *Experimental Neurology* **44**, 363-380.

Oakley, D.A. AND Russell, I.S. (1972). Neocortical lesions and Pavlovian conditioning in the rabbit. *Physiology and Behaviour* **8**, 915-926.

Oscarsson, O. (1980). Functional organization of olivary projection to the cerebellar anterior lobe. In *The Inferior Olivary Nucleus: Anatomy and Physiology*, ed. Courville, J., de Montigny, C. and Lamarre, Y., pp 279-289. Raven Press, New York.

Schmaltz, L.W. AND Theios, J. (1972). Acquisition and extinction of a classically conditioned response in hippocampectomized rabbits (Oryctolagus cuniculus). *Journal of Comparative and Physiological Psychology* **79**, 328-333.

Welsh, J.P. AND Harvey, J.A. (1989). Cerebellar lesions and the nictitating membrane reflex: performance deficits of the conditioned and unconditioned response. *Journal of Neuroscience* **9**, 299-311.

Yeo, C.H., Hardiman, M.J. AND Glickstein, M. (1985a). Classical conditioning of the nictitating membrane response of the rabbit. I. Lesions of the cerebellar nuclei. *Experimental Brain Research* **60**, 87-98.

Yeo, C.H., Hardiman, M.J. AND Glickstein, M. (1985b). Classical conditioning of the nictitating membrane response of the rabbit. II. Lesions of the cerebellar cortex. *Experimental Brain Research* **60**, 99-113.

Yeo, C.H., Hardiman, M.J. AND Glickstein, M. (1986). Classical conditioning of the nictitating membrane response of the rabbit. IV. Lesions of the inferior olive. *Experimental Brain Research* **63**, 81-92.

LOCOMOTION

CELLULAR BASES OF LOCOMOTOR BEHAVIOUR IN LAMPREY: COORDINATION AND MODULATORY CONTROL OF THE SPINAL CIRCUITRY

P. WALLÉN

Nobel Institute for Neurophysiology
Department of Neuroscience
Karolinska Institute
S-171 77 Stockholm, Sweden

SUMMARY

The rhythmic movements performed during locomotion in vertebrates are controlled by pattern-generating circuits in the spinal cord. The different network units are precisely coordinated to achieve a proper timing of the different muscle groups. Different modulatory systems act upon the circuits in order to keep the motor pattern well-adapted to external demands, or to modify the pattern to produce various forms of locomotor behaviour. The neuronal mechanisms underlying this motor control system are being investigated in the lamprey model system, in which the spinal circuitry has been characterized in considerable detail. Computer simulations are performed in conjunction with experiments in the cellular analysis of these control mechanisms. Rhythmic burst activity in the lamprey segmental network is initiated from the brainstem via activation of both NMDA- and non-NMDA types of glutamate receptors. The rate of bursting activity can be controlled by varying the relative amount of activation of the two types of receptor. Serotonergic and GABAergic systems act to modulate the activity of the spinal circuitry. Serotonin specifically reduces the slow afterhyperpolarization (sAHP) by blocking apamin-sensitive Ca-dependent K-channels, leading to a smaller contribution from sAHP summation in burst termination, and thereby burst prolongation and a lower burst rate. GABA acts both pre- and postsynaptically, via $GABA_A$- and $GABA_B$ receptors, giving differential effects on the operation of the circuitry. The intersegmental coordination of the network units along the spinal cord during the undulatory swimming movements can be accounted for by the leading segment having a higher "intrinsic burst frequency" than the other segments, which become entrained to become coordinated with a constant phase lag. The intersegmental coordination can be modified by a local control of intrinsic frequency, resulting in e.g. modifications in body curvature or backwards locomotion. Both the serotonergic and the GABAergic systems may exert such a modulatory influence on the intersegmental coordination.

INTRODUCTION

The basic design of the neural control system for locomotion in vertebrates appears similar in all studied classes, from cyclostomes to mammals, despite differences in the mode of propulsion. Thus, corresponding brainstem areas initiate tetrapod locomotion in reptiles and primates, flying in birds, and swimming in fish and cyclostomes. Spinal neuronal networks

Neural Control of Movement, Edited by W.R. Ferrell
and U. Proske, Plenum Press, New York, 1995

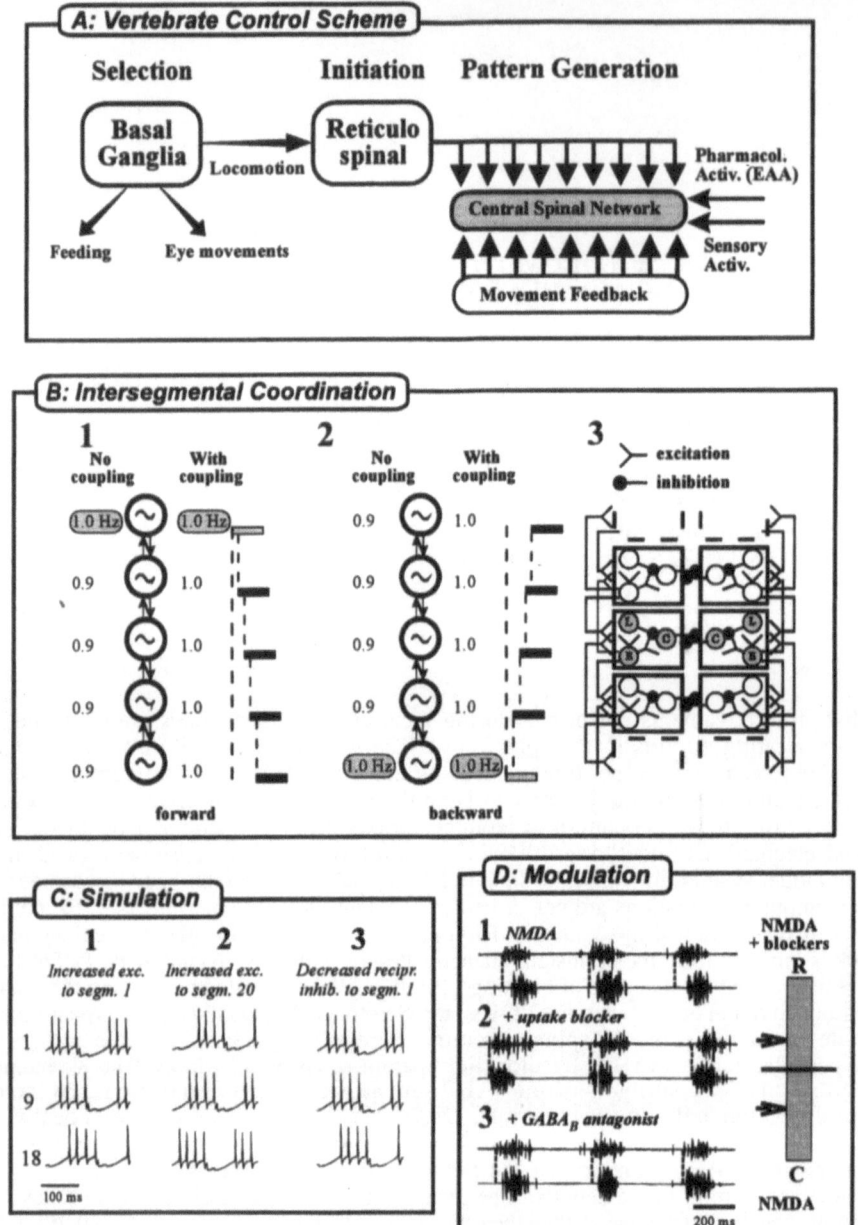

Fig.ure 1. A: General control scheme for vertebrate locomotion. The selection of different motor patterns, one of which is locomotion, can occur in the basal ganglia. The output nuclei of the basal ganglia exert a strong tonic inhibitory influence on different groups of neurones which in their turn directly or indirectly control (on the brainstem or cortical level) the neural networks generating different kinds of behaviour, like feeding, eye movements and locomotion. Once a pattern of behaviour is "selected" in the basal ganglia, the inhibition is removed allowing the specific brainstem-spinal cord network to be activated (see Hikosaka, 1991). In addition to this disinhibition, a direct excitation may occur. Locomotion is initiated by an increased activity in reticulospinal neurones, which in their turn activate the central spinal network, which then produces the locomotor pattern in close interaction with sensory, movement-related feedback. Experimentally, locomotion can also be elicited pharmacologically by administration of excitatory amino-acid agonists (EAA) and by sensory input. **B-D.** Intersegmental coordination during locomotion in the lamprey. **B:** Schematic representation of the 'trailing oscillator hypothesis'. **1-2.** The network in each segment is represented as one oscillator coupled to the other oscillators with mutual excitation. A forward phase lag (**1**) is produced when

126

(Fig. 1A) are responsible for the activation of the appropriate muscle groups with a correct timing so that the appropriate locomotor pattern is produced.

First, the CNS has to select among a multitude of different networks controlling various forms of behaviour (Fig. 1A), in order to specifically activate the locomotor system. There is growing evidence for the basal ganglia having a key role in this selection process. The basal ganglia provide strong tonic inhibition of different brainstem motor centres, and the selection of a certain motor pattern can be accomplished by a specific disinhibition of the responsible motor centre, as suggested for several forms of motor behaviour, like saccadic eye movements (Hikosaka, 1991) and locomotion (Mogenson, 1991; Noga et al.,1988; Jordan et al.,1992; Jordan, 1991).

Once the locomotor circuitry has been selected, reticulospinal neurones in the brainstem will initiate activity in the spinal networks (Fig. 1A). An increased intensity of the reticulospinal signal will result in an increase of rhythm frequency in the networks, i.e. a faster rate of locomotor movements. The spinal networks are subject to a powerful modulation from various sources. Sensory, movement-related feedback signals exert a prominent modulatory influence, providing appropriate adaptation of the locomotor pattern to changing external demands. In addition, different central mechanisms may modulate the spinal networks; notably, in the lamprey there are spinal serotonergic as well as GABAergic systems that can influence the properties of network neurones in specific ways, and thereby the locomotor output pattern.

The different network units of the spinal locomotor circuitry, each assumed to control the motor activity e.g. around a joint of a leg during walking or of a single body segment during undulatory locomotion, need to be appropriately coordinated for the particular locomotor task to be performed. Indeed, the coordination between network units is also under modulatory control, such that different modes of locomotion, e.g. forward and backward walking or swimming, can be produced (see, e.g. Grillner, 1981). This kind of pattern generator reconFig.uration has been described for several systems, and has been well analyzed on the cellular level in the crustacean stomatogastric system (Dickinson et al.,1990; Meyrand et al., 1991).

To reach an in-depth understanding of the mechanisms underlying the operation of the neuronal network controlling locomotion, a fruitful approach has proven to be the study of "simpler" preparations from lower animals, allowing a detailed cellular analysis under in vitro conditions. For the vertebrate locomotor control system, the Xenopus embryo preparation and the lamprey preparation have been used as such model systems for several years (see e.g. Roberts, 1993; Grillner et al.,1991). This paper attempts to review the major features of the lamprey model system for vertebrate locomotion, and in particular the operation of the segmental rhythm-generating network, the coordination between network units, and the various systems that can modulate the activity of the locomotor circuitry. Throughout this analysis, our desire has been to understand the cellular bases of the various aspects of the locomotor behaviour, and in this task computer modelling techniques have been extensively utilized in parallel with electrophysiological methods.

---

the oscillators are coupled and the most rostral one has a higher excitability (intrinsic frequency; example 1.0 Hz) than the remainder (0.9 Hz). A backward phase lag (2) is produced when the caudalmost oscillator has the highest excitability. 3. Chain of interconnected segmental network units, coupled by mutual excitatory connections between neighbouring segments. See text and Fig.. 2B for a description of the segmental circuitry. C: Computer simulations of 20 connected segmental oscillators have shown that a local increase of the excitability in the first segment (1) is able to produce a stable forward phase lag throughout the chain. When the extra excitation was instead given to the most caudal segment (2), the phase lag was reversed to backward coordination between all 20 segments. A forward phase lag was also obtained when the intrinsic frequency of the most rostral segment was increased by instead decreasing the amount of reciprocal inhibition (3) (Hellgren et al. 1992; cf. Grillner and Wallén, 1980). D: Modulation of intersegmental coordination by the spinal GABA system. Local application of the GABA uptake blocker nipecotic acid alters the coupling between segments in the preparation in a two-pool experiment with a partition mid-way along the spinal cord (inset to the right). NMDA and blocker(s) were added to the rostral segments, while the caudal pool was perfused with NMDA only. Under control conditions (1), the preparation exhibited fictive locomotion with a forward coordination (+1% phase lag/segment); after addition of nipecotic acid to the rostral pool (2), the phase lag shifted to a negative value (-0.8%), as during backwards locomotion. This effect of blocking GABA uptake, thereby increasing the endogenous levels of GABA, was partially counteracted by the $GABA_B$ receptor antagonist phaclofen (3), but not by the $GABA_A$ receptor antagonist bicuculline (not shown).

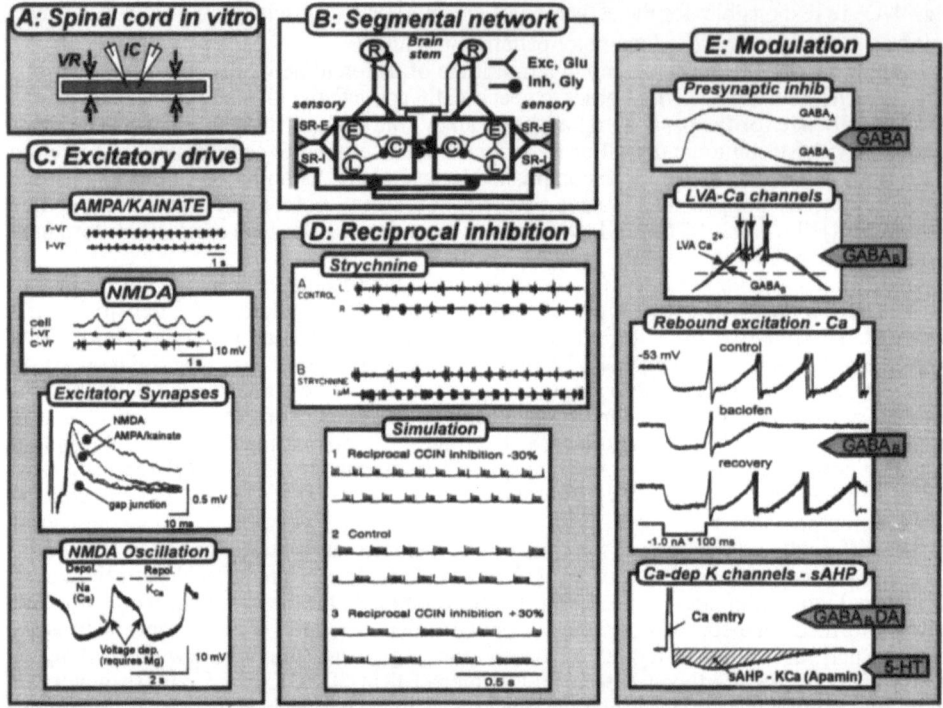

Figure 2. Cellular mechanisms involved in the pattern generation of the lamprey locomotor network. **A.** Experimental situation with a portion of the isolated lamprey spinal cord, in which fictive locomotor activity can be recorded with suction electrodes from ventral roots (VR) of the left and right sides and at different segmental levels. Simultaneous, paired intracellular recordings (IC) can be made from different classes of neurones. **B.** The lamprey segmental pattern generating network and its reticulospinal control. Locomotion is initiated by activation of the brainstem reticulospinal neurones (R) which excite all types of neurones on either side. The R neurones are subject to phasic excitatory and inhibitory feedback from the segmental circuitry. The segmental neuronal network consists of three types of interconnected interneurones. Excitatory (E) glutamate interneurones provide excitation to two types of segmental inhibitory interneurones (L and C) and to motoneurones (not shown). The C interneurone is responsible for reciprocal inhibition of all types of interneurones (E, C, L) and motoneurones on the contralateral side. The L interneurone provides inhibition to the C interneurones on the ipsilateral side. Two types of intraspinal sensory stretch receptor neurones (SR), located by the lateral margins of the spinal cord, can modulate the activity of the central neuronal network. One is excitatory (SR-E) and is connected to all ipsilateral neurones, the other is inhibitory (SE-I) and project onto all contralateral neurones. **C.** The glutamatergic excitatory drive of burst activity. AMPA/Kainate receptor activation elicits bursting at medium and high rates, while activation of NMDA receptors yields slower, stable bursting. The single glutamatergic synapse has three components: One NMDA- and one AMPA/Kainate chemical component, and an electrical, gap junction component. NMDA-receptor activation also elicits pacemaker-like membrane potential oscillations in the individual neurone. These oscillations are caused by the interplay between NMDA channels, the influx of calcium, and activation of Ca-dependent K-channels. **D.** The strength of reciprocal inhibition influences burst frequency. The synaptic transmission of inhibitory interneurones is glycinergic, and can be blocked by strychnine. Bath application of low levels of strychnine increases burst rate, while at higher levels bursting is disrupted and replaced by tonic activity. Computer simulations with alterations of the strength of reciprocal inhibition between the two sides of the segmental network, corroborate the experimental findings. **E.** Mechanisms for modulation of cellular properties. Presynaptic GABAergic inhibition of transmission among network neurones is exerted via $GABA_A$ and $GABA_B$ receptors located on interneurone axons, which respond to the respective agonist (muscimol and baclofen) in the presence of TTX. Sensory afferent transmission is also subject to presynaptic GABAergic modulation. In addition, $GABA_B$ receptor activation reduces an LVA Ca-current, which may delay the onset of discharge during locomotor driven membrane potential oscillations. The postinhibitory rebound excitation, which can be elicited by a hyperpolarizing current pulse, is also due to a Ca-current and can be modulated by $GABA_B$ receptor activation. Activation of apamin-sensitive K_{Ca} channels underlies the slow AHP following the action potential. The size of the sAHP can be modulated in several ways: $GABA_B$ receptor activation reduces Ca entry during the action potential by blocking HVA Ca channels, leading to less activation of K_{Ca} channels. Dopamine (DA), co-localized in 5-HT cells, acts through the same mechanism, whereas 5-HT appears to directly depress K_{Ca} channels without affecting Ca entry.

THE SEGMENTAL NETWORK

On the basis of experimentally established neurone types and synaptic connections, the principal organization of the segmental, rhythm-generating network has been described (Fig. 2B; see Grillner *et al.*,1991). The fast, excitatory synaptic transmission is glutamatergic, while the fast inhibitory connections utilize glycine. On each side of the segment, which here refers to a functional unit rather than a strict anatomical entity, local excitatory interneurones (E in Fig. 2B) excite two classes of inhibitory interneurones, the lateral (L) and the crossed, caudally projecting interneurones (C), and in addition the motoneurones (not indicated in Fig. 2B). Thus, the C interneurones provide reciprocal inhibition between the two sides, assuring alternating activity. Brainstem reticulospinal neurones (R) will initiate activity in the network neurones via glutamatergic transmission (cf. Fig. 2C). During rhythmic activity, the R neurones will receive excitatory as well as inhibitory, phasic feedback from the segmental networks. When one side of the network becomes active, the reciprocal connections will keep the other side silent, until the burst of activity is terminated. One way in which the burst can be terminated is via the activation of the L interneurones, which in turn will inhibit the crossing C interneurone, allowing activity on the contralateral side. However, the L interneurone effect is only one of several factors that can contribute to burst termination, as will become evident below. The capacity of a network organized in this way to produce rhythmic, alternating burst activity has been corroborated and extensively investigated in computer simulations with "semi-realistic" model neurones corresponding to the experimentally established cell types (Grillner *et al.*,1988; Wallén *et al.*,1992; Hellgren *et al.*, 1992).

The excitatory synaptic drive in the network occurs via glutamatergic transmission (Fig. 2C). On the cellular level, the glutamate EPSP may exhibit three separable components, an NMDA receptor mediated one, an AMPA/kainate component, and an electrical component (Fig. 2C). Activation of the AMPA/kainate receptor subtype, as well as of NMDA receptors, by agonist application to the perfusion solution will elicit fictive locomotor activity in the isolated spinal cord (Fig. 2C, top panels; Brodin *et al.*,1985). NMDA receptor activation results in slow to medium rate, stable rhythmic activity, while higher frequency bursting can be elicited by activation of AMPA/kainate receptors. Thus, by a combined activation of both types of receptor the whole physiological range of locomotor burst rates (0.25 Hz - 10 Hz) can be produced also in the *in vitro* preparation. The activation of NMDA receptors not only elicits slow fictive locomotor activity on the network level, but also gives rise to pacemaker-like membrane potential oscillations on the single cell level (Fig. 2C, bottom panel). The ionic mechanisms underlying these tetrodotoxin-resistant oscillations have been analyzed in detail (Wallén and Grillner, 1987), and involve an interaction between NMDA receptor channels with their voltage-dependence, and calcium-activated potassium channels (K_{Ca}), as schematically illustrated in Fig. 2C (bottom panel). These endogenous, NMDA receptor-dependent membrane potential oscillations constitute an important membrane property of network interneurones and motoneurones, and appear to be crucial for the generation of slow, stable burst activity (Wallén and Grillner, 1987; Brodin and Grillner, 1986). Indeed, in computer simulations of the segmental network it was found that the model neurones needed to be equipped with NMDA pacemaker-like properties in order to achieve slow rate bursting in the simulated network (Wallén *et al.*,1992).

By varying the level of excitatory drive, and in particular the relative involvement of NMDA and non-NMDA receptors, the rate of burst activity can thus be controlled over a wide range. Another means of regulating rhythm frequency is to vary the strength of reciprocal inhibition (Fig. 2D). With a lower amount of reciprocal inhibition, the inactive side will more rapidly reach the threshold for activation, which would give a faster burst rate. Fig. 2D shows that if the specific glycine receptor blocker strychnine is given in a low concentration, the rate of bursting does indeed increase (Grillner and Wallén, 1980). A higher concentration of strychnine will disrupt the rhythm and give tonic activity. Similarly, in network simulations it was found that a decrease of the strength of reciprocal inhibition will increase burst rate, whereas an increase instead results in a lowering of the burst rate (Fig. 2D; Hellgren *et al.*, 1992). In fact, by varying the amount of reciprocal inhibition in these simulations, the entire physiological range of burst rates could be covered.

Several different cellular properties have thus been defined which are of importance for the operation of the network. A number of these constitute factors which contribute to the onset and the termination of the burst. The *initiation of activity* on one side occurs when the reciprocal inhibition from the contralateral side stops. This disinhibition will in itself bring the membrane potential towards a more depolarized level. A further depolarization will be produced by the background excitation from reticulospinal neurones (NMDA, kainate/AMPA). As the excitatory interneurones (E) become activated they will cause a further depolarization, and activation, of their target neurones. Several additional factors may amplify the phasic depolarization and boost it near spike threshold, notably the opening of voltage dependent NMDA channels, and the type of Ca^{2+} channels activated at low threshold after a hyperpolarization (LVA Ca^{2+} channels, cf. Tsien et al., 1988); Fig. 2E). The ipsilateral excitatory stretch receptor neurones will also provide excitation in this phase of the swimcycle (Fig. 2B; see below). The maintenance of spike activity during the burst is achieved through the background excitation and is further amplified by the E interneurone activity. Again, the depolarization is due to both a kainate/AMPA and NMDA receptor activation, and the NMDA receptor induced plateau potentials provide a basis for a maintained depolarization (Fig. 2C; cf. above). For the termination of the burst, a number of cellular mechanisms contribute, like calcium dependent K^+ channels in their different roles (see below), the closing of voltage dependent NMDA channels, and circuit mechanisms, including the contralateral inhibitory stretch receptor neurones which become activated towards the end of the burst. These mechanisms all combine to end the depolarizing phase and thereby disinhibit the contralateral part of the network which will then become active.

The calcium-dependent potassium channels (K_{Ca}) and their regulation play important roles in the operation of the network. Following each action potential in the network neurones, Ca^{2+} entry will cause activation of apamin-sensitive K_{Ca} channels (Hill et al., 1992) which will produce a slow afterhyperpolarization (sAHP; cf. Fig. 2E). The sAHP is a key factor regulating the firing rate in neurones (Gustafsson, 1974), and during the burst the sAHPs will sum, which will bring the membrane potential in the hyperpolarizing direction and thereby contribute to the termination of the burst. K_{Ca} channels are also essential for the repolarization phase during NMDA-receptor induced membrane potential oscillations (cf. above), thus also in this case contributing to the termination of activity. Consequently, the addition of apamin, which selectively blocks the K_{Ca} channels, will prolong the bursts and, during slow, NMDA-induced fictive locomotion, also cause irregular activity (cf above; El Manira et al. 1994).

MODULATION OF LOCOMOTOR NETWORK ACTIVITY

During the performance of a motor act, like locomotion, the movements of the animal need to be continuously adapted to match the changing demands from the environment. A neuronal network may also be modulated to produce a different output pattern, e.g. increased movement amplitudes or backward instead of forward locomotion. The CNS contains several modulatory systems capable of altering the activity of a neuronal network, and in addition afferent systems provide movement-related feedback information that is used to modify and adapt the motor pattern. In the lamprey model system, spinal GABAergic and serotonergic systems both have a modulatory influence on the locomotor network by acting on specific ion channels, notably on different types of calcium channels and on K_{Ca} channels (cf. Fig. 2E). The mechanisms of action of these modulatory systems have been investigated in detail on the cellular level (Wallén et al., 1989; Matsushima et al., 1993), and the ability of these mechanisms to account for the effects on the network level (i.e. the burst pattern) have been for the most part corroborated in computer simulation studies (Wallén et al., 1992; Hellgren et al., 1992).

Also the mechanisms of the sensory feedback modulation have been analyzed in detail on the cellular level in the lamprey model system. Intraspinal stretch receptor neurones, located by the lateral edge on either side of the flattened spinal cord, are of two types; one which monosynaptically excites ipsilateral network neurones and another which, also monosynaptically, inhibits contralateral neurones (Fig. 2B). This pattern of connectivity, established in paired recording experiments, can account for the powerful entrainment effect on the network exerted by the stretch receptor neurones, when activated by lateral movements (Viana Di Prisco et al., 1990). Again, computer simulations have been used to verify the significance of this cellular analysis (Grillner et al., 1991; Tråvén et al., 1993).

COORDINATION BETWEEN SEGMENTAL NETWORKS ALONG THE BODY AXIS

In the lamprey, the approximately 100 segments along the body are precisely coordinated with a constant intersegmental phase lag, to produce the proper undulatory wave travelling down along the body for forward swimming. The mechanisms for this intersegmental coordination have been analyzed in detail, both experimentally, in computer simulations, and with mathematical modelling (Grillner *et al.*,1993; Matsushima and Grillner, 1992a; Wadden *et al.*, 1992; Cohen *et al.*, 1992; Sigvardt, 1993). According to the "trailing oscillator hypothesis" (Grillner *et al.*,1993; Matsushima and Grillner, 1990; Matsushima and Grillner, 1992a), the segmental network with the highest "intrinsic burst frequency" will become the leading one, and will "entrain" all other networks in the chain to follow after a certain delay (Fig. 1B). By increasing the excitatory drive to a specific segment, or by decreasing the amount of reciprocal inhibition, it will become the leader (cf Fig. 2D). If a rostrally located segment becomes the leading one, forward locomotion will be produced (Fig. 1B:1), but if instead a caudal segment is "selected" to lead, the pattern of coordination will switch to backward locomotion (Fig. 1B:2). In computer simulations of a chain of 20 segmental oscillators connected by mutual excitation (cf. Fig. 1B:3), corresponding patterns of coordination were seen when the first or last segment was given a higher excitability, respectively (Fig. 1C:1,2; (Wadden *et al.*,1992). If instead the strength of reciprocal inhibition in the first segment was decreased, thereby increasing the inherent burst frequency, a rostro-caudal coordination was again produced (Fig. 1C:3). This latter finding is in accord with recent experimental results which indicate that in the isolated spinal cord preparation, where most commonly a rostro-caudal coupling is seen, the amount of reciprocal inhibition in the most rostral segments is significantly lower than in caudal segments, due to the asymmetric projections of the descending C interneurones (Wallén *et al.*,1993).

Both the GABAergic and the serotonergic modulatory systems are capable of modifying the intersegmental coordination (Matsushima and Grillner, 1992b; Tegnér *et al.*,1993). As an example, Fig.ure 1D illustrates the effects of varying the GABAergic influence in the rostral portion of the spinal cord in a split-bath experiment (Tegnér *et al.*,1993). When the endogenous levels of GABA were raised rostrally by local application of an uptake blocker (nipecotic acid), the intersegmental coordination switched from a rostro-caudal to a caudo-rostral coupling (Fig. 1D:1,2), i.e. from a forward to a backward motor pattern. This is presumably due to a lowering of the intrinsic burst frequency in rostral segments (cf. Fig. 1B,C), since increased endogenous levels of GABA will decrease the rate of bursting (Tegnér *et al.*,1993). This effect on the intersegmental coordination is counteracted by the $GABA_B$ receptor antagonist phaclofen (Fig. 1D:3). Both pre- and postsynaptic mechanisms are presumably involved in this GABAergic modulation of the locomotor pattern (cf. above; Fig. 2E; (Tegnér *et al.*,1993).

CONCLUDING REMARKS

The aim of this paper has been to provide a brief account of the major characteristics of the control system for propulsion during locomotor behaviour in the lamprey. Due to space limitations, no attempts have been made to cover the control systems for body orientation and equilibrium, which obviously constitute an important aspect of the locomotor behaviour, and which are now being analyzed at the cellular level (Deliagina *et al.*,1992a; Deliagina *et al.*, 1992b; Orlovsky *et al.*,1992). The basic features of the rhythmic motor pattern for propulsion can be explained by the available experimental data, which is further supported by computer simulations of "semi-realistic" models of the neuronal network. A blend of cellular and network properties act in concert to produce a well-controlled and well-coordinated motor output. Furthermore, several modulatory systems operate to modify the output pattern of the circuitry in order to keep the movements well-adapted to external demands, and may even change the pattern of coordination, e.g. from forward to backward locomotion.

Although we have here dealt with the mechanisms underlying locomotion in the lamprey, several features of the control system appear general among different vertebrates. For instance, a brainstem "locomotor region" for initiation has been described for several different groups of animals, and the same applies to the rhythm-generating network in the spinal cord,

131

the sensory feedback control, etc. Thus, the principal "building blocks" for the control of locomotor behaviour appear to apply throughout the vertebrates; however, clearly a multitude of modifications of, and additions to, the control system have evolved in different species and for different types of movements.

REFERENCES

Brodin, L., Grillner, S. and Rovainen, C. M. (1985). N-Methyl-D-Aspartate (NMDA), kainate and quisqualate receptors and the generation of fictive locomotion in the lamprey spinal cord. *Brain Research* **325**, 302-306.

Brodin, L. and Grillner, S. (1986). Effects of magnesium on fictive locomotion induced by activation of N-methyl-D-aspartate (NMDA) receptors in the lamprey spinal cord in vitro. *Brain Research* **380**, 244-252.

Cohen, A. H., Ermentrout, G. B., Kiemel, T., Kopell, N., Sigvardt, K. A. and Williams, T. L. (1992). Modelling of intersegmental coordination in the lamprey central pattern generator for locomotion. *Trends in Neuroscience* **15**, 434-438.

Deliagina, T. G., Orlovsky, G. N., Grillner, S. and Wallén, P. (1992a). Vestibular control of swimming in lamprey. II. Characteristics of spatial sensitivity of reticulospinal neurones. *Experimental Brain Research* **90**, 489-498.

Deliagina, T. G., Orlovsky, G. N., Grillner, S. and Wallén, P. (1992b). Vestibular control of swimming in lamprey. III. Activity of vestibular afferents: convergence of vestibular inputs on reticulospinal neurones. *Experimental Brain Research* **90**, 499-507.

Dickinson, P. S., Mecsas, C. and Marder, E. (1990). Neuropeptide fusion of two motor-pattern generator circuits. *Nature* **344**, 155-158.

El Manira, A., Tegnér, J. and Grillner, S. (1994). Calcium-dependent potassium channels play a critical role for burst termination in the locomotor network in lamprey. *Journal of Neurophysiology* **72**, 1852-1861.

Grillner, S. (1981). Control of locomotion in bipeds, tetrapods and fish. In *Handbook of Physiology, Motor Control*, ed. Brooks, V., pp. 1179-1236. Bethesda: American Physiological Society.

Grillner, S., Buchanan, J. T. and Lansner, A. (1988). Simulation of the segmental burst generating network for locomotion in lamprey. *Neuroscience Letters* **89**, 31-35.

Grillner, S., Wallén, P., Brodin, L. and Lansner, A. (1991). Neuronal network generating locomotor behaviour in lamprey: circuitry, transmitters, membrane properties and simulation. *Annual Review of Neuroscience* **14**, 169-199.

Grillner, S., Matsushima, T., Wadden, T., Tegnér, J., El Manira, A. and Wallén, P. (1993). The neurophysiological bases of undulatory locomotion in vertebrates. *Seminars in the Neurosciences* **5**, 17-27.

Grillner, S. and Wallén, P. (1980). Does the central pattern generation for locomotion in lamprey depend on glycine inhibition?. *Acta Physiologica Scandinavica* **110**, 103-105.

Gustafsson, B. (1974). Afterhyperpolarization and the control of repetitive firing in spinal neurones of the cat. *Acta Physiologica Scandinavica Suppl.* **416**, 1-47.

Hellgren, J., Grillner, S. and Lansner, A. (1992). Computer simulation of the segmental neural network generating locomotion in lamprey by using populations of network interneurones. *Biological Cybernetics* **68**, 1-13.

Hikosaka, O. (1991). Basal ganglia--possible role in motor coordination and learning. *Current Opinion in Neurobiology* **1**, 638-643.

Hill, R., Matsushima, T., Schotland, J. and Grillner, S. (1992). Apamin blocks the slow AHP in lamprey and delays termination of locomotor bursts. *Neuroreport* **3**, 943-945.

Jordan, L. M. (1991). Brainstem and spinal cord mechanisms for the initiation of locomotion. In *Neurobiological Basis of Human Locomotion*, eds. Shimamura, M., Grillner, S. and Edgerton, V. R., pp. 3-20. Tokyo: Japan Scientific Societies Press.

Jordan, L. M., Brownstone, R. M. and Noga, B. R. (1992). Control of functional systems in the brainstem and spinal cord. *Current Opinion in Neurobiology* **2**, 794-801.

Matsushima, T., Tegnér, J., Hill, R. H. and Grillner, S. (1993). $GABA_B$ receptor activation causes a depression of low and high voltage-activated Ca^{2+}-currents, postinhibitory rebound and post-spike afterhyperpolarization in lamprey neurones. *Journal of Neurophysiology* **70**, 2606-2619.

Matsushima, T. and Grillner, S. (1990). Intersegmental co-ordination of undulatory movements - a "trailing oscillator" hypothesis. *Neuroreport* **1**, 97-100.

Matsushima, T. and Grillner, S. (1992a). Neural mechanisms of intersegmental coordination in lamprey: local excitability changes modify the phase coupling along the spinal cord. *Journal of Neurophysiology* **67**, 373-388.

Matsushima, T. and Grillner, S. (1992b). Local serotonergic modulation of calcium-dependent potassium channels controls intersegmental coordination in the lamprey spinal cord. *Journal of Neurophysiology* **67**, 1683-1690.

Meyrand, P., Simmers, J. and Moulins, M. (1991). Construction of a pattern-generating circuit with neurones of different networks. *Nature* **351**, 60-63.

Mogenson, G. J. (1991). The role of mesolimbic dopamine projections to the ventral striatum in response initiation. In *Neurobiological Basis of Human Locomotion*, eds. Shimamura, M., Grillner, S. and Edgerton, V. R., pp. 33-44. Tokyo: Japan Scientific Societies Press.

Noga, B. R., Kettler, J. and Jordan, L. M. (1988). Locomotion produced in mesencephalic cats by injections of putative transmitter substances and antagonists into the medial reticular formation and the pontomedullary locomotor strip. *Journal of Neuroscience* **8**, 2074-2086.

Orlovsky, G. N., Deliagina, T. G. and Wallén, P. (1992). Vestibular control of swimming in lamprey. I. Responses of reticulospinal neurones to roll and pitch. *Experimental Brain Research* **90**, 479-488.

Roberts, A. (1993). How does a nervous system produce behaviour? A case study in neurobiology. *Science Progress Oxford* **74**, 31-51.

Sigvardt, K. A. (1993). Intersegmental coordination in the lamprey central pattern generator for locomotion. *Seminars in the Neurosciences* **5**, 3-15.

Tegnér, J., Matsushima, T., El Manira, A. and Grillner, S. (1993). The spinal GABA system modulates burst frequency and intersegmental coordination in the lamprey: Differential effects of $GABA_A$ and $GABA_B$ receptors. *Journal of Neurophysiology* **69**, 647-657.

Tråvén, H. G. C., Brodin, L., Lansner, A., Ekeberg, Ö., Wallén, P. and Grillner, S. (1993). Computer simulations of NMDA and non-NMDA receptor-mediated synaptic drive: sensory and supraspinal modulation of neurones and small networks. *Journal of Neurophysiology* **70**, 695-709.

Tsien, R. W., Lipscombe, D., Madison, D. V., Bley, K. R. and Fox, A. P. (1988). Multiple types of neuronal calcium channels and their selective modulation. *Trends in Neuroscience* **11**, 431-438.

Viana Di Prisco, G., Wallén, P. and Grillner, S. (1990). Synaptic effects of intraspinal stretch receptor neurones mediating movement-related feedback during locomotion. *Brain Research* **530**, 161-166.

Wadden, T., Grillner, S., Matsushima, T. and Lansner, A. (1992). Realistic simulation of undulatory locomotion - a trailing oscillator hypothesis. In *Computation and Neural Systems*, eds. Eeckman, F. M. and Bower, J. M., Norwell, MA: Kluwer Academic Publishers.

Wallén, P., Buchanan, J. T., Grillner, S., Hill, R. H., Christenson, J. and Hökfelt, T. (1989). Effects of 5-hydroxytryptamine on the afterhyperpolarization, spike frequency regulation, and oscillatory membrane properties in lamprey spinal cord neurones. *Journal of Neurophysiology* **61**, 759-768.

Wallén, P., Ekeberg, Ö., Lansner, A., Brodin, L., Tråvén, H. and Grillner, S. (1992). A computer-based model for realistic simulations of neural networks. II. The segmental network generating locomotor rhythmicity in the lamprey. *Journal of Neurophysiology* **68**, 1939-1950.

Wallén, P., Shupliakov, O. and Hill, R. H. (1993). Origin of phasic synaptic inhibition in myotomal motoneurones during fictive locomotion in the lamprey. *Experimental Brain Research* **96**, 194-202.

Wallén, P. and Grillner, S. (1987). N-Methyl-D-Aspartate receptor induced, inherent oscillatory activity in neurones active during fictive locomotion in the lamprey. *Journal of Neuroscience* **7**, 2745-2755.

REFLEX REVERSAL IN THE WALKING SYSTEMS OF MAMMALS AND ARTHROPODS

K.G. PEARSON

Department of Physiology
University of Alberta
Edmonton, Canada T6G 2H7

SUMMARY

Recent investigations on cats, crabs, crayfish and insects have demonstrated that the reflex influence from some leg proprioceptors is reversed during locomotor activity. A general feature of this phenomenon is that the reversed reflex acts to reinforce the activity of motoneurones active during the stance phase of locomotion. In the cat, feedback from extensor group Ib afferents (arising from the force-sensitive Golgi tendon organs, GTOs) has an *excitatory* action on extensor motoneurones. This action of the GTOs during stance may function to regulate the level of activity in extensor motoneurones according to the load carried by the limb and/or to prevent the initiation of flexor burst activity when the extensor muscles are loaded. In arthropods, the reinforcing action of feedback from velocity sensitive afferents (chordotonal organs in crustacea and insects, and muscle receptor organs in crustacea) may regulate the speed of shortening of load bearing muscles.

INTRODUCTION

Although it is generally acknowledged that the regulation of muscle contractions during walking depends on afferent feedback from leg receptors (see reviews by Grillner 1981 and Rossignol et al.,1988), our understanding of the mechanisms of action, and the functional roles, of most groups of leg receptors remains fragmentary. For example, the function of the group Ia afferents from primary muscle spindle endings is still unclear. There are indications that they may facilitate the generation of extensor activity during stance (Dietz et al.,1979; Severin 1970; Yang et al.,1991), but whether these afferents contribute to the generation and timing of the locomotor rhythm is unknown. Even less is known about the function of the secondary spindle endings and the Golgi tendon organs (GTOs). Until recently, most proposals for their function have been speculative, based almost entirely on the patterns of convergence the afferents from these receptors make on interneurones and motoneurones (Edgley & Jankowska 1987; Lundberg et al.,1977,1978; Baldiserra et al.,1981; Jankowska 1992).

Over the past decade it has become increasingly clear that determining the role of afferent feedback in the regulation of locomotor activity requires the analysis of the characteristics of reflex pathways either in intact behaving animals, or in preparations in which the locomotor rhythm can be expressed. The reason for this is that the influence of afferents signals can vary enormously depending on the state of the system (standing, walking, running) and on the phase of the locomotor cycle (swing, stance) (Sillar 1989,1991; Rossignol et al.,1988; Duysens et al., 1992). The most striking manifestation of reflex modulation is the reversal in the sign of reflexes evoked by stimulation of specific groups of afferents. In this article I review the occurrence of reflex reversals that occur at the onset of locomotor activity in mammals and arthropods, and discuss some of the implications of reflex reversals for our understanding of the afferent regulation of stepping. Not considered in this article are phase-dependent reflex reversals since these have been described in a number of recent reviews (Rossignol *et al.*, 1988; Sillar 1991; Duysens & Tax 1994).

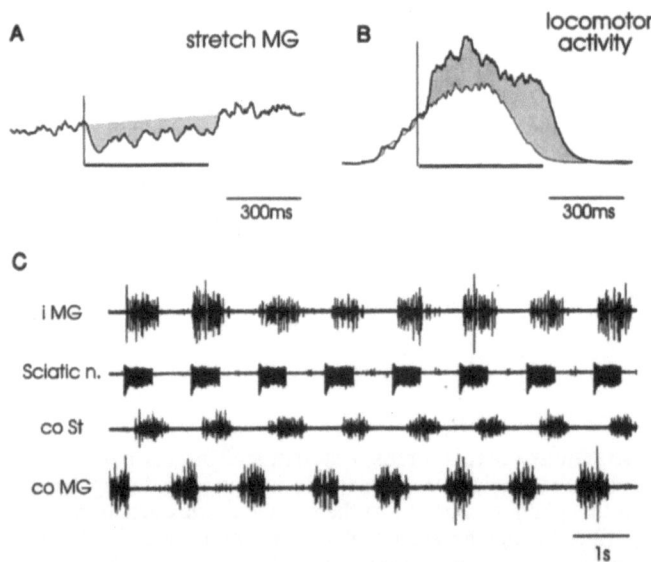

Figure 1. A and **B**: Reversal of the influence of group I afferents from plantaris on activity in MG muscle. Records show rectified, filtered and averaged EMGs in MG muscle during ramps stretches of the MG muscle (A), and locomotor activity (B). In the absence of locomotor activity (A) stimulation of the plantaris nerve (solid bar; 150Hz, 1.4xT) reduced ongoing, stretch-evoked activity. During locomotor activity (B) the same stimulus presented during MG bursts increased the magnitude of the bursts (thick trace). The thin trace in B shows the MG bursts in the absence of plantaris nerve stimulation. The stippled area in each panel indicates the influence of the stimulus train. **C**: Entrainment of the locomotor rhythm by repetitive stimulation of the plantaris nerve. The stimulus trains (150Hz, 1.6xT) were monitored by a recording from the sciatic nerve (sciatic n.). Note that during entrainment bursts in the ipsilateral MG (i MG) occur in-phase with the stimulus trains. The panel also shows coordinated bursting in the contralateral semitendinosus (co St) and MG (co MG) muscles. Clonidine treated chronic spinal cat. (Modified from Pearson & Collins, 1993).

REFLEX REVERSALS ASSOCIATED WITH LOCOMOTOR ACTIVITY

In the walking system of the cat the influence of input from the Golgi tendon organs (GTOs) in extensor muscles onto extensor motoneurones is reversed from inhibitory to excitatory at the onset of locomotor activity (Figs. 1,2; Conway *et al.*,1987; Pearson *et al.*,1992; Gossard *et al.* 1994; Pearson & Collins 1993). This has been demonstrated most clearly for the pathway from plantaris group Ib afferents to medial gastrocnemius (MG) motoneurones [the absence of significant monosynaptic connections from plantaris group Ia afferents to MG motoneurones (Eccles *et al.*,1957a) is advantageous in visualizing the effects for inputs from group Ib afferents when the plantaris nerve is stimulated electrically]. In the absence of locomotor activity, electrical stimulation of group I afferents in the plantaris nerve produces inhibition of any ongoing activity in MG motoneurones (Fig. 1A). This inhibitory action is most likely due to activation of the group Ib afferents in the plantaris nerve (Eccles *et al.*, 1957b). During locomotor activity, or under conditions that induce locomotor activity such as the administration of L-DOPA to spinal cats, the same stimulus produces excitation of MG motoneurones (Fig. 1B; Gossard *et al.*,1994). A number of observations have indicated that the excitatory pathway is via the system of interneurones (not yet defined) that centrally generates the extensor bursts, i.e. via the extensor half-centre: 1) the locomotor rhythm can be entrained and reset by stimulation of plantaris group I afferents, 2) during entrainment extensor bursts occur in-phase with the stimulus trains (Fig. 1C), 3) the rhythm is reset by an excitation of extensor motoneurones (Conway *et al.*,1987), and 4) the EPSPs evoked in the MG motoneurones are modulated in a phase-dependent manner such that they are minimal during extensor activity (Gossard *et al.*,1994). The latter is considered to be due to occlusion produced by activity in the extensor half-centre

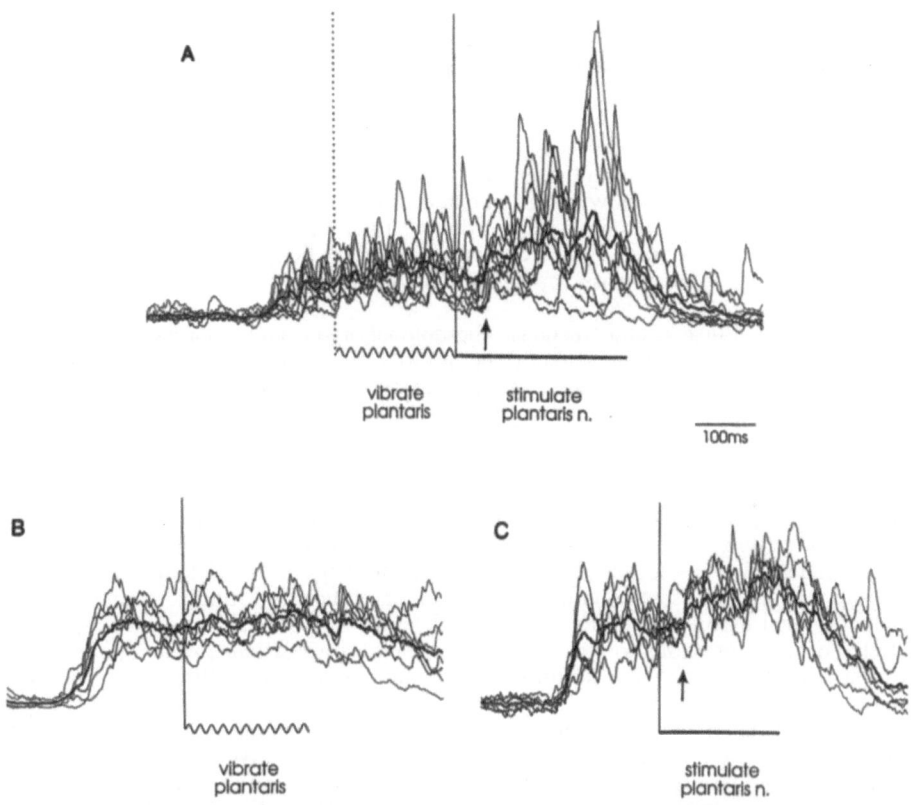

Figure 2. Input from group Ib afferents from plantaris excites MG motoneurones during fictive locomotor activity. Burst activity was recorded in the MG nerve. Either the plantaris muscle was vibrated (150Hz), or the plantaris nerve was electrically stimulated (150Hz, 2xT), at a preset delay following the onset of MG burst activity. Thick traces - averaged rectified and filtered bursts, thin traces - individual trials. In **A**) the nerve stimulation immediately followed the vibration of the muscle. Note that vibration did not increase the magnitude of the MG bursts but electrical stimulation increased the magnitude with a latency of about 50ms (onset of increase indicated by arrow). **B**) and **C**): In another preparation the muscle vibration and nerve stimulation were presented at the same time relative to the onset of MG bursts. Again note that only the electrical stimulation increased the magnitude of the MG bursts. (Modified from Pearson & Collins, 1993).

An important issue is whether the excitatory action produced in MG motoneurones during locomotor activity is the result of stimulating group Ia afferents, group Ib afferents, or a combination of both groups of afferents. The strongest evidence that input from group Ib afferents from the plantaris muscle have an excitatory action on MG motoneurones during locomotion is that selective activation of the group Ia afferents by vibration of the plantaris muscle has no influence on the magnitude of the MG bursts (Fig. 2 A,B), whereas electrical stimulation of group Ia and Ib afferents in the plantaris nerve excites MG motoneurones (Fig. 2 A,C). This observation does not exclude the possibility that input from group Ia afferents could facilitate the transmission in the excitatory Ib pathway by a subthreshold excitatory action of common interneurones. Indeed, this must be considered likely since all other known pathways transmitting information from group Ib afferents have been found to be facilitated by group Ia afferents (Jankowska & McCrea 1983; McCrea 1992). Another indication that input from group Ib afferents of extensor muscles has an excitatory action on extensor motoneurones during locomotion is that selective activation of group Ia afferents by vibration does not entrain or reset the locomotor rhythm (Conway et al.,1987; Pearson et al.,1992). It follows, therefore, that the resetting and entrainment of the rhythm by electrical stimulation of group I afferents must, in part, be produced by an excitation of the extensor half-centre by group Ib afferents. The final observation implicating the group Ib afferents is that the excitatory effects on the amplitude of the MG bursts, and on the timing of the locomotor rhythm, are usually only apparent when the plantaris nerve is stimulated at strengths (greater than 1.3xT) where the inhibitory synaptic actions are observed in the absence of locomotor activity.

Although recent research has focused on examining the excitatory pathway from plantaris group Ib afferents to the MG motoneurones (Gossard et al., 1994; Pearson & Collins 1993), there are reasons for believing that group Ib afferents from other extensor muscles also have an excitatory action on extensor motoneurones during locomotor activity. Electrical stimulation of group I afferents from quadriceps and triceps surae muscles during locomotor activity evoke slow EPSPs in extensor motoneurones similar to those evoked by stimulation of the plantaris nerve (Gossard et al.,1994), and entrain and reset the locomotor rhythm by eliciting extensor bursts (Conway et al.,1987; Pearson et al.,1992). To date, these phenomena have not been observed by stimulation of group I afferents from flexor muscles (Conway et al.,1987).

In the legs of insects and crustacea, chordotonal organs are important proprioceptors for the detection of movements about the joints. Activation of these proprioceptors is responsible for eliciting resistance reflexes to imposed leg movements in passive animals. Since these receptors are not regulated by efferent motor signals, they have similar response properties regardless of whether movements are externally imposed or internally generated by muscle contractions. It is not surprising, therefore, that the resistance reflexes evoked from these receptors are reduced (Barnes et al., 1972; Head & Bush 1991) or reversed (Bässler 1988; DiCaprio et al.,1981; El Manira et al.,1991; Head & Bush 1991) during locomotor activity, otherwise the reflexes would antagonize active movements.

A well studied case of reflex reversal is in the forelegs of the stick insect (Bässler 1988). During the stance phase of walking there is flexion of the femur-tibia joint and this stretches a chordotonal organ located in the proximal femur (Fig. 3 A,B). In passive preparations stretch of the chordotonal organ (tibial flexion) elicits a resistance reflex, i.e. activation of extensor motoneurones and inhibition of flexor motoneurones (Fig. 3C). This reflex is reversed in active preparations; the same stretch of the chordotonal organ now elicits excitation of flexors and inhibition of extensors (Fig. 3D). This response has been termed the "active-reaction". The active-reaction only occurs at relatively slow rates of stretching of the chordotonal organ. With rapid stretches a resistance reflex is elicited. Furthermore, even with slow stretches the positive feedback action of the chordotonal organ onto flexor motoneurones reverts to a resistance reflex near the end of the stretch. Thus, the chordotonal organ has two actions during walking: 1) to reinforce flexor activity during the early part of the stance phase provided the rate of flexion is not high, and 2) to terminate flexor activity during the latter part of stance and thus initiate the transition from stance to swing. These two actions are considered to depend on input from velocity and position sensitive afferents, respectively (Bässler 1988).

A very similar situation exists in the walking systems of crayfish and crabs. In these animals an important proprioceptor regulating leg movements during walking is the thoracic-coxal muscle receptor organ (TCMRO) (Sillar et al.,1986; Head & Bush 1991). The TCMRO consists of two non-spiking afferent neurones, the T and S fibres which are velocity and position sensitive, respectively. Rhythmically stretching the TCMRO can entrain the fictive locomotor rhythm, and depolarization of the T-fibre can trigger bursts in remotor motoneurones during locomotor activity (Sillar et al.,1986). In the absence of the locomotor rhythm, depolarization of the T fibre excites the promotor motoneurones. The remotor and promotor motoneurones are normally active during the stance and swing phases of walking, respectively. Since the T fibre is depolarized by remotion of the coxa, it follows that input from this velocity sensitive receptor acts to reinforce ongoing activity in the remotor motoneurones during the stance in a walking animal. When the animal is not walking, the T-fibre depolarization produced by remotion would contribute to the afferent signal for a resistance reflex. The action of the position sensitive S fibre does not reverse during locomotor activity. In both passive and active animals it excites promotor motoneurones and appears to be involved in regulating the transitions from remotion (stance) to promotion (swing) (Sillar et al.,1986).

Finally, the action of input from some of the chordotonal organs in the legs of crayfish and crabs has been found to be reversed during locomotor activity (El Manira *et al.*,1991; Head & Bush 1991)). Here the input from a chordotonal organ spanning the coxal-basal joint (CBCO) is activated by leg depression. Stretch of this receptor in passive animals excites leg elevator motoneurones and inhibits depressor motoneurones thus evoking a resistance reflex. During the expression of the fictive locomotor rhythm, however, the same stretch excites depressor motoneurones. Thus during locomotion input from the CBCO would reinforce the generation of activity in motoneurones active during the stance phase of walking, i.e. the depressors.

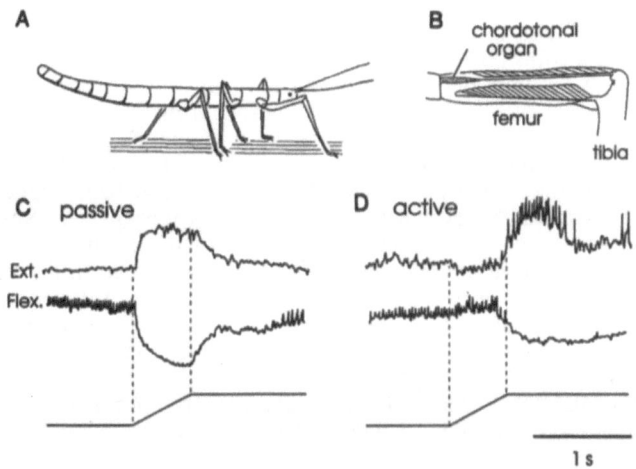

Figure 3: Reversal of the influence of input from the foreleg chordotonal organ to tibial extensor and flexor motoneurones in the stick insect. **A** and **B**: Diagrams showing stick insect and location of the chordotonal organ (CO) in the foreleg. **C**: Ramp stretch of the CO in a passive animal elicits a depolarization in an extensor tibia motoneurone (Ext.) and inhibition in a flexor tibia motoneurone (Flex.). These responses produce a pattern of muscle activity that resists imposed flexion movements of the tibia. **D**: The same stretch of the CO in an active animal causes the opposite effects in the extensor and flexor tibia motoneurones - extensor inhibition and flexor excitation. In this situation the CO input assists the contraction of the flexor muscle. Note that near the end of the stretch the assistive action reverts to a resistive action. (Modified from Bässler, 1988).

FUNCTIONAL IMPLICATIONS OF REFLEX REVERSAL

Currently, we can only speculate about the function of the reflex reversal from the GTOs during walking in the cat. This reflex reversal has only been characterized in reduced (often immobilized) preparations and has not yet been shown to occur in intact walking animals. Nevertheless, a two plausible hypotheses have been suggested that are subject to experimental verification in normal animals.

The first of these is that positive feedback from the GTOs contributes to the generation of extensor activity during the stance phase of walking. We know that afferent feedback contributes to the generation of extensor activity during the stance in humans (Yang *et al.,* 1991) and in decerebrate cats (Severin 1970). Until now this facilitatory action has been attributed to feedback from the primary muscle spindle afferents. However, the recent findings in reduced spinal preparations indicates that positive feedback from the GTOs may also contribute. If this does occur then the GTOs may function to provide a mechanism for automatically compensating for variations in the loads carried by the extensor muscles. For example, the load increase when an animal begins to walk up an incline would increase the feedback from GTOs, thus automatically increasing the level of activity in extensor motoneurones. Load compensation by positive feedback from force sensing afferents has also been proposed in the walking system of the cockroach (Pearson 1972).

Another proposal for the function of reflex reversal from the GTOs is that it allows these receptors to control the transition from stance to swing. In addition to increasing the magnitude of extensor burst activity, input from the GTOs can also prolong the duration of the extensor bursts (Fig. 1B) and inhibit, or delay, the onset of flexor bursts (Duysens & Pearson 1980; Conway *et al.,*1987; Pearson *et al.,* 1992). These observations have suggested that feedback from the GTOs acts to prevent the initiation of the flexor burst activity until the load carried by the extensor muscles falls below a critical level. The decline in GTO input does not necessarily trigger the stance-to-swing transition, since other biomechanical conditions may by required [such as an adequate extension at the hip (Grillner & Rossignol 1978)]. Unloading of the legs is also necessary for the initiation of the swing phase during walking in insects (Pearson 1972; Bässler 1988).

The normal function of the chordotonal organs in the legs of arthropods has also been extrapolated from observations in non-walking preparations, so again no firm conclusions regarding function can be drawn at present. However, it is clear that if the reinforcing action of input from the chordotonal organs does occur in intact animals then this does not function for load compensation. This follows from the facts that these receptors signal the velocity of joint movements and joint position, and they receive no efferent innervation. Thus any additional loading of the limb during stance would reduce input from these receptors and hence reduce the level of facilitation to the motoneurones active during stance. An alternative function may be to regulate the velocity of muscle shortening, as has been proposed for the chordotonal organs in the legs of the stick insect (Bässler 1988; Cruse 1985). Regulation of the velocity of limb retraction has also been suggested to be the function of the T-fibre of the TCMRO in the crayfish (Skorupski & Sillar 1986). The detailed mechanisms for the regulation of the velocity of muscle shortening by proprioceptive feedback have not been established in any animal. In the stick insect it is proposed that at the beginning of stance the reinforcing action of the chordotonal organs quickly leads to an appropriate rate of muscle shortening, and that the velocity is limited by a velocity-dependent reduction in the gain of the positive feedback pathway.

ACKNOWLEDGEMENTS. I thank Gordon Hiebert for his comments on a draft of this paper. Supported by a grant from the Canadian Medical Research Council.

REFERENCES

Baldissera, F., Hultborn, H. & Illert, M. (1981) Integration in spinal neuronal systems. In *Handbook of Physiology. The Nervous System*, Sect.1, Vol. 2. Edited by J.M. Brookhart, V.B. Mountcastle, V.B. Brooks and S.R. Geiger. American Physiological Society, Bethesda. pp. 509-595

Barnes, W.J., Spirito, C.P. & Evoy, W.H. (1972) Nervous control of walking in the crab *Cardisoma guanhumi*. II. Role of resistance reflexes in walking. *Zeitschrift vergl Physiologie* **76**,16-31

Bässler, U. (1988) Functional principles of pattern generation for walking movements of stick insect forelegs: the role of femoral chordotonal organ afferences. *Journal of Experimental Biology* **136**, 125-147

Conway, B.A., Hultborn, H. & Kiehn, O. (1987) Proprioceptive input resets central locomotor rhythm in the spinal cat. *Experimental Brain Research* **68**, 643-656

Cruse, H. (1985) Which parameters control the leg movement of a walking insect? I. Velocity control during the stance phase. *Journal of Experimental Biology* **116**, 343-355

DiCaprio, R.A. & Clarac, F. (1981) Reversal of a walking leg reflex elicited by a muscle receptor. *Journal of Experimental Biology* **90**, 197-203

Dietz, V., Schmidtbleicher, H.R. & Noth, J. (1979) Neuronal mechanisms of human locomotion. *Journal of Neurophysiology* **42**, 1212-1223

Duysens, J.D. & Pearson, K.G. (1980) Inhibition of flexor burst generation by loading ankle extensor muscles in walking cats. *Experimental Brain Research* **187**, 31-332

Duysens, J.D. & Tax, A.A.M. (1994) Interlimb reflexes during gait in cat and man. In: *Interlimb Coordination: Neural, Dynamical, and Cognitive Constrains*. Edited by S.P. Swinnen, H, Heuer, J. Massion and P. Casaer. Academic Press, New York. In press.

Duysens, J.D., Tax A.A.M., Trippel, M. & Dietz, V. (1992) Phase-dependent reversal of reflexly induced movements during human gait. *Experimental Brain Research* **90**, 404-414

Eccles, J.C., Eccles, R.M. & Lundberg, A. (1957a) The convergence of monosynaptic excitatory afferents on to many different species of alpha motoneurones. *Journal of Physiology* **137**, 22-50

Eccles, J.C., Eccles, R.M. & Lundberg, A. (1957b) Synaptic actions on motoneurones caused by impulses in golgi tendon organ afferents. *Journal of Physiology* **138**, 227-252

Edgley, S.A. & Jankowska, E. (1987) An interneuronal relay for group I and II muscle afferents in the midlumbar segments of the cat spinal cord. *Journal of Physiology* **389**, 647-674

El Manira, A., DiCaprio, R.A., Cattaert, D. & Clarac, F. (1991) Monosynaptic interjoint reflexes and their central modulation during fictive locomotion in crayfish. *European Journal of Neuroscience* **3**, 1219-1231

Gossard, J.P., Brownstone, R.M., Barajon, I. & Hultborn, H. (1994) Transmission in a locomotor-related group Ib pathway from hindlimb extensor muscles in the cat. *Experimental Brain Research* **98,** 213-228.

Grillner, S. (1981) Control of locomotion in bipeds, tetrapods and fish. In *Handbook of Physiology, Sect.1, Vol. 2. The Nervous System, Motor Control*, Edited by V.B. Brooks, American Physiological Society, Bethesda. pp. 1179-1236

Grillner, S. & Rossignol, S. (1978) On the initiation of the swing phase of locomotion in chronic spinal cats. *Brain Research* **144,** 411-414

Head, S.I. & Bush, B.M.H. (1991) Proprioceptive reflex interactions with central motor rhythms in the isolated thoracic ganglion of the shore crab. *Journal of Comparative Physiology* **168,** 445-459

Jankowska, E. (1992) Interneuronal relay in spinal pathways from proprioceptors. *Progress in Neurobiology* **38,** 335-378

Jankowska, E. & McCrea, D.A. (1983) Shared reflex pathways from Ib tendon organ afferents and Ia muscle spindle afferents in the cat. *Journal of Physiology* **338,** 99-111

Lundberg, A., Malmgren, K. & Schomburg, E.D. (1977) Cutaneous facilitation of transmission in reflex pathways from Ib afferents to motoneurones. *Journal of Physiology* **265,** 763-780

Lundberg, A., Malmgren, K. & Schomburg, E.D. (1978) The role of joint afferents in motor control exemplified by effects on reflex pathways from Ib afferents. *Journal of Physiology* **284,** 327-343

McCrea, D.A. (1992) Can sense be made of spinal interneurone circuits? *Behavioral and Brain Sciences* **15,** 633-643

Pearson, K.G. (1972) Central programming and reflex control of walking in the cockroach. *Journal of Experimental Biology* **56,** 321-330

Pearson, K.G. & Collins, D.F. (1993) Reversal of the influence of group Ib afferents from plantaris on activity in medial gastrocnemius muscle during locomotor activity. *Journal of Neurophysiology* **70,** 1009-1017

Pearson, K.G., Ramirez, J.M. & Jiang, W. (1992) Entrainment of the locomotor rhythm by group Ib afferents from ankle extensor muscles in spinal cats. *Experimental Brain Research* **90,** 557-566

Rossignol, S., Lund, J.P. & Drew, T. (1988) The role of sensory inputs regulating pattern of rhythmical movements in higher vertebrates. In *Neural Control of Rhythmic Movements in Vertebrates*, Edited by A. Cohen, S. Rossignol and S. Grillner. John Wiley & Sons, New York. pp. 201-283

Severin, F.V. (1970) The role of gamma motor system in the activation of the extensor alpha-motoneurones during controlled locomotion. *Biophysics* **14,** 1138-1145

Sillar, K.T. (1989) Synaptic modulation of cutaneous pathways in the vertebrate spinal cord. *Seminars in Neuroscience* **1,** 45-54

Sillar, K.T. (1991) Spinal pattern generation and sensory Gating mechanisms. *Current Opinion in Neurobiology* **1,** 583-589

Sillar, K.T., Skorupski, P., Elson, R.C. & Bush, B.M.H. (1986) Two identified afferent neurones entrain a central locomotor rhythm generator. *Nature* **323,** 440-443

Skorupski, P. & Sillar, K.T. (1986) Phase-dependent reversal of reflexes mediated by the thoracocoxal muscle receptor organ in the crayfish, *Pacifastacus leniusculus. Journal of Neurophysiology* **55,** 689-695

Yang, J.F., Stein, R.B. & James, K.B. (1991) Contribution of peripheral afferents to the activation of the soleus muscle during walking in humans. *Experimental Brain Research* **87,** 679-687.

A LOCOMOTOR-RELATED "AUTOGENETIC" IB EXCITATION OF HIND-LIMB EXTENSOR MUSCLES IN THE CAT

H. HULTBORN, R.M. BROWNSTONE and J.-P. GOSSARD

Department of Medical Physiology
The Panum Institute
University of Copenhagen, Blegdamsvej 3
DK-2200 Copenhagen N, Denmark

SUMMARY

In this Chapter we review our findings that stimulation of group I afferents from ankle and knee extensor muscles may reset and/or entrain the intrinsic spinal rhythm of "fictive" locomotion in the cat; these afferents are thus acting on the motoneurones through the spinal rhythm generators. A train of group I volleys delivered during a flexor burst would abruptly terminate the flexor activity and initiate an extensor burst. The same stimulus given during an extensor burst prolonged the extensor activity, while delaying the appearance of the following flexor burst. It has been shown that the major part of these effects originates from Golgi tendon organ Ib afferents. Intracellular recordings from extensor motoneurones during locomotion revealed that stimulation of group I afferents produces oligosynaptic excitation in this situation, rather than the "classical" Ib inhibition. We also discuss the possible functions of this proprioceptive control as well as the present understanding of the organization of the rhythm generator for locomotion in the cat.

INTRODUCTION

It is now well established that spinal networks are capable of generating locomotor-like activity in the absence of afferent signals (for recent reviews see Grillner 1981; Gossard and Hultborn 1991). However, in order to compensate for unexpected postural disturbances, or changes in the terrain, effective locomotor behaviour must utilize sensory feedback. Sensory input may influence the locomotor movements via reflex pathways in which transmission is either phasically modulated during the locomotor cycle, or acts "directly" upon the rhythm generators themselves. Experiments on a variety of species have revealed that proprioceptive inputs indeed have access to the locomotor rhythm generators. In the cat it has been shown that two types of afferents are critical for the initiation of the swing phase in the hindlimb: (1) the hip must reach a certain extended position (Grillner and Rossignol 1978, Andersson and Grillner 1981, 1983) and (2) the extensor muscles need to be unloaded (Duysens and Pearson 1980, Pearson et al.,1992). In Copenhagen we have recently studied the pathway related to extensor loading during fictive locomotion (Conway et al., 1987, Gossard et al.,1994). The aim of this work, to be reviewed here, is to use the sensory control of the spinal rhythm generator as a tool for investigating the intrinsic organization of the rhythm-generating circuits in the cat's spinal cord and, finally, to be able to identify the interneurones which are part of this network.

Neural Control of Movement, Edited by W.R. Ferrell
and U. Proske, Plenum Press, New York, 1995

Resetting and Entrainment of the Locomotor Rhythm

In the studies described here the spinal locomotor networks were activated either by intravenous administration of nialamide (a monoamine oxidase inhibitor) and l-DOPA to the acute spinal cat (Jankowska et al., 1967a,b; Grillner and Zangger 1979) or by stimulation of the mesencephalic locomotor region (MLR) in the decerebrate cat (Shik et al., 1966). The cats were paralyzed by neuromuscular blockade ("fictive" locomotion) and the locomotor activity was monitored by recording the activity in several nerves to flexor and extensor muscles of both hindlimbs. In this preparation there is no movement-related sensory feedback.

Figure 1 illustrates the basic finding that short trains of group I volleys from knee or ankle extensors were effective in resetting the activity in a coordinated fashion for all recorded nerves (Conway et al., 1987). When given during the flexor phase (Fig. 1A) a train of stimuli to the Plantaris nerve at group I strength was able to reset the locomotor rhythm by abruptly terminating the ipsilateral flexor activity and initiating a new extensor burst. On the contralateral side the opposite effect was observed; the extensor activity ceased and a new flexor phase was initiated. The same stimulus given during an extensor burst (Fig.1B) prolonged the extensor activity while delaying the appearance of the following flexor burst on the ipsilateral side.

When trains were given at frequencies in the same range as the spontaneous rhythm it was seen that the locomotor rhythm could be entrained. When the stimulation frequency was

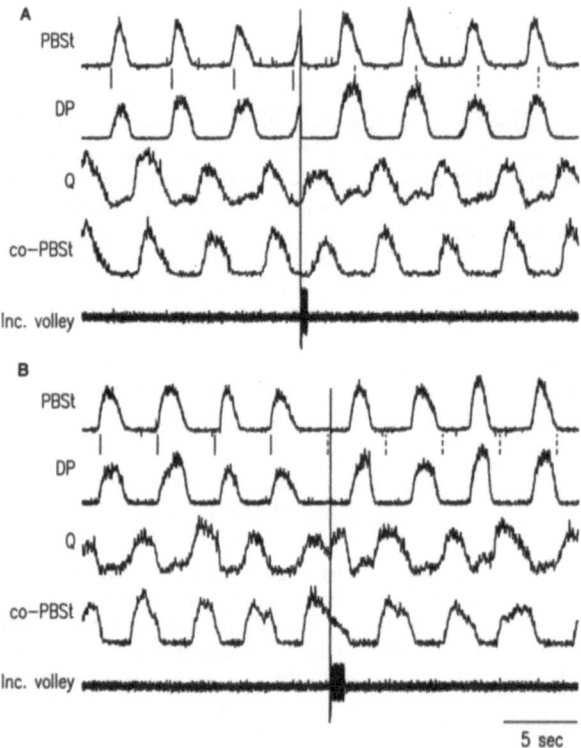

Figure 1. Resetting of locomotor rhythm by trains of group I stimuli to the plantaris nerve. Acute spinal (C1) preparation administered nialamide and l-DOPA. Upper traces are electroneurograms from a variety of ipsilateral and contralateral nerves to flexors (posterior biceps and semitendinosus, PBSt; tibial anterior and extensor digitorum longus - the deep peroneal nerve, DP) and extensors (quadriceps Q). The lowermost trace shows the cord dorsum potential. The timing of the plantaris train is shown by the long vertical lines. The uninterrupted vertical markers before the plantaris train indicate the onset of the PBSt activity. Comparisons of onsets of flexor activity after the plantaris train with the interrupted markers, indicating the *expected* onsets of ipsilateral flexor bursts had the rhythm been unperturbed, clearly demonstrates resetting of the rhythmicity. The examples in A and B show the effects of the train of stimuli to the plantaris nerve when given during the ipsilateral flexor and extensor phase respectively. (Unpublished material from Schomburg et al.. 1993).

decreased beyond a certain level the endogenous rhythm "escaped", while the rhythm failed to follow 1:1 when the imposed frequency increased.

This resetting and entrainment by group I afferents was seen following stimulation of toe, ankle and knee extensor nerves, but not from hip extensors and never from flexors at any of the hindlimb joints (Conway *et al.*, 1987, Gossard *et al.*, 1994). From a functional point of view it is of importance to elucidate the receptor origin of this group I effect. The major effect seems to be caused by Ib afferents from Golgi tendon organs, as muscle contraction evoked by intramuscular stimulation (activating Golgi tendon organs, while silencing many muscle spindle afferents) reproduces the effect of electrical nerve stimulation, while vibration (activating muscle spindle afferents) does not (Conway *et al.*, 1987). This does not exclude, however, the possibility that Ia afferents contribute to some convergent excitation, which alone is not sufficient to activate this pathway.

The demonstration that load receptors in extensors can alter the timing of the central rhythm has important functional implications. As discussed by Pearson and collaborators (Pearson *et al.*, 1992, Duysens and Pearson 1980) and Conway *et al.*, (1987) the function is most likely related to a variation of load signals during the stance phase. An increase in load on the hindlimbs following foot contact with the ground, and events preventing the release of load at the end of the stance phase, would reinforce extensor activity and prolong the extensor phase. Conversely, if a limb fails to meet a surface during the extensor phase of the step, or if the surface suddenly gives way, the lack of activation of these load receptors may provoke an early flexor phase ipsilaterally and a prolonged extensor phase contralaterally.

Although trains of extensor Ib impulses can terminate the flexor activity and trigger a new extensor burst under experimental conditions (cf. Fig. 1A) we regard this phenomenon as an experimental artefact, albeit a very useful one for the analysis of the locomotor network. Firstly, it is difficult to imagine a situation in which a powerful activation of extensor Ib afferents could occur during the swing phase. Secondly, even *if* such an activity has occurred (e.g. experimentally by electrical stimulation) it seems likely that the remaining proprioceptive information from the ipsilateral and contralateral limb would give conflicting information and likely prevent the resetting. This is exemplified in the following two results: (1) The resetting illustrated in Fig. 1A was obtained during fictive locomotion when other competing (phasic) sensory information was excluded. In this case the new ipsilateral rhythm was imposed on the contralateral limb so that the interlimb coordination remained after the resetting. (2) The experiments of Duysens and Pearson (1980) were performed during treadmill locomotion in decerebrate cats. Stretch of the ankle extensors caused maintained extensor activity of the ipsilateral limb which was restrained and partly denervated, while the contralateral limb continued with unperturbed stepping. The most likely explanation is that the sensory information from the contralateral (walking) limb was overriding the "erroneous" information from the ipsilateral limb causing an "uncoupling" of the rhythm generators of the two hindlimbs.

Transmission in the Locomotor-related Excitatory Group Ib Pathway to Extensors

In the acute spinal cat, stimulation of group Ib afferents from extensor muscles generally evokes inhibition among extensor motoneurones ("classical" Ib inhibition, see Baldissera *et al.*,1981 and Jankowska 1992 for reviews). The resetting described in the preceding section, however, would require the opposite pattern of effects. The effects of trains of group I stimulation were therefore investigated by intracellular recording of extensor motoneurones as l-DOPA was injected in acute spinal cats, as well as in relation to stimulation of the mesencephalic locomotor region (MLR) in decerebrate cats. Figure 2A illustrates the response to a group I train before, 17min and 30min after the start of l-DOPA injection. Before injection the response is dominated by the summating waves of IPSPs that are typical for classical Ib inhibition. During the injection a new excitatory wave (group I EPSP) progressively grows in response to the same stimulation. The appearance of the dominating excitatory response actually occurs before the emergence of spontaneous locomotion and can thus be studied on a quiescent background. This is of considerable experimental advantage.

When spontaneous locomotion appears the Ib EPSPs can be seen during both the flexor phase and the active extensor phase. When the stimulus train during the flexor phase is long enough, this EPSP simply merges with the new advanced extensor phase. With shorter trains, however, distinct stimulus-related EPSPs can be seen in both phases. The amplitude was usually smaller when evoked during the active extensor phase. The reduced amplitude during this phase is interpreted as due to an *occlusion* at interneuronal level, as the Ib EPSP is thought to be mediated by the very same interneurones that depolarize the motoneurones during the active extensor phase (the extensor half-centre; see diagram in Fig. 3).

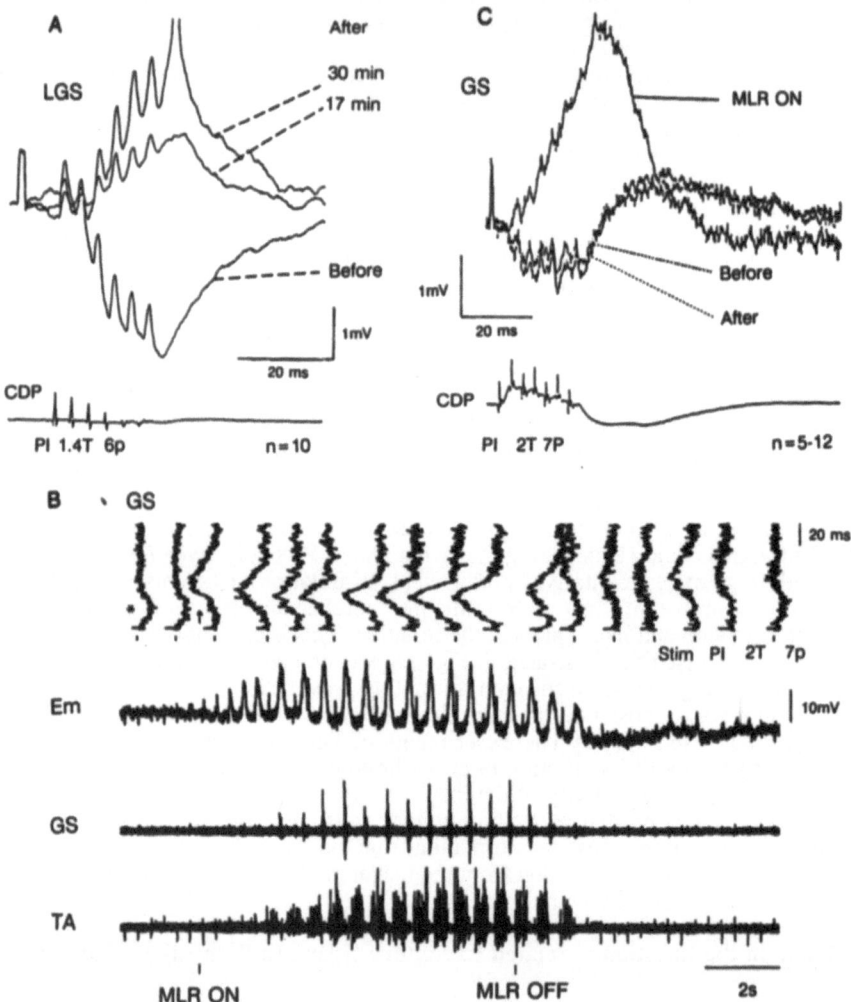

Figure 2. Emergence of group I EPSPs in two gastrocnemius motoneurones with administration of DOPA (*A*) and during MLR-induced fictive locomotion (*B-C*). **A.** Upper superimposed traces are averaged intracellular records whereas the lower trace shows a sample of the cord dorsum potentials. The intracellular traces show the response to a train of stimuli of group I fibers of the plantaris nerve (Pl 1.4 xT), before, 17 min and 30 min after injection of l-DOPA. **B.** From top to bottom: (1) tilted vertically are the high gain intracellular responses to group I stimulation of the plantaris nerve. These are expanded periods (100 ms) obtained from (2) the slow low gain intracellular record displaying the locomotor drive potentials, and the electroneurograms from (3) the gastrocnemius-soleus nerves and (4) the tibialis anterior nerve. The periods of the slow time base recordings, which are expanded in the upper vertical traces, are indicated by markers above the continuous recording. The group I stimulation coincided with the beginning of the fast vertical traces (as indicated in C). The beginning and the end of a period of continuous MLR-stimulation is marked at bottom. **C.** Upper traces: superimposition of averaged responses of the fast vertical traces from the sequence in B, obtained before, during and after the locomotor period. The lowermost trace is the cord dorsum potential aligned with the averaged intracellular responses. (Rearranged from Gossard *et al.*,1994).

As shown in Fig. 2B stimulation of extensor group I fibers before, during and after periods of MLR-evoked locomotor activity gave very similar results. Before and after the locomotor period the group I train evokes a small inhibition (the "classical" Ib inhibition), but during the period of locomotor activity a large EPSP appears following the same stimulus. Averages of the responses before, during and after the locomotor period is shown in Fig. 2C.

Intracellular recording from flexor motoneurones during locomotor activity has shown that extensor group I trains do not elicit any response during the extensor phase, but that they cause a hyperpolarization during the active flexor phase. By the use of Cl⁻ electrodes, this hyperpolarization was shown to be a *disfacilitation* rather than inhibition. As indicated in Fig. 3 the inhibition of the interneurones driving the flexor motoneurones (the flexor "half-centre") was thought to occur secondarily to the activation of the extensor "half-centre".

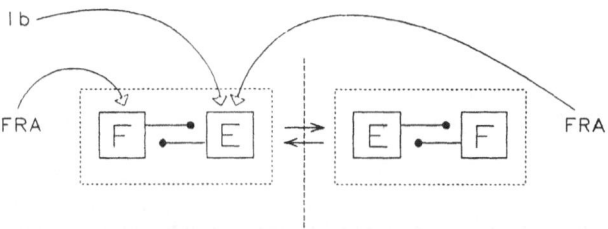

Figure 3. Schematic diagram summarizing results regarding the locomotor-related Ib responses in relation to locomotor centres and FRA inputs. The vertical dashed line separates the ipsi- and contralateral sides. We have adopted the "half-centre" model and assume that the core of the rhythm generator is composed of flexor (F) and extensor (E) half-centres with reciprocal inhibitory connections. The Ib afferent pathways are projecting to the extensor half-centre.

The central latency of the Ib EPSPs was in the order of 3.5 - 4.0ms. In the spinal preparation, in which locomotion was induced by l-DOPA, it was found that the locomotor-related Ib EPSPs could still be evoked following lesions at the level of the caudal L6 segment. The pathway of the locomotor-related Ib excitation to ankle extensors - which must encompass the (extensor part of the) rhythm generator - thus resides in the L7 - S1 segments.

Convergence in the locomotor-related Ib pathway

Stimulation of the MLR produces locomotion as well as short-latency PSPs in motoneurones. These PSPs are modulated through the step cycle such that extensor motoneurones receive EPSPs during the active extensor phase and IPSPs during the flexor phase (Shefchyk and Jordan 1985). It has been proposed that these PSPs are mediated via the spinal rhythm-generating circuits (Jordan 1991) and it would therefore be expected that the pathways for both the MLR-evoked EPSPs and the Ib EPSPs in extensor motoneurones would involve common interneurones. Recent experiments indeed support this by demonstrating a spatial facilitation between the two systems while recording the response in extensor motoneurones (Brownstone et al.,1992).

Administration of l-DOPA to acute spinal preparations reveals a pair of long-latency, long-lasting reflexes evoked from stimulation of "flexor reflex afferents" (FRA) (Jankowska et al., 1967 ab). Stimulation of the ipsilateral FRA results in excitation of flexor motoneurones, while stimulation of the contralateral side causes excitation of extensors.

These two pathways are organized with reciprocally inhibitory interactions; transmission in one of them prevents simultaneous transmission in the other. When pre-treated with nialamide (which potentiates the DOPA effects) single trains of FRA stimulation trigger brief periods of alternating activity in flexor and extensor nerves (Lundberg 1979). These bursts are likely to be forerunners of fictive locomotion, and in fact are thought to involve the same spinal networks as locomotion (see Lundberg 1979). As would be required from this hypothesis, it has recently been demonstrated that FRA stimulation can reset the cycle of the spontaneous DOPA-induced spinal locomotor rhythm (Schomburg et al.,1993). As would be expected from this scheme spatial facilitation has also been demonstrated for Ib EPSPs and contralateral FRA EPSPs in extensor motoneurones (Conway et al., 1987; Gossard et al.,1994).

The block diagram of Fig. 3 summarizes the results regarding the locomotor-related Ib excitation of extensor motoneurones. Candidate interneurones for mediating these actions have recently been described in lamina VII (Gossard *et al.*, 1994). Responses to peripheral nerve stimulation (group I and FRA) were in concordance with the convergence studies in motoneurones. The interneurones were also rhythmically active in the appropriate phases of the locomotor cycle, as predicted by their response pattern. However, none of these neurones was monosynaptically excited from group I afferents, neither were their target cells identified. A current goal is to obtain a full description of this locomotor-related Ib pathway as it will encompass (part of) the extensor "half-centre" and thus provide an opening for studying the central network for locomotion at an interneuronal level.

Concluding Remarks on the Spinal "Central Pattern Generator" (CPG) for Locomotion in Higher Mammals

For a long period of time, reference to CPGs for various types of motor behaviours, including locomotion, was little more than a convenient way to designate the capability of *central networks* to generate rhythmic patterns of activity. Now the CPGs for rhythmic activity in several invertebrates and even for swimming in lower vertebrates (lamprey, Grillner *et al.*, 1991; *Xenopus* embryos, Roberts 1990; Arshavsky *et al.*, 1993)have been reasonably well described. Mechanisms underlying rhythm generation are diverse, but often involve both intrinsic properties in individual neurones (endogenous rhythm generators) and synaptic interconnections within a network of neurones.

One enduring scheme for CPG organization emphasises reciprocal inhibitory connections between two groups of interneurones that control the motoneurones for antagonistic pairs of muscles with alternating activity. This was first detailed by Graham Brown (1911, 1914, 1924), who proposed a simple neuronal organization for locomotion in such terms. He envisaged that there was two "half-centres" (pools of interneurones) - one responsible for activation of flexor motoneurones, the other for extensor motoneurones. These two half-centres may receive a common excitatory drive. Furthermore, he assumed that the two half-centres were connected with mutual reciprocal inhibition, which would ensure that when one is active, the other is suppressed. Switching of activity from one half-centre to the other was thought to result from accumulating fatigue in the active half-centre leading to a decrease in the reciprocal inhibition and finally the other half-centre taking over. Fifty years after Graham Brown proposed this "half-centre hypothesis", Lundberg and co-workers described the spinal network subserving the long-latency, long-lasting FRA responses in acute spinal cats following l-DOPA administration. As described above, this network is thought to involve the same spinal networks as locomotion. Indeed, the properties of this organization, including the mutual inhibition between the excitatory centres to flexor and extensor motoneurones, are similar to what was predicted in the half-centre hypothesis. Furthermore, the activation of this network by l-DOPA in acute spinal cats closely parallels the development of spontaneous fictive locomotion and stimulation of the FRA indeed resets spontaneous locomotor rhythm in this preparation (for refs. see above). On the basis of the half-centre hypothesis and his own findings on the organization of the late DOPA reflexes, Lundberg (1969) proposed that the centrally generated pattern consisted of a simple alternating excitation of all flexors and all extensors. He hypothesized that this simple central pattern was then "sculptured" by proprioceptive reflex activity resulting in the refined and individualized pattern of muscular activity seen during normal locomotion. However, it has since been positively demonstrated that many details of the individual muscle patterns can be seen during fictive locomotion (i.e. in the absence of phasic movement-related sensory feedback) in the spinal cat (see e.g. Pearson and Rossignol 1991). Furthermore there is an asymmetric structure of the flexor/extensor activation patterns in real locomotion, such that with increasing speed (decrease in the cycle period) the extensor phase (stance) decreases much more than the flexor phase. This is also often seen in fictive spinal locomotion and would not be expected from a symmetric half-centre model. It is therefore obvious that the original , simplistic version of the half-centre hypothesis is inadequate in explaining many findings concerning spinal locomotor activity.

Can the individualized pattern of muscular activity, and the asymmetric structure of the step cycle be explained under the general framework of a half-centre model of the CPG? The individualized patterns could be accomplished by one of several schemes. Firstly, there may be an additional layer of "output interneurones" following the *rhythm*-generating circuit. Indeed, recent studies in respiration (Feldman and Smith 1989) and mastication (Donga and Lund 1991) now suggests a multilayer organization of CPGs (see also Lennard and Hermanson 1985). A second possibility is that several half-centres, each pair controlling antagonist pairs of muscles around an individual joint, are coupled together in a flexible and controllable manner. Such a scheme was proposed by Grillner (1981) in his hypothesis on a mosaic of coupled "unit burst generators", and would certainly allow for a flexible combination of activity patterns. There is also experimental evidence on coupled segmental rhythm generators in the lamprey spinal cord (see Grillner et al.,1991). A further analysis of the pathway(s) of the locomotor-related Ib EPSPs - and its possible fractionation into parallel pathways to extensors at different joints - may actually help to elucidate these questions.

As for the specific time relations between extensor and flexor phases of the step cycle, an *asymmetric* drive to the two symmetric half-centres would likely be enough to explain this behaviour. At present, however, nothing excludes either an asymmetric connectivity or differences in the intrinsic properties of the interneurones of the two half-centres. It seems that this question may have to await direct recordings from identified interneurones.

It is relevant to refer to the two examples of CPGs in (lower) vertebrates that are presently best understood, i.e. the networks underlying the swimming behaviour in the lamprey (Grillner et al.,1991) and in the *Xenopus* embryo (Roberts 1990, Arshavsky et al., 1993). In these simple systems, alternating activity between the two sides seems to be the rule, and in fact the underlying interneuronal organization is reminiscent of Graham Brown's original half-centre hypothesis for alternating flexor/extensor activity in cat locomotion. On the basis of the above considerations we find that an eclectic use of the half-centre terminology is reasonable and helpful both in relating present findings with Lundberg's original work (Jankowska et al.,1967 a,b; see Lundberg 1979) and in describing new findings such as the convergence outlined in Fig. 3. As for all models, this cartoon will be modified and expanded, and must not serve to "straightjacket" the interpretation of new data.

ACKNOWLEDGEMENTS. Our research has been supported mainly by grants from the Danish Medical Research Council, the Lundbeck Foundation and the Human Frontiers Science Program Organization.

REFERENCES

Andersson, O. & Grillner, S. (1981) Peripheral control of the cat's step cycle. I. Phase dependent effects of ramp-movements of the hip during "fictive locomotion". *Acta Physiologica Scandinavica* **113,** 89-101

Andersson, O. & Grillner, S. (1983) Peripheral control of the cat's step cycle. II. Entrainment of the central pattern generators for locomotion by sinusiodal hip movements during "fictive locomotion". *Acta Physiologica Scandinavica* **118,** 229-239

Arshavsky, Y.I., Orlovsky, G.N., Panchin, Y.V., Roberts, A. & Soffe, S.R. (1993) Neuronal control of swimming locomotion: analysis of the pteropod mollusc Clione and the embryos of the amphibian Xenopus. *Trends in Neurosciences* **16,** 22-233

Baldissera, F., Hultborn, H. & Illert, I. (1981) Integration in spinal neuronal systems. In: *Handbook of Physiology,* Motor Control, section 1, vol. II, part 1, ed Brooks V., pp 508-595. Bethesda, MD, U.S.A.: American Physiological Society

Brown, T.G. (1911) The intrinsic factors in the act of progression in the mammal. *Proceedings of the Royal Society* (London) **84,** 308-319

Brown, T.G. (1914) On the nature of the fundamental activity of the nervous centres; together with an analysis of the conditioning of rhythmic activity in progression, and a theory of evolution of function in the nervous system. *Journal of Physiology* (London) **48,** 18-46

Brown, T.G. (1924) Studies in the physiology of the nervous system. XXVIII: absence of algebraic equality between the magnitudes of central excitation and effective central inhibition given in the reflex centre of a single limb by the same reflex stimulus. *Quarterly Journal of Experimental Biology* **14,** 1-23

Brownstone, R.M., Noga B.R. & Jordan, L.M. (1992) Convergence of excitatory Ib and descending locomotor pathways in the cat. *Society for Neurosciences* Abstracts **18,** 315

Conway, B.A., Hultborn, H. & Kiehn, O. (1987) Proprioceptive input resets central locomotor rhythm in the spinal cat. *Experimental Brain Research* **68,** 643-656

Donga, R. & Lund, J.P. (1991) Discharge patterns of trigeminal commissural last-order interneurones during fictive mastication in the rabbit. *Journal of Neurophysiology* **66,** 1564-1578

Duysens, J. & Pearson, K.G. (1980) Inhibition of flexor burst generator by loading ankle extensor muscles in walking cats. *Brain Research* **187,** 321-332

Feldman, J.L. & Smith, J.C. (1989) Cellular mechanisms underlying modulation of breathing pattern in mammals. *Annals of New York Academy of Science* **563,** 114-130

Gossard, J.-P., Brownstone, R.M., Barajon, I. & Hultborn, H. (1994) Transmission in a locomotor-related group Ib pathway from hindlimb extensor muscles in the cat. *Experimental Brain Research* In Press

Gossard, J.-P. & Hultborn, H. (1991) The organization of the spinal rhythm generation in locomotion. In: *Plasticity of motoneuronal connections.* ed. Wernig, A. pp 385-403, Elsevier Science Publishers

Grillner, S. (1981) Control of locomotion in bipeds, tetrapods, and fish. In: *Handbook of Physiology,* Motor control, section 1, vol. II, part 1, ed. Brooks, V., pp 1179-1236. Bethesda, MD, U.S.A.: American Physiological Society

Grillner, S. & Rossignol, S. (1978) On the initiation of the swing phase of locomotion in the chronic spinal cats. *Brain Research* **146,** 269-277

Grillner, S., Wallén, P., Brodin, L. & Lansner, A. (1991) Neuronal network generating locomotor behaviour in lamprey: Circuitry, transmitters, membrane properties and simulation. *Annual Review of Neuroscience* **14,** 169 - 199

Grillner, S. & Zangger, P. (1979) On the central generation of locomotion in the low spinal cat. *Experimental Brain Research* **34,** 241-261

Jankowska, E. (1992) Interneuronal relay in spinal pathways from proprioceptors. *Progress in Neurobiology* **38,** 335-378

Jankowska, E., Jukes, M.G.M., Lund, S. & Lundberg, A. (1967a) The effect of DOPA on the spinal cord. V. Reciprocal organization of pathways transmitting excitatory actions to alpha motoneurones of flexors and extensors. *Acta Physiologica Scandinavica* **70,** 369-388

Jankowska, E., Jukes, M.G.M., Lund, S. & Lundberg, A. (1967b) The effect of DOPA on the spinal cord. VI. Half-centre organization of interneurones transmitting effects from the flexor reflex afferents. *Acta Physiologica Scandinavica* **70,** 389-403

Jordan, L.M. (1991) Brain stem and spinal cord mechanisms for the initiation of locomotion. In: *Neurobiological basis of human locomotion,* ed Shimamura, M., Grillner, S. & Edgerton,V.R., pp 3 -20. Tokyo: Japan Scientific Societies Press.

Lennard, P.R. & Hermansson, J.W. (1985) Central reflex modulation during locomotion. *Trends in Neurosciences* **8,** 483-486

Lundberg, A. (1969) Reflex control of stepping. *The Nansen memorial lecture V,* Universitetsforlaget, Oslo, pp 5-42

Lundberg, A. (1979) Multisensory control of spinal reflex pathways. In: Reflex control of posture and movement, ed Pompeiano, O. *Progress of Brain Research* **50,** 11-28

Pearson, K.G. & Rossignol, S. (1991) Fictive motor patterns in chronic spinal cats. *Journal of Neurophysiology* **66,** 1874-1887

Pearson, K.G., Ramirez, J.M. & Jiang, W. (1992) Entrainment of the locomotor rhythm by group Ib afferents from ankle extensor muscles in spinal cats. *Experimental Brain Research* **90,** 557-566

Roberts, A. (1990) How does a nervous system produce behaviour? A case study in neurobiology. *Science Progress* **74,** 31-51

Schomburg, E.D., Petersen, N., Barajon, I. & Hultborn, H. (1993) Flexor reflex afferents (FRA) reset the step cycle during fictive locomotion in the cat. *Acta Physiologica Scandinavica* **149,** 22A

Shefchyk, S.J. & Jordan, L.M. (1985) Excitatory and inhibitory postsynaptic potentials in alpha-motoneurones produced during fictive locomotion by stimulation of the mesencephalic locomotor region. *Journal of Neurophysiology* **53,** 1345-1355

Shik, M.I., Severin, F.V. & Orlovskii, G.N. (1966) Control of walking and running by means of electrical stimulation of the mid-brain. *Biofizika* **11,** 659-666

MODULATION OF STRETCH REFLEXES DURING BEHAVIOUR

R.B. STEIN, S.J. DESERRES and *R.E. KEARNEY

Division of Neuroscience
University of Alberta
Edmonton, Canada T6G 2S2

*Department of Biomedical Engineering
McGill University
Montreal, Canada

SUMMARY

Far from being the simple, stereotyped tendon jerk tested by physicians, stretch reflexes are highly modulated within a given motor task and the type of modulation is task-dependent. There are several sources of this modulation. Static and dynamic γ motoneurones are modulated differently during cat locomotion in a way that could permit the primary and secondary muscle spindle afferents to play distinct roles. Central modulation also occurs in both the cat and the human, since electrical stimulation of the afferent nerve (H-reflex) produces different responses at different times in the step cycle. Moreover, the reflex response is smaller in humans running compared to walking on a treadmill, even when the speed and EMG levels are matched. Mechanical inputs produce reflex responses with marked amplitude and time-dependent nonlinearities, which are being investigated in further cat and human experiments to provide better estimates of the contribution of reflexes to ongoing movements.

Despite being studied for seventy years, the role of stretch reflexes in the control of movement remains something of a mystery (Liddell and Sherrington, 1924; Matthews, 1990). This is even more surprising when one considers that there is a strong monosynaptic connection from primary muscle spindle afferents to α-motoneurones. Inputs and outputs can be easily monitored with no intervening interneurones, a simplification that is found nowhere else in the nervous system.

With this ease of recording, one might anticipate that a wealth of data would be available from which logical, consistent theories could be derived with broad predictive power. Instead, a number of theories, such as the length servo theory of Merton (1953) and the stiffness regulation hypothesis of Houk (1979) have been tested against experimental data and found to be wanting. Others such as the equilibrium point hypothesis (Levin et al., 1992) have been repeatedly updated to take account of the complexities of real systems.

There are other complexities, such as the longer latency components of the stretch reflex, including those that involve pathways from the limbs to the brain and back. Indeed, Melvill Jones and Watt (1971) named these longer latency pathways the "functional stretch reflex" and suggested that the monosynaptic component was relatively weak and stereotyped (see also Dietz, 1992).

However, even the monosynaptic component can be modulated so that it may be important under one condition and insignificant under another. Three classes of modulation can be considered: 1) fusimotor, 2) synaptic and 3) mechanical. A discussion of these types of modulation forms the body of this paper.

Fusimotor Modulation

Activity in fusimotor neurones is well known to modify the properties of the muscle spindle afferent and the detailed actions of fusimotor neurones occupied much of the working life of Ian Boyd, in whose honour this symposium is being held. This topic will be considered in other papers presented here, so I will only give a single example taken from our own work.

The research was motivated by the desire to find out how the fusimotor system was actually used during a behaviour such as locomotion. For this purpose we recorded from fusimotor neurones to the triceps surae muscles of high decerebrate cats which will walk spontaneously if suspended over a moving treadmill (Shik and Orlovsky, 1976). Two distinct patterns of activity were observed: one type of fusimotor neurone (Fig. 1A) had a low resting rate of nerve impulses, gradually increased this rate when the treadmill was turned on (arrow labelled 1) and <u>tonically</u> maintained a high rate during the period of locomotion until the treadmill was turned off (arrow labelled 2). The other type had a higher resting rate, which often decreased on average during the period of locomotion and showed a strong <u>phasic</u> modulation during each step.

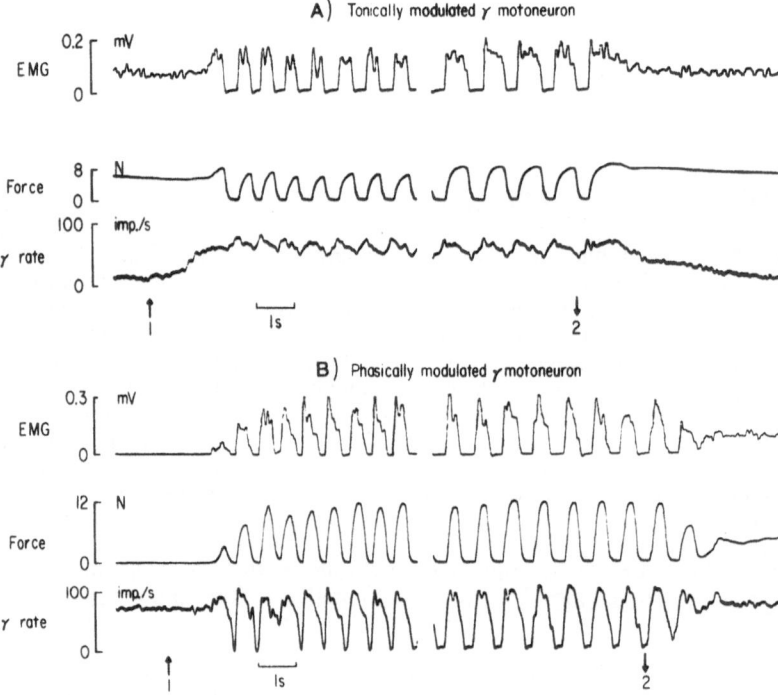

Figure 1. A: Tonically modulated γ-motoneurones increased their impulse rate from a low resting value when the treadmill was turned on (arrow 1) and maintained it throughout the period of walking until the treadmill was turned off (arrow 2). The period of walking is indicated by rhythmic bursts of EMG and cyclic force changes. Only a few steps at the beginning and end of a period of walking are shown. **B:** Phasically modulated γ-motoneurones had a high impulse rate at rest which was modulated with each step. From Murphy et al., 1984.

We were able to show that the tonically firing neurones were static fusimotor fibres and that the phasically firing neurones were dynamic fusimotor fibres in two ways: 1) in a few instances we could record from fusimotor fibres in continuity to observe their patterns during locomotion. Then, we could stimulate these fibres and determine their effects on muscle spindle afferents in order to classify them in the usual ways (Matthews, 1972). 2) In addition, we recorded from muscle spindle afferents during locomotion (Taylor et al., 1985). Secondary afferents which only receive major inputs from static fusimotor fibres also fired tonically when held at a steady length. In contrast, primary afferents, which also receive inputs from dynamic fusimotor neurones, showed large changes in their rate of impulses and their sensitivity during the step cycle. This was consistent with the patterns observed, if the phasic fibres were dynamic fusimotor fibres.

These changes in sensitivity were also consistent with the changes in stretch reflexes observed during locomotion in the same preparation (Akazawa *et al.*, 1982), although other factors can modify the overall reflex, as described below. More recently, my colleague Arthur Prochazka recorded from muscle spindle afferents under a variety of conditions and inferred the pattern of fusimotor input using a simulation technique developed by Hulliger *et al.*, (1987). Depending on conditions he found considerable variation in the levels of fusimotor activity, which was termed "fusimotor set" (Prochazka, 1989). Certainly, a large number of sensory and descending inputs can affect fusimotor neurones and Murphy and Hammond (1991) showed that cutaneous inputs to fusimotor neurones are gated by the locomotor pattern.

We conclude at present that there is a basic pattern of fusimotor activity during locomotion that modulates the sensitivity of the primary muscle spindles afferents appropriately for their reflex role. The effects, for example of cutaneous inputs onto fusimotor neurones, that are expected during locomotion may be gated out, but other factors such as the speed or difficulty of the task may modify the basic pattern of fusimotor activity and "set" it appropriately for the task.

Synaptic Modulation

To the extent that the monosynaptic component of the stretch reflex predominates, one might expect that the opportunity for synaptic modulation would be limited, because of the single synapse from muscle spindle afferents onto α–motoneurones. In fact, dramatic effects are observed from three different sources: 1) potentiation and depression of the monosynaptic EPSP depending on the pattern of input, 2) modulation of the percentage of motoneurones firing due· to other synaptic inputs to the motoneurone pool and 3) presynaptic inhibition of transmission under a variety of conditions. Each of these factors will be considered briefly in turn.

The EPSP's recorded in α-motoneurones from Ia inputs can be greatly potentiated by tetanic inputs to the point that virtually the whole motoneurone pool will discharge to a single Ia volley (Clamann *et al.*, 1974). More recently, Koerber and Mendell (1991) have shown that the pattern of input, normally observed in Ia fibres during walking, modulates the size of the EPSP significantly.

Ten thousand or more synapses from a number of sources may end on a single α-motoneurone, which serves as "the final common pathway" for the motor system, to use Sherrington's term. All of these inputs could, in principle, modify the synaptic effect of the spindle input. The most common way to study these effects is to produce a Ia volley by electrical stimuli and record the Hoffmann or H-wave in an appropriate muscle that results under a variety of conditions. The advantage of this technique is that it can be used in normal human subjects doing a variety of tasks. As shown in Fig. 2, the size of the H-wave during walking varies considerably depending on the level of EMG activity in the soleus muscle.

Similar variation has been seen in other muscles with natural (length changes) as well as electrical stimuli. Matthews (1986) referred to this variation as an "automatic gain control," since a priori one would expect, if other synaptic inputs depolarize motoneurones closer to threshold, more would fire in response to the Ia volley. The situation is more complex, as shown in Fig. 2. When the same subject was asked to generate a range of EMG levels tonically, the variation in the H-reflex was quite different. The relation between the H-wave and EMG activity is also different in running (Edamura *et al.*, 1991) and cycling (Brooke *et al.*, 1992).

Evidence from experimental and theoretical studies support the idea that this task-dependent variation in H-waves is due to presynaptic inhibition. If the levels of post-synaptic excitation and inhibition are varied in models of the motoneurone pool so as to produce the same net activity, the size of the H-reflex is identical under a fairly broad range of conditions (Capaday and Stein, 1987). This prediction was verified experimentally by varying the level of post-synaptic inhibition from Renshaw cells in decerebrate cats (Capaday and Stein, 1989). Conversely, the model indicated that the relation between the H-wave and the level of activity in the motor pool could be dramatically changed by presynaptic inhibition. Experimentally, Baclofen, an agonist of the presynaptic transmitter GABA, clearly altered the relation in decerebrate cats (Stein *et al.*, 1993).

In conclusion, the efficacy of the synapses from Ia fibres to α-motoneurones can vary widely depending on the pattern of activity in these fibres or others which synapse onto the motoneurones. Functionally, the two most important modes of varying transmission appear to be: 1) covariation of reflexes with the net level of activity in the motoneurone pool (automatic gain control) and 2) task-dependent variation of the reflex at a given level of activity in the motoneurone pool by means of presynaptic inhibition.

Mechanical Factors

By concentrating on the electrical activity of muscle spindles or H-reflexes the mechanical factors which shape reflexes are often ignored. The magnitude of these factors has only recently become obvious as a result of experiments with mechanical inputs to cat and human muscles. Fig. 3 shows the effect of a brief 3 mm stretch to soleus muscle of a decerebrate cat. In Fig. 3A the muscle was at rest and the response was a large EMG and force; i.e., the typical "tendon jerk" familiar to clinicians and patients alike. In this example, the muscle was attached to a puller which received length and velocity feedback so that it behaved like a spring of stiffness 5 N/mm. The effect of the reflex force was therefore to shorten the muscle against the puller. Clinically, it is this shortening of the muscles after the tendon tap that is observed and used in assessing the integrity and magnitude of reflexes. The larger the reflex shortening the higher is the apparent "gain" of the reflex.

In Fig. 3B the same command signal was applied to the stretcher when the muscle was generating a high force, as a result of eliciting a crossed extensor reflex. Because of the increased intrinsic stiffness of the muscle due to the contraction, the length change was smaller and the immediate force larger. Nonetheless, a comparable reflex EMG was produced. Surprisingly, no extra reflex force or reflex shortening is apparent. In fact, the stretch broke many existing bonds and the force would have dropped considerably (not shown), if these bonds were not replaced by the reflex contraction.

Is the "reflex gain" low under these conditions because one does not see a large reflex force and shortening or is it high, because it almost precisely cancels the slow deviation in force and position that occurs when the reflex is abolished? Clearly, terms such as gain that apply to linear systems are not appropriate here, since the effect of extra force is far from additive; i.e., the system is highly nonlinear. Nonetheless, during contraction the reflex is effective in preventing slow deviations that would arise from the breakage of bonds by external perturbations.

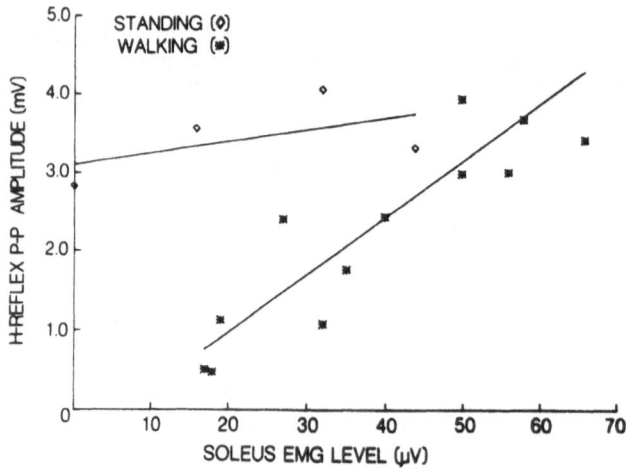

Figure 2. H-reflexes in soleus muscle of a normal human subject during walking (*) and standing (◊) as a function of EMG level. Note the marked difference in the slopes and y-intercepts of the straight lines, which were computed to minimize the mean-square errors. From Capaday and Stein, 1986.

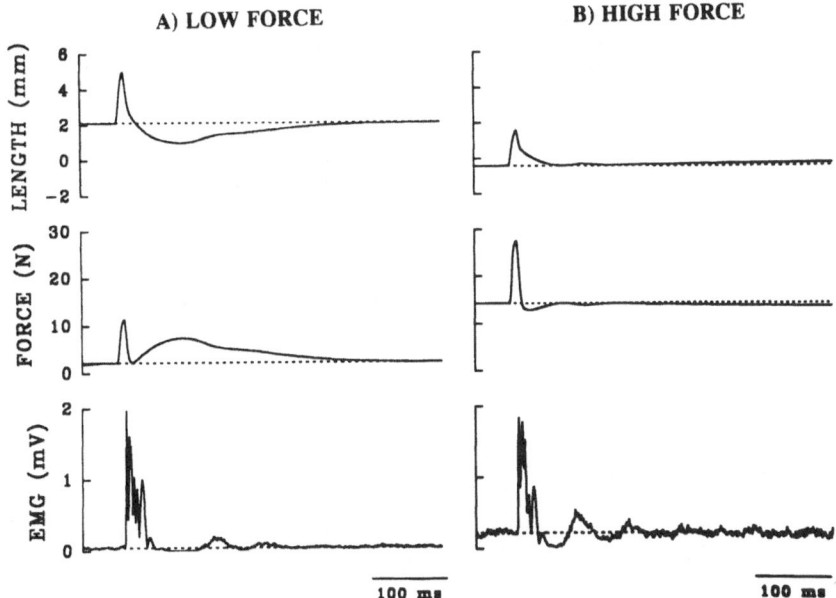

Figure 3. A: Approximately 3mm stretches were applied to soleus muscle in a decerebrate cat. The muscle was attached to a puller that had feedback so that it behaved like a spring of stiffness 5 N/mm. At rest the stretch produced some force due to the intrinsic stiffness of the muscle. This was followed by EMG activity (which has been rectified and lightly filtered) and reflex force. **B:** The same command to the puller when the muscle was generating higher levels of force produced approximately the same level of EMG, but negligible reflex force. The dashed lines show the average levels prior to the stretch. Further explanation in the text.

These results are reminiscent of the experiments of Nichols and Houk (1976) on "yielding" during ramp and hold stretches which gave rise to the stiffness regulation hypothesis. The muscle is clearly behaving much more like a spring of constant stiffness when the reflexes are present in Fig. 3, but the point we wish to make is that factors such as the level of background force can modify the reflex force and length changes dramatically by mechanical means, even when the reflex EMG is comparable.

The final example is taken from experiments done on the human ankle joint. Fig. 4 shows the average effect of small length changes produced by a hydraulic actuator. The early oscillatory torque associated with the flexion of the ankle (solid lines) is due to the inertia of the foot and actuator attachment and intrinsic muscle mechanics. Slightly later, a brisk EMG and reflex torque are observed at latencies consistent with a monosynaptic reflex in human subjects. An upward movement (flexion) produces a downward (extensor) torque, as expected for a resistance reflex. In fact, these small perturbations (0.035 radians or approximately 2^O) produced torques that were often 15-20% of the subjects' maximum voluntary contraction (MVC). Thus, the reflex under these conditions was remarkably potent.

Also shown in Fig. 4 is the averaged response to ankle extensions (dotted lines) of the same amplitude and velocity as the flexions. The early, inertial component of the torque is reversed, but a delayed reflex EMG and torque are observed which are in the same directions as during flexions. The extension did not itself produce a reflex response in the ankle extensor muscles or the antagonist muscles which flex the ankle joint. Thus, the reflex behaves as a half-wave rectifier, responding preferentially to one direction of input. Indeed, triceps surae EMG has been shown to be well described as a non-linear system with unidirectional rate sensitivity (Kearney and Hunter, 1988).

Reflex EMG and torques are observed, but at latencies which suggest that they arise from the restretch of the ankle extensor muscles to their original length. This suggestion was confirmed by using a variety of pulse widths. However, the response to the restretch of the ankle extensors, following the release, is much smaller than the response to the same size of stretch. The amplitude difference was observed with pulse widths up to 200ms, when the subjects were exerting steady, voluntary torques, as in Fig. 4. Thus, the initial extension can have a fairly major, long-lasting effect, even though it does not itself elicit a reflex. An even longer-lasting effect (up to 1 s) was observed in subjects who were relaxed. This suggests that it may be a mechanical effect, depending on the rate of bonds cycling in the muscle. Whether the mechanical effect is in the intrafusal muscle and affects the spindle response or in the extrafusal muscle remains to be determined. However, these examples show that the magnitude of human reflexes can clearly be affected in a major way by mechanical factors, as well as fusimotor and synaptic effects.

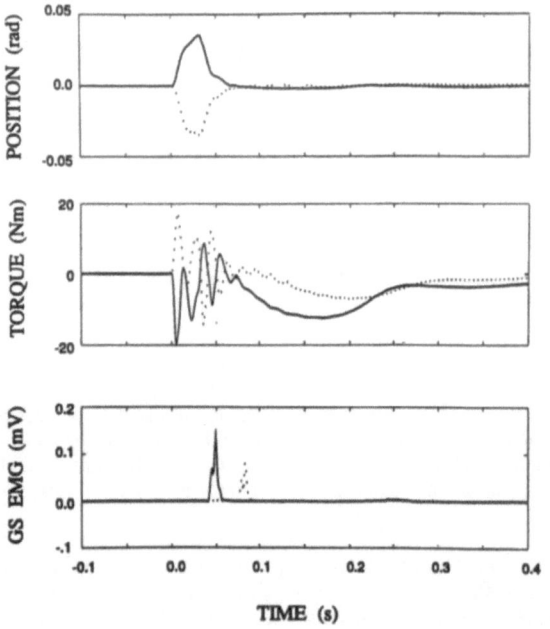

Figure 4. Brief flexions (solid lines) and extensions (dotted lines) of the ankle joint of a human subject produced EMG responses (rectified and lightly filtered records from the gastrocnemius and soleus (GS) muscles) and torque. Note that the reflex produced extensor torque (downward) for both directions of perturbation. The subject was generating a constant, voluntary extensor torque of 5Nm at a position of 0.1 radian of ankle flexion with respect to a right angle. These steady values have been subtracted from the torque and position records. Further explanation in the text. From Stein and Kearney, 1993.

DISCUSSION

The stretch reflex clearly contains a large number of nonlinearities, only some of which have been described here. These include: 1) the muscle spindle afferents have a small amplitude linear region, but are highly nonlinear for stretches larger than a fraction of a mm (Matthews and Stein, 1969). 2) Fusimotor modulation, such as shown in Fig. 1, is well known to increase and decrease the sensitivity of the afferents to stretch. 3) As discussed in the section on synaptic modulation, the EPSPs produced by each impulse will depend on the pattern of activity. 4) Each motoneurone will have a threshold, one or more linear ranges (Kernell, 1965) and will adapt to maintained inputs over a period of time. These will all affect the response to a given pattern of afferent input. 5) The percentage of motoneurones firing is a function of various post-synaptic inputs to the motoneurone pool and this leads to the "automatic gain control" discussed in relation to Fig. 2. 6) A variety of inputs can presynaptically inhibit transmission to the motoneurones. A dramatic example of this was studied recently when we applied small, broad band-width inputs. Inputs with amplitudes less than 1[o] could dramatically reduce the stretch reflex of human muscles (Kearney and Stein, 1993), presumably by means of presynaptic inhibition. 7) Generation of muscle force involves formation of cross-bridges between the myofilaments, which can be mechanically broken by stretch. Thus, the reflex force produced by a stretch will vary with the force level, as shown in Fig. 3, even when the EMG produced is similar. 8) The rate at which cross-bridges cycle in the muscle is relatively slow, so the delay for a muscle to take up slack after it is shortened can be substantial. As shown in Fig. 4, this can affect the response to stretch for several hundred ms. 9) The number of cross-bridges that can be formed and the spontaneous activity in muscle spindle afferents vary with muscle length, so the reflex force will depend on muscle length (Gordon *et al.*, 1966). 10) In addition geometric factors vary with the angle of a joint which will affect the mechanical advantage and hence force output.

This list of nonlinearities could no doubt be extended further, but it is obvious that one can not simply speak of the "gain" of the stretch reflex. In a half-wave rectifier the gain would switch between zero and non-zero values. For some nonlinearities (*e.g.*, the unidirectional velocity sensitivity of Kearney and Hunter, 1988) it may not even be possible to define an "effective" gain sensibly. For others, an equivalent or effective gain can be defined by linearizing about the operating point. However, with time varying conditions such as occur during the execution of movements, the effective gain can vary continuously over a substantial range on the time scale of the movements. We have in the last few years begun to understand how the reflex is varied adaptively to suit the requirements of particular voluntary tasks. Much work remains to be done before we fully understand the many roles of this simplest of reflexes in the coordination and control of behaviour.

References

Akazawa, K., Aldridge, J.W., Steeves, J.D., Stein, R.B. (1982) Modulation of stretch reflexes during locomotion in the mesencephalic cat. *Journal of Physiology* **329:** 553-567.

Brooke, J.D., McIlroy, W.E., Collins, D.F. (1992) Movement features and H-reflex modulation: I. Pedalling versus matched controls. Brain Res. **582:** 78-84.

Capaday, C., Stein, R.B. (1986) Amplitude modulation of the soleus H-reflex in the human during walking and standing. *Journal of Neuroscience* **6:** 1308-1313.

Capaday, C., Stein, R.B. (1987) A method for simulating the reflex output of a motoneurone pool. *Journal of Neuroscience Methods* **21:** 91-104.

Capaday, C., Stein, R.B. (1989) The effects of postsynaptic inhibition on the monosynaptic reflex of the cat at different levels of motoneurone pool activity. *Experimental Brain Research* **77:** 577-584.

Clamann, H.P., Gillies, J.D., Skinner, R.D., Henneman, E. (1974) Quantitative measures of output of a motoneurone pool during monosynaptic reflexes. *Journal of Neurophysiology* **37:** 1328-1337.

Dietz, V. (1992) Human neuronal control of automatic functional movements: interaction between central programs and afferent input. *Physiological Reviews* **72:** 33-50.

Edamura, M., Yang, J.F., Stein, R.B. Factors that determine the magnitude and time course of human H-reflexes in locomotion. *Journal of Neuroscience* (1991) 11: 420-427.

Gordon, A.M., Huxley, A.F., Julian, F.J. (1966) The variation in isometric tension with sarcomere length in vertebrate muscle fibres. *Journal of Physiology* **184:** 170-192.

Houk, J.C. (1979) Regulation of stiffness by skeletomotor reflexes. *Annual Review of Physiology* **41:** 99-114.

Hulliger, M., Horber, F., Medved, A., Prochazka, A. (1987) An experimental simulation method for iterative and interactive reconstruction of unknown (fusimotor) inputs contributing to known (spindle afferent) response. *Journal of Neuroscience Methods* **21:** 225-238.

Kearney, R.E., Hunter, I.W. (1988) Nonlinear identification of stretch reflex dynamics. *Annals of Biomedical Engineering* **16:** 79-94.

Kearney, R.E., Stein, R.B. (1993) Influence of perturbation properties on the identification of stretch reflexes at the human ankle joint. *International Conference IEEE Engineering in Medicine and Biology Society*, San Diego pp. 1169-1170.

Kernell, D. (1965) High-frequency repetitive firing of cat lumbosacral motoneurones stimulated by long-lasting injected currents. *Acta Physiologica Scandinavica* **64:** 75-86.

Koerber, H.R., Mendell, L.M. (1991) Modulation of synaptic transmission at Ia-afferent connections on motoneurones during high-frequency afferent stimulation: dependence on motor task. *Journal of Neurophysiology* **65:** 1313-1320.

Levin, M.F., Feldman, A.G., Milner, T.E., Lamarre, Y. (1992) Reciprocal and coactivation commands for fast wrist movements. *Experimental Brain Research* **89:** 669-677.

Liddell, E.G.T., Sherrington, C. (1924) Reflexes in response to stretch (myotatic reflexes). *Proceedings of the Royal Society* **B96:** 212-242.

Matthews, P.B.C. (1972). *Mammalian Muscle Receptors and their Central Actions.* London: Arnold.

Matthews, P.B.C. (1986) Observations on the automatic compensation of reflex gain on varying the pre-existing level of motor discharge in man. *Journal of Physiology* **374:** 73-90.

Matthews, P.B.C. (1990) The knee jerk: still an enigma? *Canadian Journal of Physiology and Pharmacology* **68:** 347-354.

Matthews, P.B.C., Stein, R.B. (1969) The sensitivity of muscle spindle afferents to small sinusoidal changes in length. *Journal of Physiology* **200:** 723-743.

Melvill Jones, G., Watt, D.G.D. (1971) Observations on the control of stepping and hopping movements in man. *Journal of Physiology* **219:** 709-727.

Merton, P.A. (1953) Speculations on the servo-control of movement. In Wolstenholme, G.E.W., ed. *The Spinal Cord,* CIBA Found. Symp. London: Churchill 247-255.

Murphy, P.R., Hammond, G.R. (1991) The role of cutaneous afferents in the control of γ-motoneurones during locomotion in the decerebrate cat. *Journal of Physiology* **434:** 529-547.

Murphy, P.R., Stein, R.B., Taylor, J. (1984) Phasic and tonic modulation of impulse rates in γ-motoneurones during locomotion in premammillary cats. *Journal of Neurophysiology* **52:** 228-243.

Nichols, T.R., Houk, J.C. (1976) The improvement in linearity and the regulation of stiffness that results from the actions of the stretch reflex. *Journal of Neurophysiology* **39:** 119-142.

Prochazka, A. (1989) Sensorimotor gain control: a basic strategy of motor systems? *Progress in Neurobiology* **33:** 281-307.

Shik, M.L., Orlovsky, G.N. (1976) Neurophysiology of locomotor automatism. Physiological Reviews **56:** 465-501.

Stein, R.B. and Kearney, R.E. Nonlinear behavior of stretch reflexes at the human ankle joint. *International Conference IEEE Engineering in Medicine and Biology Society,* San Diego (1993) pp. 1167-1168.

Stein, R.B., Yang, J.F., Belanger, M., Pearson, K.G. (1993) Modification of reflexes in normal and abnormal movements. *Progress in Brain Research* **97:** 189-196.

Taylor, J., Stein, R.B., Murphy, P.R. (1985) Impulse rates and sensitivity to stretch of soleus muscle spindle afferent fibers during locomotion in the premammillary cat. *Journal of Neurophysiology* **53:** 341-360.

A MULTIPLE-LEVEL APPROACH TO MOTOR PATTERN GENERATION

P.S.G. STEIN

Department of Biology
Washington University
St. Louis, MO. 63130 U.S.A.

SUMMARY

Motor control is studied at multiple levels, e.g., the movement level, the network/circuit level, and the cellular/synaptic/molecular level. The movement level requires the nervous system as well as the muscular system and the structural support system, e.g., skeletal system. The network/circuit level occurs strictly within the central nervous system (CNS). The cellular/synaptic/molecular level has its focus on each individual neurone. Understanding each level illuminates properties at other levels. A levels analysis is used to approach the comparative physiology of motor pattern generation. Specific examples from work on the physiology of turtle scratch reflex are used (Stein, 1989).

INTRODUCTION

Churchland and Sejnowski (1988) stress the importance of a multilevel analysis of the nervous system and behaviour. The levels they describe are CNS, systems, maps, networks, neurones, synapses, and molecules. Bunge (1989) also describes a multilevel approach to the nervous system control of behaviour. He emphasizes the importance of a levels analysis that is balanced, i.e., neither strictly reductionistic nor strictly holistic. Bunge also notes that at a specific level (1) certain properties are properties of lower levels and (2) still other properties, termed "emergent properties," are present at that specific level and not at lower levels. The term "emergent property" has connotations for some that may exceed those required for the present paper. A neutral term "levels property" will be used here to describe the set of properties at a given level. This discussion will describe the levels that exhibit a given property and, in addition, will note the lowest level at which a given property is present.

Many classic experiments in the field of the comparative physiology of motor pattern generation are designed to reveal the lowest level at which a given property, e.g., "rhythm," is present (Stein, 1984). Issues from these classic experiments can be expressed clearly from the perspective of a levels analysis. For example, is a motor rhythm present only at the movement level, i.e., only when movement-related feedback is present? Can a motor rhythm be expressed at a network level in the absence of movement-related feedback? If a motor rhythm is present at a network level, is it also present at a cellular level? The answers to these questions have been revealed in many biological systems (Delcomyn, 1980; Stein, 1984); the levels perspective can be used to illuminate important generalizations that apply to several phyla.

Neural Control of Movement, Edited by W.R. Ferrell
and U. Proske, Plenum Press, New York, 1995

GENERAL CONCEPTS AT THE MOVEMENT LEVEL

The nervous system, the muscular system, and the structural support system are all required for function at the movement level. Several concepts assist analyses at the movement level. First, a classification scheme for tasks and the forms of a task can be helpful. Second, descriptions using each of several coordinate systems are also useful. Classification and coordinate concepts will be presented in this section. In the following section, these concepts will be applied to a specific behaviour, namely, scratching in a spinal vertebrate, the spinal turtle.

Classification of Movements: Tasks and the Forms of a Task

Movements may be classified according to their function or "goal" (Stein, Mortin and Robertson, 1986). A movement with a particular function is termed a task. For example, scratching functions to generate force against a site on the body surface; locomotion functions to move an organism's centre-of-mass from one location to another. A given task may be performed using each of several movement strategies. A person can scratch the side of one's thorax with either an elbow or a hand; a horse can walk, trot, or gallop; a person can walk forward or backward. Each movement strategy used to perform a task is termed a form of that task (Stein *et al.*,1986). For each task, discriminators must be developed to classify each of the forms of that task.

The concept of a "descriptor space" is helpful in analyzing the forms of a task (Stein *et al.*,1986). A descriptor space for overground locomotion of a horse can be the set of velocity vectors for the horse's centre-of-mass. In a pure-form region of this descriptor space, only a single form is expressed, e.g., there is a set of velocities always associated with a gallop. In a transition-zone region of this descriptor space, either of two forms may be expressed, e.g., there is a set of velocities slower than the fastest trot and faster than the slowest gallop in which the horse may trot or gallop. Specific application of the concept of the descriptor space to the receptive field for turtle scratch reflex will be detailed in a later section.

Coordinate Systems for Movements

First, the rectilinear "Cartesian" coordinate system, with its three orthogonal x, y and z coordinates, is useful in describing the space in which movement occurs. A Cartesian coordinate system may be also useful in defining the "descriptor space" for the forms of a task (see above section). Other coordinate systems such as polar or spherical coordinates may also be useful.

Second, the "body degree-of-freedom" coordinate system is useful in describing the actual movements performed by an organism. In a vertebrate, this coordinate system has been termed a "joint-angle" coordinate system. Some joints, e.g., knee, have only one degree of freedom; other joints, e.g., hip, have three degrees of freedom, and therefore require three dimensions to describe its actions. The dimension of time is also required for the full description of movements in joint-angle space. When a rhythmic movement is analyzed, rhythmicity is a property of the movement level that can be observed in each of several joint angles. The movement pattern, e.g., a regulated timing in one dimension of joint-angle space with respect to a rhythm in another dimension can also be a movement-level property in joint-angle space.

Third, the muscle/motor-pool coordinate system is useful in describing the actual output of the nervous system during movement. In this coordinate system, each muscle is viewed as a dimension. A measure of the intensity of muscular action, e.g., the full-wave rectified electromyogram, can be used to describe the magnitude along that muscle's dimension. Of course, other ways of measuring the magnitude of a muscle's activation, e.g., a weighted sum of the firing frequencies of all motor-units/motor-neurones, may be used instead. The dimension of time is also required to describe motor output during actual movements. The term "motor pattern" is used to describe the time course of a set of muscle/motor-pool activities. During a rhythmic movement, rhythmicity can be observed in each of several muscles. In addition, the motor pattern, the regulated timing of one muscle's activity with respect to the rhythm of another muscle's activity, can be described in this coordinate system.

MOVEMENT-LEVEL ANALYSIS OF SCRATCH REFLEX IN A SPINAL VERTEBRATE

The function of scratching is to generate force against a specific site on the body surface that has received a tactile stimulus. Limbed vertebrates may utilize the movements of a nearby limb to generate this force. The present discussion will focus on the scratch reflex produced by a limbed vertebrate with a complete transection of the spinal cord, i.e., a spinal vertebrate, in response to a tactile stimulus delivered to a site in a dermatome innervated by a spinal segment caudal to the complete transection. Scratch reflex has been demonstrated in a number of spinal vertebrates, e.g., cat, dog, turtle, frog (Stein, 1983).

Classification of the Forms of Hindlimb Scratching in the Spinal Turtle

In the spinal turtle, three distinct movement strategies have been described for hindlimb scratching (Mortin, Keifer and Stein, 1985). The strategy that is expressed is determined by which portion of the limb is used to rub against the specific site on the body surface that has received the tactile stimulus. For each form of the scratch, a distinct portion of the limb is used to exert force against the stimulated site.

Cartesian Coordinates for the Receptive Fields for Scratch Reflex

The Cartesian coordinate system that describes the body surface can be used to characterize the receptive field for the hindlimb scratch in the turtle (Mortin et al., 1985, Stein et al., 1986). The scratch receptive field can be viewed as a "descriptor space" for the scratch reflex. This field is a continuous surface that may be divided into five regions, three pure-form regions and two transition-zone regions. Stimulation of each site in a pure-form region evokes scratching characterized by only one form. A transition zone is a region that separates two nearby pure-form regions. Stimulation of each site in a transition zone evokes either of two possible pure-form responses or a blended response (either a switch or a hybrid). In a switch, several cycles of one form are followed immediately by several cycles of the other form; in a hybrid, in each of several successive cycles, each of two movement strategies is used. The concept of "motor equivalence" may be used to describe the several strategies exhibited in response to stimulation of a single site in a transition zone (Berkinblit, Feldman and Fukson, 1986).

Joint-Angle Analysis of Hindlimb Scratching in the Spinal Turtle

For each of the three forms of hindlimb scratching in the spinal turtle, (1) there is rhythmic alternation between hip flexion (protraction) and hip extension (retraction) and (2) the knee extends during the rub against the stimulated site (Mortin et al., 1985). Distinct for each form of the scratch is the timing of knee extension in the cycle of hip flexion and hip extension. Thus each movement strategy is characterized by a distinct regulation in the timing of a distal joint's movement within the cycle of a proximal joint's movement; each movement strategy has a distinct movement pattern.

Muscle/Motor-Pool Analysis of Hindlimb Scratching: the Motor Pattern

The activity of a muscle can be monitored by recording the EMG (electromyographic activity) of that muscle; this recording is also a measure of the activities of the motor units in that muscle and thus also a measure of the motor pool, i.e., the set of motor neurones that innervate that muscle. The activity of the motor pool for a muscle can also be monitored by recording the ENG (electroneurographic activity) of the nerve innervating that specific muscle. The characteristics of each muscle's EMG are similar to those of the ENG of that muscle's specific nerve.

The pattern of activity recorded either as EMGs or ENGs is termed the "motor pattern." Common to the motor patterns for all three forms of the scratch in the spinal turtle is rhythmic alternation between hip flexor (protractor) muscle activity and hip extensor (retractor) muscle activity (Robertson, Mortin, Keifer and Stein, 1985). For each form of scratch, the monoarticular knee extensor muscle is active during the rub against the stimulated site. Distinct for each form of the scratch is the timing of knee extensor muscle activity in the cycle of hip muscle activity. Thus, there is a regulated timing of one muscle's activation in the cycle of another muscle's activation, i.e., there is a distinct motor pattern for each scratch form.

Properties at the Movement Level for Scratch Reflex in the Spinal Turtle Observed in Joint-Angle Space and in Muscle/Motor-Pool Space

During scratching produced by the spinal turtle, "rhythm," "pattern," and "selection" are several properties at the movement level that can be observed either in joint-angle space or in muscle/motor-pool space. First, rhythm is a characteristic of each of several dimensions of joint-angle space as well as each of several dimensions of muscle/motor-pool space. Second, pattern is a characteristic of these two spaces as well. Correlates of the pattern in movement space can be observed in muscle/motor-unit space. For example, the statement in movement space that there is regulated timing of knee extension in the cycle of hip flexion and extension has a correlate in muscle/motor-pool space; namely, there is regulated timing of monoarticular knee extensor muscle activity in the cycle of hip flexor muscle and hip extensor muscle activities. Third, since each of several forms of a task is produced by the spinal turtle, the spinal cord and the associated muscular-skeletal system exhibits the property of selection among movement strategies; supraspinal structures are not required for this motor strategy selection.

Characteristics Present Only at the Movement Level and Characteristics Present at Both the Movement Level and the Network/Circuit Level

By definition, the movements of joints occur at the movement level and not at a lower level; thus, movement patterns can be observed at the movement level and not at a lower level. The motor pattern is present at the movement level. It is an experimental question to determine if the motor pattern is also present at the network/circuit level. The following section describes the experiments that establish that the motor pattern is also present at the network/circuit level. Thus motor patterns are a powerful tool in motor physiology since they may be analyzed at both the movement level and the network/circuit level.

THE NETWORK/CIRCUIT LEVEL

The nervous system control of motor pattern can be studied in the absence of actual movements. The motor pattern is recorded as ENG activities from each of several specific nerves; each nerve contains the axons of a motor pool that innervates a specific muscle. The ENG motor pattern produced in the absence of movement has been termed a "fictive" motor pattern since it is observed when there is no "real" movement (Stein, 1984).

Several types of preparations are studied at the network/circuit level. First, in the *in vitro* preparation, all muscles can be removed to eliminate movement. Second, in the neuromuscular blockade preparation, a neuromuscular blocking agent, e.g., an acetylcholine receptor antagonist in a vertebrate, can be used to prevent movement. In either of these preparations, all movement-related feedback is prevented. For many of these preparations, the motor pattern recorded in the absence of movement-related feedback, termed the "central motor pattern," is an excellent replica of the motor pattern recorded during actual movements (Stein, 1984). Thus the motor pattern is a property of the network/circuit level as well as the movement level.

The set of CNS neurones responsible for generating a central motor pattern is termed the "central pattern generator" or CPG. It is an experimental issue to locate the CNS region that contains the CPG for a given motor pattern as well as the individual neurones that comprise the CPG.

In a spinal turtle immobilized with a neuromuscular blocking agent, scratch reflex ENG motor patterns can be recorded (Robertson *et al.,* 1985). The rhythmic motor pattern characteristic of each scratch form is elicited by the stimulation of that form's receptive field. Common to the patterns for all three scratch forms is rhythmic alternation between hip flexor motor-pool activity and hip extensor motor-pool activity. Distinct for each form of the scratch is the timing of monoarticular knee extensor motor-pool activity in the cycle of hip motor-pool activity. These results in a spinal, immobilized turtle establish that rhythm, pattern, and selection are properties of the spinal cord at the network/circuit level. Similar conclusions have been obtained in *Xenopus* for struggling and swimming (Soffe,

1993) and in lamprey for forward swimming and backward swimming (Matsushima and Grillner, 1992).

The first segment of the turtle hindlimb enlargement, when isolated from the remainder of the CNS, contains sufficient neural circuitry to generate a scratch motor rhythm; the first three segments of the turtle hindlimb enlargement contain sufficient circuitry to generate a rhythmic scratch motor pattern (Mortin and Stein, 1989). The importance of anterior segments of the hindlimb enlargement for rhythmogenesis has also been demonstrated for chick embryo (Ho and O'Donovan, 1993) and cat (Deliagina, Orlovsky and Pavlova, 1983).

Rhythm, pattern, and selection have also been well studied at the network/circuit level in the crustacean stomatogastric nervous system (Harris-Warrick and Marder, 1991; Harris-Warrick, Marder, Selverston and Moulins, 1992). When this system is studied *in vitro*, it can generate each of several rhythmic motor patterns in response to each of several neuromodulators. Thus, a specific anatomical network of neurones can be configured into each of several physiological circuits depending upon the responses of each of the network elements to specific neuromodulatory inputs (see also Getting, 1989). The result obtained with PS neurone stimulation in the stomatogastric system is an important example of network reconfiguration (Meyrand, Simmers and Moulins, 1994). A vertebrate example of major changes in network properties under the influence of a neuromodulator is the result obtained with serotonin on *Xenopus* swim motor pattern (Sillar, Wedderburn, Woolston and Simmers, 1993).

It is an experimental question to reveal what aspects of the properties of "rhythm," "pattern," and "selection" observed at the network/circuit level are also observed at the cellular/synaptic/molecular level. The next section will describe aspects of rhythm that are present at the single-neurone level; however, other aspects of rhythm clearly require the network/circuit level, e.g., the contribution of reciprocal inhibition to rhythmogenesis (Roberts and Tunstall, 1990). Aspects of pattern are present at the single-neurone level, e.g., rhythmic bursts of action potentials; however, those aspects of pattern that relate the timing of one motor pool to that of another require a network/circuit. The activity of a single neurone can correlate with motor pattern selection (Wine and Krasne, 1972); that neurone must work in conjunction with a network, however. Motor pattern selection may also be strictly a network property, however; selection may result from the summed activity of a population of broadly-tuned neurones (Berkowitz and Stein, 1994a,b).

THE CELLULAR/SYNAPTIC/MOLECULAR LEVEL

The levels properties at the cellular/synaptic/molecular level serve as "building blocks" for the network/circuit level (Getting, 1989; Harris-Warrick and Marder, 1991; Pearson, 1993). These building blocks can be specific molecules, e.g., specific channels and/or receptors in the membrane. Particular combinations of building blocks may result in rhythmicity at the single-neurone level, e.g., NMDA channels that are calcium permeable that interact with calcium-dependent potassium channels to act as a single-neurone rhythm generator (Grillner, Wallen, Brodin and Lansner, 1991).

An important finding at the cellular level is the powerful influence of neuromodulatory agents on the properties of the cellular/synaptic/molecular building blocks. For example, the AB neurone in the crustacean stomatogastric ganglion is a conditional bursting neurone, i.e., it will fire bursts of action potentials in the absence of synaptic inputs and in the presence of each of several specific neuromodulators (Harris-Warrick and Flamm, 1987); the characteristics of the bursting pattern are dramatically different according to which neuromodulator activates the neurone.

A major thrust of work at the cellular/synaptic/molecular level is to determine the characteristics of the building blocks that can be utilized in the construction of networks/circuits. For example, temporal summation plays an important role in activation of scratch reflex (Currie and Stein, 1990); both NMDA receptors (Currie and Stein, 1992; Daw, Stein and Fox, 1993) and calcium channels (Russo and Hounsgaard, 1993) contribute to temporal summation.

CONCLUSIONS

Future work at the cellular/synaptic/molecular level will contribute to our understanding at the network/circuit level. A full understanding at the network/circuit level is required to elucidate the mechanisms that act at the movement level. Further work at the movement level is also required. Additional work at each level will provide important insights to improve understanding of properties at all levels.

ACKNOWLEDGEMENTS. NIH Grant NS30786 provides support for research in the PSGS laboratory. I thank Edelle Field for her editorial assistance.

REFERENCES

Berkinblit, M.B., Feldman, A.G. and Fukson, O.I. (1986). Adaptability of innate motor patterns and motor control mechanisms. *Behavioral and Brain Sciences* **9**, 585-599.

Berkowitz, A. and Stein, P.S.G. (1994a). Activity of descending propriospinal axons in the turtle hindlimb enlargement during two forms of fictive scratching: broad tuning to regions of the body surface. *Journal of Neuroscience* **14**, 5089-5104.

Berkowitz, A. and Stein, P.S.G. (1994b). Activity of descending propriospinal axons in the turtle hindlimb enlargement during two forms of fictive scratching: phase analyses. *Journal of Neuroscience* **14**, 5105-5119.

Bunge, M. (1989). From neurone to mind. *News In Physiological Sciences* **4**, 206-209.

Currie, S.N. and Stein, P.S.G. (1990). Cutaneous stimulation evokes long-lasting excitation of spinal interneurons in the turtle. *Journal of Neurophysiology* **64**, 1134-1148.

Currie, S.N. and Stein, P.S.G. (1992). Glutamate antagonists applied to midbody spinal cord segments reduce the excitability of the fictive rostral scratch reflex in the turtle. *Brain Research* **581**, 91-100.

Churchland, P.S. and Sejnowski, T.J. (1988). Perspectives on cognitive neuroscience. *Science* **242**, 741-745.

Daw, N., Stein, P.S.G. and Fox, K. (1993). The role of NMDA receptors in information processing. *Annual Review of Neuroscience* **16**, 207-222.

Delcomyn, F. (1980). Neural basis of rhythmic behaviour in animals. *Science* **210**, 492-498.

Deliagina, T.G., Orlovsky, G.N. and Pavlova, G.A. (1983). The capacity for generation of rhythmic oscillations is distributed in the lumbosacral spinal cord of the cat. *Experimental Brain Research* **53**, 81-90.

Getting, P.A. (1989). Emerging principles governing the operation of neural networks. *Annual Review of Neuroscience* **12**, 185-204.

Grillner, S., Wallen, P., Brodin, L. and Lansner, A. (1991). Neuronal network generating locomotor behaviour in lamprey: circuitry, transmitters, membrane properties, and simulation. *Annual Review of Neuroscience* **14**, 169-199.

Harris-Warrick, R.M. and Flamm, R.E. (1987). Multiple mechanisms of bursting in a conditional bursting neurone. *Journal of Neuroscience* **7**, 2113-2128.

Harris-Warrick, R.M. and Marder, E. (1991). Modulation of neural networks for behaviour. *Annual Review of Neuroscience* **14**, 39-57.

Harris-Warrick, R.M., Marder, E., Selverston, A.I. and Moulins, M. eds. (1992). *Dynamic Biological Networks: The Stomatogastric Nervous System*, Cambridge, Massachusetts: MIT Press.

Ho, S. and O'Donovan, M.J. (1993). Regionalization and intersegmental coordination of rhythm-generating networks in the spinal cord of chick embryos. *Journal of Neuroscience* **13**, 1354-1371.

Matsushima, T. and Grillner, S. (1992). Neural mechanisms of intersegmental coordination in lamprey: local excitability changes modify the phase coupling along the spinal cord. *Journal of Neurophysiology* **67**, 373-388.

Meyrand, P., Simmers, J. and Moulins, M. (1994). Dynamic construction of a neural network from multiple pattern generators in the lobster stomatogastric nervous system. *Journal of Neuroscience* **14**, 630-644.

Mortin, L.I. and Stein, P.S.G. (1989). Spinal cord segments containing key elements of the central pattern generators for three forms of scratch reflex in the turtle. *Journal of Neuroscience* **9**, 2285-2296.

Mortin, L.I., Keifer, J. and Stein, P.S.G. (1985). Three forms of the scratch reflex in the spinal turtle: movement analyses. *Journal of Neurophysiology* **53**, 1501-1516.

Pearson, K.G. (1993). Common principles of motor control in vertebrates and invertebrates. *Annual Review of Neuroscience* **16**, 265-297.

Roberts, A. and Tunstall, M.J. (1990). Mutual re-excitation with post-inhibitory rebound: a simulation study on the mechanisms for locomotor rhythm generation in the spinal cord of *Xenopus* embryos. *European Journal of Neuroscience* **2**, 11-23.

Robertson, G.A., Mortin, L.I., Keifer, J. and Stein, P.S.G. (1985). Three forms of the scratch reflex in the spinal turtle: central generation of motor patterns. *Journal of Neurophysiology* **53**, 1517-1534.

Russo, R.E. and Hounsgaard, J. (1993). Wind up mediated by L-type calcium channels in turtle dorsal horn neurones. *Society for Neuroscience Abstracts* **19**, 1196.

Sillar, K.T., Wedderburn, J.F.S., Woolston, A.M. and Simmers, A.J. (1993). Control of locomotor movements during vertebrate development. *News in Physiological Sciences* **8**, 107-111.

Soffe, S.R. (1993). Two distinct rhythmic motor patterns are driven by common premotor and motor neurones in a simple vertebrate spinal cord. *Journal of Neuroscience* **13**, 4456-4469.

Stein, P.S.G. (1983). The vertebrate scratch reflex. *Symposia of the Society for Experimental Biology* **37**, 383-403.

Stein, P.S.G. (1984). Central pattern generators in the spinal cord, In *Handbook of the Spinal Cord, Vols. 2 and 3: Anatomy and Physiology*, ed. Davidoff, R.A., pp. 647-672. New York: Marcel Dekker.

Stein, P.S.G. (1989). Spinal cord circuits for motor pattern selection in the turtle. *Annals of the New York Academy of Sciences* **563**, 1-10.

Stein, P.S.G., Mortin, L.I. and Robertson, G.A. (1986). The forms of a task and their blends, In *Neurobiology of Vertebrate Locomotion*, eds. Grillner, S., Stein, P.S.G., Stuart, D.G., Forssberg, H. and Herman, R.M., pp. 201-216. London: Macmillan Press.

Wine, J.J. and Krasne, F.B. (1972). The organization of escape behaviour in the crayfish. *Journal of Experimental Biology* **56**, 1-18.

THE STATUS OF THE PREMOTOR AREAS: EVIDENCE FROM PET SCANNING

R.E. PASSINGHAM

Department of Experimental Psychology
University of Oxford
South Parks Road
Oxford OX1 3UD, U.K.

SUMMARY

There has been no agreement as to the role of the premotor areas including the supplementary motor cortex. Some believe them to play a role in the execution of movement; others believe them to play a role in the selection of movement. This paper suggests a resolution. Recent evidence from PET studies indicates that these areas should be divided into anterior and posterior sectors. Evidence is provided that the posterior sectors play a role in the execution of movement and that the anterior sectors play a role in the selection of movement.

INTRODUCTION

For many years there has been a controversy concerning the status of the premotor areas. The posterior parts of the lateral and medial area 6 are premotor areas in the strict sense that they send projections to the motor cortex (Muakkassa and Strick, 1979). Woolsey *et al.*, (1952) suggested that the medial premotor cortex (supplementary motor area) should be called MII, because electrical stimulation revealed a somatotopic map that was independent of the map in motor cortex (MI). Wise (1985) has suggested that the lateral and medial area 6 be referred to as 'non-primary motor' areas.

The use of this term can be justified by pointing to both physiological and anatomical evidence. It has recently been shown using microstimulation that there are somatotopic maps in the medial convexity of area 6 (SMA) and in the cingulate sulcus (Mitz and Wise, 1987; Luppino *et al.*, 1991; Dum and Strick, 1993). Furthermore it has also been shown by using microstimulation that there is a somatotopic map in the lateral premotor cortex (area 6) (Kurata, 1989, Godschalk *et al.*, 1990).

The existence of motor maps in the premotor areas is further confirmed by recent anatomical studies. Dum and Strick (1993) have shown that there are pyramidal projections to the spinal cord from the medial convexity of area 6 (SMA) as well as from the upper and lower banks of the cingulate sulcus. On the lateral surface there are also projections to the cord from the posterior part of the lateral premotor cortex, including the region termed 'the arcuate premotor area' by Schell and Strick (1984) that lies around the bow of the arcuate sulcus (Dum and Strick, 1991). These projections are organized in a somatotopic fashion (He *et al.*, 1993).

These physiological and anatomical studies point to the existence of multiple mapped motor areas in area 6 and the cingulate cortex that lies under the medial part of area 6. We do not know the particular contribution of each of these areas, but it is reasonable to suppose that they may be specialized for different aspects of the execution of movement.

Behavioural studies have, however, pointed to a different role for the premotor areas. If the lateral area 6 is removed bilaterally in macaque monkeys, the animals are very impaired at relearning tasks on which they must select between two movements on the basis of visual cues (Halsband and Passingham, 1982, 1985; Petrides 1982, 1987). The evidence is reviewed by Passingham (1993), and he concludes that the lateral premotor cortex is involved in the process by which movements are selected on the basis of external contexts.

Similarly, if the medial area 6 (SMA) is removed bilaterally in macaque monkeys, the animals are very impaired at learning sequence tasks on which they have to learn to perform either two movements (Chen *et al.*, in press, Passingham 1993) or three movements (Halsband, 1987; Passingham 1987) in a particular order. These and other studies are reviewed by Passingham (1993), and he concludes that the medial premotor cortex is involved in the process by which movements are selected on the basis of an internal context.

Thus, there are two different views of the functions of area 6. The first holds that it has executive functions, and the second that it plays a role in the selection of movements. There is a need to reconcile these views. This paper suggests a resolution on the basis of studies using the PET scanner.

EVIDENCE FROM PET STUDIES

Execution

We have scanned ten subjects while they execute a series of repetitive movements (Playford, unpublished data). These subjects include the six subjects described by Playford *et al.*,(1992).

Every time a tone sounds, the subjects push a joystick forwards with their right hand. We call this the 'repetitive' condition. Activation of the motor areas is measured by changes in regional cerebral blood flow compared with a rest condition in which the subjects hear the tones but make no movements.

There is significant activation in the motor cortex, lateral premotor cortex, medial premotor cortex (SMA) and the posterior cingulate cortex lying under the medial premotor cortex.

Figure 1 gives the coordinates for the peak of maximum significance in the right lateral premotor cortex and the peak in the left medial premotor cortex. For the lateral premotor cortex the peak is taken on the right; the activation for the left lateral premotor cortex is continuous with that for the left motor cortex, and it is difficult to give an accurate estimate of the peak of maximal significance for the lateral premotor area on the left. The coordinates are given in the space used by the atlas of Talairach and Tournoux (1988). It is defined with reference to the line through the anterior and posterior commissure; the VCA line is the line drawn vertically through the anterior commissure, at $90°$ to the AC-PC line. The filled circles gives the peaks of maximum significance. It will be seen that in both premotor areas these peaks lie behind the VCA line.

In monkeys most of the pyramidal fibres from the premotor areas come from tissue behind the level of the anterior commissure (Dum and Strick, 1991). The anterior commissure lies roughly at the level of the back of the bow of the arcuate sulcus. Braak (1976) has described an area in the cingulate sulcus with gigantopyramidal cells; this also lies behind the level of the anterior commissure.

Selection

Now compare the activation when subjects must select which movement to make each time that the tone sounds. In this 'free-selection' condition the subjects are required to decide on each trial whether to move the joystick forwards, backwards, left or right. Thus they make a new decision on each trial. The activation for this condition is measured by comparing it with the activation for the 'repetitive' condition, described above. This comparison shows the areas that are more active when subjects select a movement compared with the condition in which they always push the joystick forwards.

The triangles in Figure 1 gives the peaks of maximum significance for this comparison in the right lateral premotor cortex and left medial premotor cortex (SMA). In the right lateral premotor cortex two positions are marked by triangles. The posterior triangle shows the peak of maximum significance in the study by Playford (unpublished data). The anterior triangle shows the peak of maximum significance for the eight subjects we tested in an earlier study (Deiber *et al.*, 1991). It will be seen that when subjects select between movements, and a comparison is made with the 'repetitive' condition, the peak of activation lies anterior to the VCA line in both the lateral and medial premotor cortex.

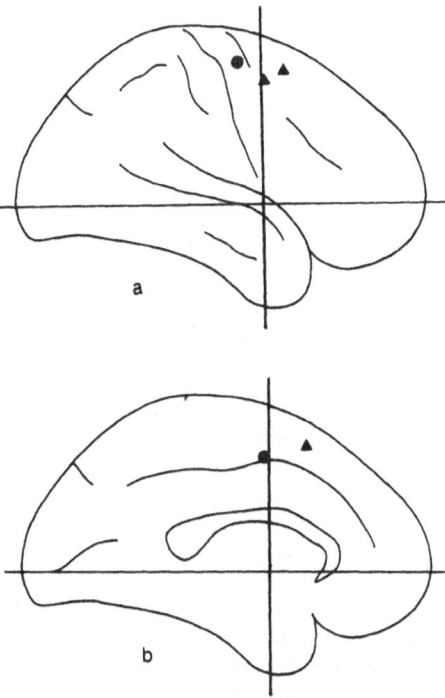

Figure 1. Peaks of activation in the right lateral premotor cortex above) and the left medial premotor cortex (below). The circles shows the peaks of maximal significance comparing the 'repetitive' task with a rest condition. The triangles show the peaks of maximal significance comparing the 'free-selection' task with the 'repetitive' task. Of the two triangles shown for the anterior part of the lateral premotor cortex, the back one is for the data of Playford (unpublished data) and the front one for the data of Deiber *et al.*,(1991).

Learning

If the anterior part of the premotor areas is activated when subjects select between movements, then it should also be activated while subjects learn what movements to make. We have scanned subjects while they learn a sequence of 8 finger movements with their right hand. We have also scanned the same subjects while they perform another sequence of 8 movements on which they were trained before scanning; they were trained for 75 minutes until they could perform the sequence automatically (Jenkins *et al.*, 1994). While learning a new sequence the subjects must make decisions as to which movement to try; but when they perform a pre-learned sequence there are no new decisions to make, since the order of the movements is now fixed and unvarying.

Figure 2 shows the activation in the premotor areas. Figure 2a shows the site at which activation if greater during learning compared with performance of the pre-learned sequence, and figure 2b the site at which activation is greater during performance of the pre-learned sequence compared with new learning. It will be seen that during new learning there is greater activation in the lateral premotor cortex, anterior to the VCA line, and that during performance of the pre-learned sequence there is greater activation in the medial premotor cortex, posterior to the VCA line.

Comparison of Macaque and Human Brain

The argument would be more secure if it could be shown that the anterior part of area 6 in the human brain is functionally similar to the anterior part in the macaque brain. We have preliminary evidence that this may be so. Matsuzaka *et al.*,(1992) compared cells in the anterior and posterior parts of the medial premotor cortex; they found more cells in the anterior part that were active when monkeys prepared to make movements. We have tested human subjects in two conditions. In both they have to raise the forefinger when they hear a tone. In one condition the tone is fairly regular, with a mean of one every three seconds or so; in the other condition the tone is unpredictable, varying from one every two seconds to one every seven seconds. We find activation of the more anterior part of the medial premotor cortex when subjects can prepare than when they cannot (Passingham, in press).

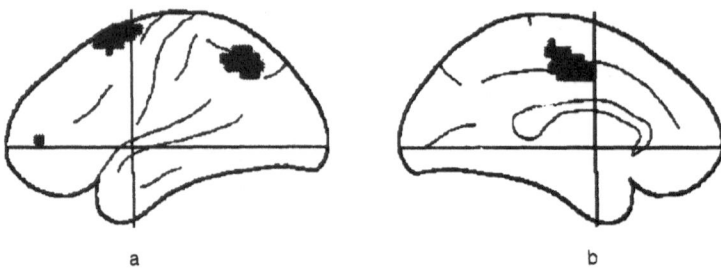

a   b

Figure 2. Areas in which there is a significant difference in activation. (a) shows the areas on the lateral surface that are significant comparing new learning with performance of the pre-learned task. (b) shows an area on the medial surface that is significant comparing performance of the pre-learned task with new learning. Data from Jenkins *et al.*, 1994.

DISCUSSION

These results suggest a resolution of the debate on the status of the premotor areas. The debate has polarized around two views, the first that the premotor areas are concerned with higher aspects of the execution of movement, the second that the premotor areas are concerned with the learned selection of movements.

The resolution suggested here is that the premotor areas should be divided into posterior and anterior regions, and that the first view correctly characterizes the functions of the posterior regions and that the second view correctly characterizes the functions of the anterior regions.

Further evidence for this view comes from the electrophysiological study reported by Matsuzaka *et al.*,(1992). The found that in the caudal part of the medial premotor cortex (SMA) many of the cells were time-locked to the onset of movement, but that in the rostral part of the medial premotor cortex many cells responded when cues informed the animal of the movement to be selected.

Recent architectural studies using cytochrome oxidase have indicated that the premotor areas should be subdivided into anterior and posterior regions. Figure 3 shows the map suggested by the studies of Matelli *et al.*,(1985, 1991). The dorsal lateral premotor cortex is divided into regions termed F2 and F7, and the medial premotor cortex into regions termed F6 and F3. The area F7 corresponds to the anterior part of the lateral premotor cortex as described in this paper, and the area F6 to the anterior part of the SMA.

Cingulate cortex

The cingulate premotor areas are not discussed in detail in this paper. However, it may be possible to divide these also into anterior and posterior sectors. Shima *et al.*,(1991) divide the anterior cingulate cortex into anterior an posterior sectors; they use as the boundary the back of the arcuate sulcus, which roughly corresponds to the position of the anterior commissure. They report that on a self-paced task more cells fire early before movement in the anterior than in the posterior sector.

Anterior commissure

The position of the anterior commissure is a useful landmark for dividing anterior and posterior sectors on the medial surface. However, the division between the anterior and posterior sectors of the lateral premotor cortex cannot be made using this landmark. In the macaque monkey it is possible to delineate anterior and posterior sectors, but most of area 6 lies in front of the anterior commissure.

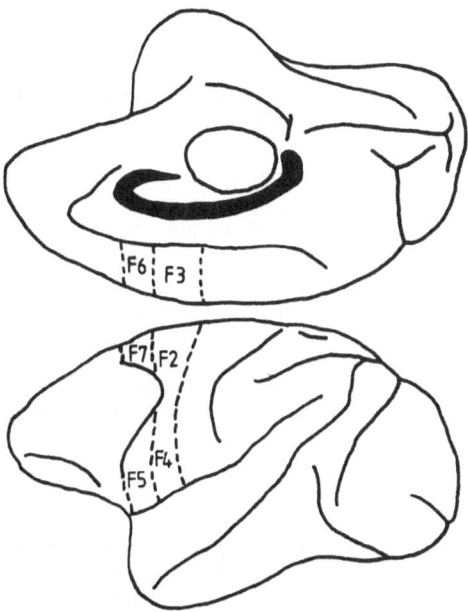

Figure 3. Subdivisions of the premotor areas on the basis of staining with cytochrome oxidase. Redrawn from Matelli *et al.*,(1985, 1991).

Discrete lesions

There are two predictions concerning the effects of removing these areas separately.

1) First, lesions which remove the anterior part of the premotor areas alone should lead to impairments in the selection of movements. There is preliminary evidence that this may be so. We have studied the effects of removing the medial premotor cortex on the ability to learn three movements in sequence. We placed lesions in the medial premotor cortex in three monkeys, but in one animal the lesion was placed more anteriorly than intended; it included the anterior part of the medial premotor cortex together with some medial frontal tissue further forwards (Passingham, 1987). Like the animals with more complete lesions this animal was also impaired on learning the motor sequence.

2) The second prediction is that if both the anterior and posterior regions are removed the animals should be impaired both in the execution and selection of movements. We have removed all of the lateral area 6, except that we deliberately left some of the tissue dorsal to the precentral sulcus so as not to encroach on the medial premotor cortex (Passingham, 1985). The monkeys were impaired at relearning a conditional task on which they had to select between movements; but they also had minor impairments in the execution of movements in the weeks after surgery. They were clumsy at first and inaccurate at locating small holes with their fingers (Passingham, 1985). Traverse and Latto (1986) have also noticed executive problems after they had made a similar removal. Rizzolatti *et al.,*(1983) made small unilateral lesions that included the posterior bank of the ventral limb of the arcuate sulcus (the arcuate premotor area); and in the first weeks after surgery their animals were slow to respond to stimuli in the contralateral half-field.

We have also removed both the anterior and posterior regions of the medial premotor cortex in monkeys (Thaler and Passingham, 1989; Passingham *et al.*, 1989). In the first few days after surgery the monkeys tended to take food by mouth rather than hand (Thaler,

1988). At first their gait was also abnormal, but it is possible that this was due to oedema in the leg area on the medial convexity of the motor cortex (Thaler, 1988).

Our papers on the effects of lesions in lateral or medial premotor cortex have stressed the effects on the selection of movement. The issue was not whether premotor lesions have effects on movement, but whether, at the time at which the animals were tested, the animals could make the movements required for the task; and we were able to show that they could (Passingham, 1993).

But it is also clear that lesions in the premotor areas can have effects on the execution of movement. Freund and Hummelsheim (1985) have also argued that patients can suffer impairments in their axial musculature when lesions are apparently confined to area 6.

We now know that there are multiple mapped motor areas in the posterior part of the lateral and medial area 6, and it is a challenge to work out the specialized contributions of each area. Removal of area 4 leads to paralysis in monkeys in the first few weeks after surgery (Gilman *et al.*, 1974), whereas monkeys are neither paralysed nor akinetic after removal of either the lateral (Passingham, 1985) or medial parts of area 6 (Thaler and Passingham, 1989; Thaler *et al.*, in press). These results suggest that there is a hierarchical difference in the contribution to movement of areas 4 and 6.

Cortico-cortical connections

It is clear that we should divide the frontal cortex into four strips and not three. Figure 4 shows this in the form of a simplified diagram.

Premotor area 8 is omitted from this diagram, because this paper only considers the areas concerned with movements of the limbs. The diagram is also oversimplified in that it fails to show all the connections between these areas.

The dorsal prefrontal cortex (areas 9 and 46) project to the anterior division of the lateral (Barbas and Pandya, 1987, 1989) and medial (Bates and Goldman-Rakic, 1993, Lu *et al.*, 1994) premotor cortex. In turn there are projections from the anterior part of the medial premotor cortex to the posterior part (Luppino *et al.*, 1990), and from the anterior part of the dorsal part of the lateral premotor cortex to the posterior part (Pandya and Barnes, 1987). Finally there are projections from the posterior regions of the lateral and medial premotor cortex to the motor cortex (area 4) (Muakkassa and Strick, 1979; Dum and Strick, 1991). Both the posterior premotor areas and the motor cortex sends projections to the cord through the pyramidal tract (Dum and Strick, 1991).

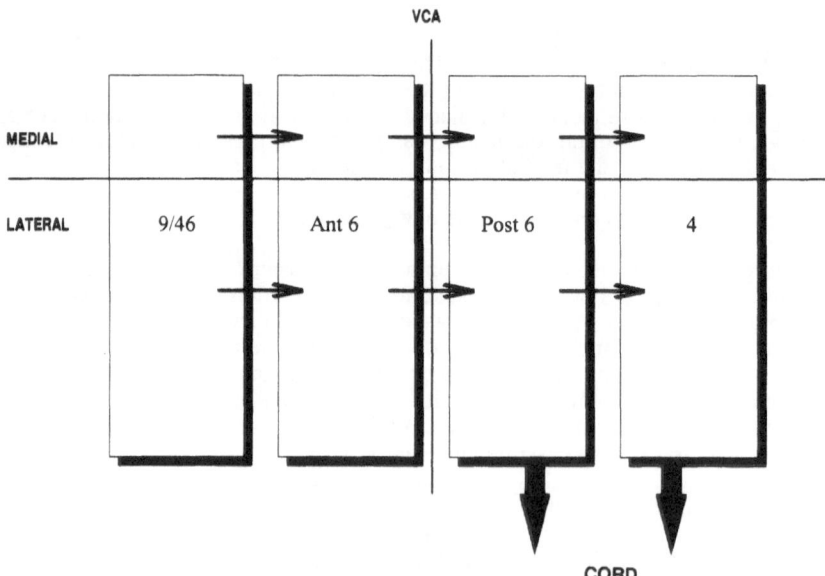

Figure 4. Diagram of four strips of cortex with some of the connections between them. Border of the lateral and medial surface is shown by the horizontal line. Thick arrows show pyramidal projections to the spinal cord.

Subcortical-cortical connections

The division of the premotor areas into anterior and posterior regions further clarifies the debate about the inputs to the premotor areas from the ventral thalamus. Hoover and Strick (1993) have shown that there is a linkage between the putamen, globus pallidus pars internal, the VLo nucleus of the ventral thalamus, and the medial premotor cortex and arcuate premotor area in the lateral premotor cortex. Schell and Strick (1984) have also stressed that the cerebellum can influence the arcuate premotor cortex via projections from the cerebellar nuclei to the arcuate premotor area.

It also appears, however, that the VA nucleus sends projections to the anterior part of the medial premotor cortex (Wiesendanger and Wiesendanger, 1985) and the anterior part of the dorsal lateral premotor cortex (Miyata and Sasaki, 1983; Matelli *et al.,* 1989; Shook *et al.,* 1991). The caudate nucleus sends projections to the dorsal part of the globus pallidus pars interna (Smith and Parent, 1986; Selemon and Goldman-Rakic, 1990), and there are suggestions that the dorsomedial globus pallidus may project to the VA nucleus of the ventral thalamus (Kim *et al.,* 1976; DeVito and Anderson, 1982; Ilinsky and Kultas-Ilinsky, 1987).

These connections form a discernible pattern. The putamen has inputs from sensorimotor areas (Percheron *et al.,* 1984) and the caudate from prefrontal areas (Selemon and Goldman-Rakic, 1985). In turn the posterior regions of the premotor cortex receive their subcortical influence from the putamen and the anterior regions from the caudate. The premotor areas project both to lateral putamen and lateral caudate (Selemon and Goldman-Rakic, 1985). It would be interesting to establish whether the posterior regions have stronger projections to the putamen, and the anterior regions stronger projections to the caudate.

CONCLUSIONS

The question of the status of the premotor areas can be resolved if it is appreciated that both the lateral and medial premotor areas should be further subdivided. Wiesendanger and Wiesendanger (1984) reviewed the evidence available at that time on the functions of the medial premotor cortex, and concluded that it had both 'higher' and 'lower' functions in the control of movement. The resolution is that the posterior regions play some role in the execution of movements, and the anterior regions some role in the selection of movements.

There is clearly a need for a new terminology. The term 'non-primary motor areas' as suggested by Wise (1985) is apt for the posterior regions. The term 'Pre-SMA' has been suggested for the anterior part of the medial premotor cortex (Matsuzaka *et al.,* 1992), but the term 'Pre-premotor cortex' would be inelegant, though anatomically correct. It might be better to call on the vague term 'association areas', since this already widely used for other areas that are neither sensory nor motor. The anterior regions could thus be called 'the premotor association areas'.

ACKNOWLEDGEMENTS. I thank Matthew Rushworth for commenting on an early draft of this paper. The research was supported by grants from the Wellcome Trust. The experiments on functional imaging of the human brain were carried out at the MRC Cyclotron Unit, Hammersmith Hospital, London. The were performed in collaboration with Professor R.S.J. Frackwick and Professor D.J. Brooks.

REFERENCES

Barbas, H. and Pandya, D.N. (1987) Architecture and frontal cortical connections of the premotor cortex (area 6) in the rhesus monkey. *Journal of Comparative Neurology* **256**, 211-228.

Barbas, H. and Pandya, D.N. (1989) Architecture and intrinsic connections of the prefrontal cortex in the rhesus monkey. *Journal of Comparative Neurology* **286**, 353-375.

Bates, J.F. and Goldman-Rakic, P.S. (1993) Prefrontal connections of medial motor areas in the rhesus monkey. *Journal of Comparative Neurology* **335**, 1-18.

Braak, H. (1976) A primitive gigantopyramidal field buried in the depth of the cingulate sulcus of the human brain. *Brain Research* **109**, 219-233.

Chen, Y-C, Thaler, D., Nixon, P.D., Stern, C. and Passingham, R.E. (in press) The functions of the medial premotor cortex (SMA). II. The timing and selection of learned movements. *Experimental Brain Research.*

Deiber, M.-P., Passingham, R.E., Colebatch, J.G., Friston, K.J., Nixon, P.D. and Frackowiak R.S.J. (1991) Cortical areas and the selection of movement: a study with positron emission tomography. *Experimental Brain Research* **84**, 393-402.

DeVito, J.L. and Anderson, M.E. (1982) An autoradiographic study of efferent connections of the globus pallidus in *Macaca mulatta*. *Experimental Brain Research* **46**, 107-117.

Dum, R. P. and Strick, P.L. (1991) The origin of corticospinal projections from the premotor areas in the frontal lobe. *Journal of Neuroscience* **11**, 667-689.

Dum, R. P. and Strick, P.L. (1993) Cingulate motor areas. In *Neurobiology of Cingulate Cortex and Limbic Thalamus,* eds Vogt B.A. and Gabriel M., 415-441, Birkhauser, Boston.

Freund, H.-J. and Hummelsheim, H. (1985) Lesions of premotor cortex in man. *Brain* **108**, 697-734.

Gilman, P.F.C., Lieberman, T.S. and Marco, L.A. (1974) Spinal mechanisms underlying the effects of unilateral ablation of area 4 and 6 in monkeys. *Brain* **97**, 49-64.

Godschalk, M., Mitz, A.R., van der Burg, J. and van Duin, B. (1990) Microstimulation map of the monkey premotor cortex. *Society of Neuroscience Abstracts* **16**, 1133.

Halsband, U. (1982) *Higher Movement Disorders in Monkeys,* Unpublished D.Phil. thesis, Oxford University

Halsband, U. (1987) Higher disturbances of movement in monkeys (Macaca mulatta). In *Motor Control* eds Gantchev G.N., Dimitev B. and Gatev P.C. 79-85, Plenum, New York.

Halsband, U. and Passingham, R.E. (1982) The role of the premotor and parietal cortex in the direction of action. *Brain Research* **240**, 368-372.

Halsband, U. and Passingham, R.E. (1985) Premotor cortex and the conditions for movement in monkeys (*Macaca mulatta*). *Behavioural Brain Research* **18**, 269-276.

He, S.Q., Dum, R.P. and Strick, P.L. (1993) Topographic organization of corticospinal projections from the frontal lobe: motor areas on the lateral surface of the hemisphere. *Journal of Neuroscience* **13**, 952-980.

Hoover, J.E. and Strick, P.L. (1993) Multiple output channels in the basal ganglia. *Science* **259**, 819-821.

Ilinsky, I. A. and Kultas-Ilinsky, K. (1987) Sagittal cytoarchitectonic maps of the *Macaca mulatta* thalamus with a revised nomenclature of the motor-related nuclei validated by observations of their connectivity. *Journal of Comparative Neurology* **262**, 331-364.

Jenkins, I. H., Brooks, D.J., Nixon, P.D., Frackowiak, R.S.J. and Passingham, R.E. (in press) Motor sequence learning: a study with positron emission tomography. *Journal of Neuroscience.*

Kim, R., Nakano, A., Jayaraman, A. and Carpenter, M.B. (1976) Projections of the globus pallidus and adjacent structures: an autoradiographic study in the monkey. *Journal of Comparative Neurology* **169**, 263-290.

Kurata, K. (1989) Distribution of neurons with set- and movement-related activity before hand and foot movements in the premotor cortex of rhesus monkeys. Experimental Brain Research **77**, 245-256.

Lu, M.T., Preston J.B. and Strick, P.L. (1994). Interconnection between the prefrontal cortex and the premotor areas in the frontal lobe. *Journal of Comparative Neurology* **341**, 375-392.

Luppino, G., Matelli, M., Camarda, R.M., Gallese, V. and Rizzolatti, G. (1991) Multiple representations of body movements in mesial area 6 and the adjacent cingulate cortex: an intracortical microstimulation study in the macaque monkey. *Journal of Comparative Neurology* **311**, 463-482.

Luppino, G., Matelli, M. and Rizzolatti, G. (1990) Cortico-cortical connections of two electrophysiologically identified arm representations in the mesial agranular frontal cortex. *Experimental Brain Research* **82**, 214-218.

Matelli, M., Luppino, G., Fogassi, L. and Rizzolatti, G. (1989) Thalamic input to inferior area 6 and area 4 in the macaque monkey. *Journal of Comparative Neurology* **280**, 448-458.

Matelli, M., Luppino, G. and Rizzolatti, G. (1991) Architecture of superior and mesial area 6 and the adjacent cingulate cortex in the macaque monkey. *Journal of Comparative Neurology* **311**, 445-462.

Matelli, W., Luppino, G. and Rizzolatti, G. (1985) Pattern of cytochrome oxidase activity in frontal agranular cortex of the macaque monkey. *Behavioural Brain Research* **18**, 125-136.

Matsuzaka, Y., Aizawa, H. and Tanji, J. (1992) Motor area rostral to the supplementary motor area (presupplementary motor area) in the monkey: neuronal activity during a learned motor task. *Journal of Neurophysiology* **68**, 653-662.

Mitz, A.R. and Wise, S.P. (1987) The somatotopic organization of the supplementary motor areas: intracortical microstimulation mapping. *Journal of Neuroscience* **7**, 1010-1021

Miyata, M. and Sasaki, K. (1983) HRP studies on thalamocortical neurons related to the cerebellocerebral projection in the monkey. *Brain Research* **274**, 213-224.

Muakkassa, K. F. and Strick, P.L. (1979) Frontal lobe inputs to primate motor cortex: evidence for four somatotopically organized "premotor areas". *Brain Research* **177**, 176-182.

Pandya, D. N., and Barnes, C. (1987) Architecture and connections of the frontal lobe. In *The Frontal Lobes Revisited* ed Perecman E., 41-72, IBRN Press, New York.

Passingham, R.E. (1985) Cortical mechanisms and cues for action. *Philosophical Transactions of the Royal Society of London B* **308**, 101-111.

Passingham, R.E. (1987) Two cortical systems for directing movements. In CIBA Symposium. 132: *Motor Areas of the Cerebral Cortex,* ed Porter R., 151-164, Wiley, Chichester.

Passingham, R.E. (1993) *The Frontal Lobes and Voluntary Action,* Oxford University Press, Oxford.

174

Passingham, R.E. (in press). Function al specialization of the SMA in monkey and man. In *The Supplementary Sensorimotor Area* ed. Lüders, H.O., Raven Press, New York.

Passingham, R.E., Thaler, D.E. and Chen, Y. (1989) Supplementary motor cortex and self-initiated movement. In *Neural Programming,* ed. Ito M., 13-24, Karger, Basel.

Percheron, G., Yelnik, J. and Francois, C. (1984) A Golgi analysis of the primate globus pallidus: III Spatial organization of the striato-pallidal complex. *Journal of Comparative Neurology* **227**, 214-227.

Petrides, M. (1982) Motor conditional associative-learning after selective prefrontal lesions in the monkey. *Behavioural Brain Research* **5**, 407-413.

Petrides, M. (1987) Conditional learning and primate frontal lobes. In *The Frontal Lobes Revisited,* ed. Perecman E., 91-108, IBRN press, New York.

Playford, E.D., Jenkins, I.H., Passingham, R.E., Nutt, J., Frackowiak, R.S.J. and Brooks, D.J. (1992) Impaired mesial frontal and putamen activation in Parkinson's disease: a positron emission tomography study. *Annals of Neurology* **32**, 151-161.

Rizzolatti, G., Matelli, M. and Pavesi, G. (1983) Deficits in attention and movement following the removal of postarcuate (area 6) and prearcuate (area 8) cortex in macaque monkeys. *Brain* **106**, 655-673.

Schell, G.R. and Strick, P.L. (1984) The origin of thalamic inputs to the arcuate premotor and supplementary motor areas. *Journal of Neuroscience* **4**, 539-560.

Selemon, L.D. and Goldman-Rakic, P.S. (1985) Longitudinal topography and interdigitation of corticostriatal projections in the rhesus monkey. *Journal of Neuroscience* **5**, 776-794.

Shima, K., Aya, K., Mushiake, H., Inase, M., Aizawa, H. and Tanji, J. (1991) Two movement-related foci in the primate cingulate cortex observed in signal-triggered and self-paced forelimb movements. *Journal of Neurophysiology* **65**, 188-202.

Shook, B.L., Schlag-Rey, M. and Schlag, J. (1991) Primate supplementary eye field. II Comparative aspects of connections with the thalamus, corpus striatum, and related forebrain nuclei. *Journal of Comparative Neurology* **307**, 562-583.

Smith, Y. and Parent, A. (1986) Differential connections of caudate nucleus and putamen in the squirrel monkey. *Neuroscience* **18**, 347-371.

Talairach, J. and Tournoux, P. (1988) *Co-Planar Stereotaxic Atlas of the Human Brain,* Thieme, Stuttgart.

Thaler, D.E. (1988) *Supplementary motor cortex and the control of action,* Unpublished D.Phil. thesis, Oxford University

Thaler, D.E. and Passingham, R.E. (1989) The supplementary motor cortex and internally directed movement. In *Neural Mechanisms in Disorders of Movement,* ed. A. R. Crossman, 175-181, Libbey, London.

Thaler, D., Chen, Y-C., Nixon, P.D., Stern, C. and Passingham, R.E. (in press) The functions of the medial premotor cortex (SMA). I. Simple learned movements. *Experimental Brain Research.*

Traverse, J. and Latto, R. (1986) Impairments in route negotiation through a maze after dorsolateral frontal, inferior parietal or premotor lesions in cynomolgous monkeys. *Behavioural Brain Research* **20**, 203-215.

Wiesendanger, M. and Wiesendanger, R. (1984) The supplementary motor cortex in the light of recent investigations. *Experimental Brain Research* Suppl. **9**, 382-392.

Wiesendanger, R. and Wiesendanger, M. (1985) The thalamic connections with medial area 6 (supplementary motor cortex) in the monkey (Macaca fascicularis). *Experimental Brain Research* **59**, 91-104.

Wise, S.P. (1985) The primate premotor cortex: past, present and preparatory. *Annual Review of Neuroscience* **8**, 1-20.

Woolsey, C.N., Settlage, P.H., Meyer, D.R., Sencer, W., Pinto-Hamuy, T. and Travis, M. (1952) Pattern of localization in precentral and "supplementary" motor area and their relation to the concept of a premotor cortex. *Association for Research in Nervous and Mental Diseases* **30**, 238-264.

DEVELOPMENT

ORGANIZATION OF SPINAL LOCOMOTOR NETWORKS AND THEIR AFFERENT CONTROL IN THE NEONATAL RAT

O. KIEHN and O. KJAERULFF

Division of Neurophysiology
Department of Medical Physiology
The Panum Institute, Blegdamsvej 3
Copenhagen N, Denmark

SUMMARY

In this short review we discuss recent findings on the details of the transmitter-induced hindlimb motor pattern, the afferent control of the centrally generated rhythm and the localization of putative rhythm generating spinal neurones in neonatal rats aged 0-4 days. With regard to the motor output, it is concluded that multiple transmitters (like 5-HT and dopamine) control different aspects of both the rhythm and pattern in the neonatal rat. Like in adult animals the centrally generated rhythm can be perturbed by peripheral muscle afferents in an organized fashion, although the resetting from low threshold extensor muscle afferents in the neonatal rat (P0-P3) is to a new flexor phase and not to a new extensor phase as in the adult cat. We suggest, however, that these findings can provide a tool for identification of spinal neurones engaged in rhythm generation. Moreover, based on the use of sulforhodamine 101, a newly identified activity marker, we suggest that cells located in lamina VII and around the central canal are engaged in generating locomotor behaviour in the neonatal rat.

INTRODUCTION

In almost all species studied so far rhythmic movements, such as walking, swimming and breathing can be generated in the absence of afferent input (Delcomyn, 1980). The neuronal networks which generate the rhythmic motor patterns are termed central pattern generators (CPGs). Considerable details of the interplay between cellular and network mechanisms in generating motor programs are revealed in invertebrates and also in some simple vertebrates (Harris-Warrick and Marder, 1991; Roberts, Soffe and Dale, 1986; Grillner and Matsushima, 1991). Much less is known about these mechanisms for both spinal and supraspinal motor CPGs in mammalians. Since there is no *a priori* reason to believe that the cellular and network properties of mammalian motor CPGs can be inferred from studies of motor CPGs in more simple nervous systems there is an inherent need to study the cellular basis for motor behaviour in mammalians. For a long time the adult cat has been the classical preparation for studying such behaviour (Grillner, 1981). Recently, the *in vitro* neonatal rat spinal cord preparation has appeared on the scene (Kudo and Yamada, 1987; Smith and Feldman, 1987). This preparation seems well suited for studying motor behaviour in a four-legged animal. Not only can the spinal cord, with the brainstem or an appropriate part of the periphery attached, survive *in vitro* for a long time, but spinal locomotor activity can readily be induced by external drug application (Kudo and Yamada, 1987; Smith and Feldman, 1987).

Furthermore, the neonatal rat preparation has successfully been used to unravel some of the basic features of the respiratory network (Feldman, Smith and Liu, 1991).

In this short review we will discuss recent findings on the details of the transmitter-induced hindlimb motor pattern, the afferent control of the centrally generated rhythm and the localization of putative rhythm generating spinal neurones (Kiehn, 1993).

CHEMICALLY INDUCED MOTOR OUTPUT

In the neonatal rat, like in all other animals studied so far, spinal locomotor networks can be activated chemically. Early studies using the isolated spinal cord preparation demonstrated that excitatory amino acid receptor activation, especially of NMDA receptors, effectively initiates locomotor activity (Kudo and Yamada, 1987; Smith and Feldman 1987). Other neurochemicals, including dopamine (DA), acetylcholine and substance P were also effective (Smith, Feldman and Schmidt, 1988). More recent studies have shown that in addition to these neurochemicals non-NMDA receptor agonists and serotonin (5-HT) can initiate locomotor activity in the neonatal rat (Cazalets, Sqalli-Houssaini and Clarac, 1992). It therefore appears that several transmitter systems can play instructive roles in initiation of locomotor activity in the neonatal rat. While this knowledge, from an experimental point of view, is practical (many substances can be used to turn on the CPG) the CPG is still treated as a black box. Therefore, we can gain knowledge about the CPG by studing the details of the transmitter-induced motor output. Moreover, it is necessary to clarify whether the activity pattern has reached its final form at birth or if changes are still occuring in the time window we usually study (postnatal rats aged 0-4 days).

As a first step to approach some of these questions we have performed a detailed EMG study of transmitter-induced locomotor activity in hindlimb-attached spinal cord preparations from rats aged 0-4 days (Kjærulff and Kiehn, 1993). We have used 5-HT (10-100 μM) and dopamine (0.1-1 mM) to induce the rhythm.

5-HT and DA initiated different EMG patterns in all concentration ranges. Characteristically, 5-HT induced a regular rhythmic EMG pattern (Fig. 1A) while dopamine induced an irregular EMG pattern (Fig. 1B). The irregularity of the dopamine pattern was most pronounced at birth and shortly after (P0-P1; Fig. 1B), but it remained irregular in all ages investigated.

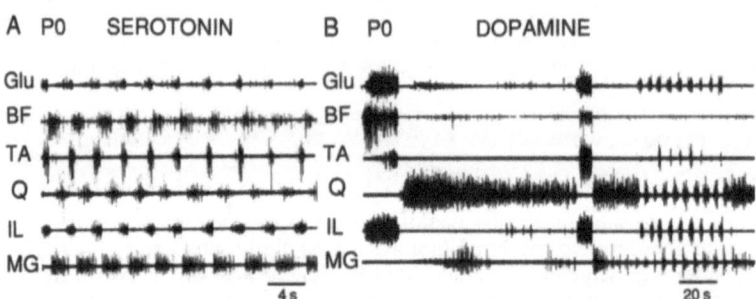

Figure 1. Transmitter-induced EMG pattern in the neonatal hindlimb-attached spinal cord preparation. The patterns of EMG activity of different hindlimb muscles vary with the transmitter used to induce spinal locomotor activity. Serotonin (5-HT) initiates a fast regular rhythmic EMG pattern (**A**) while dopamine (DA) induces an irregular slow EMG pattern (**B**). Note that biceps femoris (BF) shifts phase from A to B. Data from a P0 rat. Drug concentrations were 10μM in A and 0.7mM in B. BF: biceps femoris (ant/post); Glu: gluteus maximus; IL: iliopsoas; Q: quadriceps; MG: medial gastrocnemius; TA: tibialis anterior.Previously unpublished data.

When phase-diagrams were composed it could be observed that in some muscles, e.g. the biceps femoris and semitendinosus muscles, were active in the different phases of the locomotor cycle dependent on the transmitter (see Fig. 1A-B). The rectus femoris, vastus lateralis and vastus medialis muscles were also active in different phases with serotonin and dopamine (not shown).

The differences in the 5-HT and DA-induced patterns were also reflected in the ranges for the cycle durations. Over the effective concentration ranges 5-HT induced cycle periods of 1.6 ± 0.2 to 5.1 ± 0.6s (n=15) while DA induced cycle periods of 2.2 ± 0.6 to 30.3 ± 17.1s (n=12). The range for the 5-HT cycle period was similar in all ages while the very long cycle periods in DA mainly was found in P0-P1 rats.

Thus, the main conclusions from this study is that multiple transmitters control different aspects of both the rhythm and pattern in the neonatal rat.

From invertebrate studies it is known that different transmitters can produce specific motor output by selectively controlling the single cell properties and interneuronal connections thereby *re-configuring* a given network (Harris-Warrick, 1988; Harris-Warrick and Marder, 1991). To what an extent the same is true for the locomotor network in the neonatal rat is at present unknown. We suggest, however, that it is desirable to 'fix' the spinal locomotor CPG in a reproducible activity-state when looking at its neuronal details.

Moreover our study has sorted the muscle in flexors and extensors. This might seem trivial. But, although there are many similarities between for example the 5-HT-induced neonatal EMG pattern and the EMG pattern in adult *locomoting* rats (Grunner and Altman, 1980; Nicolopoulos-Stournaras and Iles, 1984; Goudard, Orsal and Cabelguen; 1992) there are also some clear differences. Most notably is that the semitendinosus muscle always behaved as and extensor in the 5-HT induced rhythm and only shifted to flexor-phase when dopamine was applied. Several other muscles behaved either as flexors or extensors dependent on the transmitter.

Along this line we have also discovered that the ventral root burst in different roots cannot be ascribed purely to flexor and extensor activity as previously has been suggested (Cazaletz *et al.*, 1992). For example the L2 ventral roots burst was found to be in phase with the main flexors and out of phase with several main extensors. Since the L2 burst during locomotor activity is in phase with the main L3 burst and out of phase with the L5 burst the later cannot be a flexor burst as suggested by Cazalets *et al.*, (1992).

AFFERENT PERTURBATION OF THE ONGOING RHYTHM

Although it is generally accepted that spinal rhythmic motor behaviour can be generated in the absence of input from supraspinal brain structures or the periphery, the actual coordination of the motor behaviour depend on a interplay between intrinsic spinal structures, descending command signals and afferent inputs. In the adult animal several descending command signals are known to interfere with the locomotor rhythm (Gossard and Hultborn, 1991; Pearson, 1993). Likewise, afferents for example from the hip joint (Andersson and Grillner, 1983) regulate phase-transitions and load receptors reinforce ongoing muscle activity in the extensor phase of the step cycle (Duysens and Pearson, 1980; Conway, Hultborn and Kiehn, 1987). From a functional point of view this kind of interplay might not be so interesting in the neonatal rat preparation, since the animal is not performing overground locomotion before 12-13 days after birth. However, information on how and when afferents establish connections to spinal CPGs may provide important insights into the structural organization of CPGs during development. Furthermore, the ability to activate afferent inputs, which have access to the CPG, is a necessary tool for an unambiguous identification of neurones engaged in rhythm generation. For these reasons we have studied the afferent perturbation of the spinal locomotor rhythm in the neonatal rat 0-3 days old.

At the outset we, in collaboration with Kudo's laboratory, investigated effects on the centrally generated rhythm from low threshold afferents in an isolated spinal cord preparation (Kiehn, Iizuka and Kudo, 1992; Kiehn, 1993). In the adult cat these afferents reset the locomotor rhythm in a coordinated fashion (Duysens and Pearson, 1980; Conway et al.,1987; Gossard and Hultborn, 1991). Alternating rhythmicity was induced by application of NMDA or 5-HT. Under these conditions the L3 and L5 ventral root bursts reciprocate and are out of phase with the corresponding roots on the opposite side reflecting an underlying locomotor activity. During such locomotor activity a brief train in low threshold afferents of the L3 (or L2) dorsal root (1.7 times threshold (xT) for the incoming volley) caused an effective resetting of the ongoing rhythm. This was due to an abrupt excitation of the L3 ventral root discharges accompanied by a suppression of the L5 ventral root discharge on the ipsilateral side and mirrored by reverse effects on the contralateral side. In contrast to these pronounced effects L4 and L5 dorsal root stimulation usually caused weak or no effects in this group of rats (Kiehn et al.,1992)

Stimulating low threshold afferents in the quadriceps nerve caused similar effects on the ventral root discharges as L2 and L3 dorsal root stimulation (Kiehn et al.,1992). The threshold for resetting was 1.2-1.3 x T and strong effects were seen around 1.5-2.0 x T.

In the adult cat a strong resetting from low threshold afferents appear to be conveyed through extensor Ib afferents (Duysens and Pearson, 1980; Conway et al.,1987; Gossard and Hultborn, 1991). Stimulation of these afferents characteristically suppress ongoing flexor activity with a simultaneous activation of extensors. To evaluate if a similar resetting pattern is present in the neonatal rat we have recorded from identified flexor and extensor muscles in the hindlimb-attached preparation during chemically induced locomotor activity while stimulating different peripheral nerves. An example from such an experiment, where graded stimulation was applied to the quadriceps-nerve, is shown in Fig. 2A-B. In contrast to the adult cat low threshold afferent stimulation of the quadriceps nerve caused resetting to a new *flexor* phase (Fig. 2B) and the resetting pattern remained constant when high threshold afferents were recruited (Fig. 2A). Flexor resetting, although weaker than from the quadriceps nerve, was also observed from stimulating other extensor nerves. In addition strong flexor resetting occurred from stimulation of both high and low threshold afferents in the saphenous and suralis nerves.

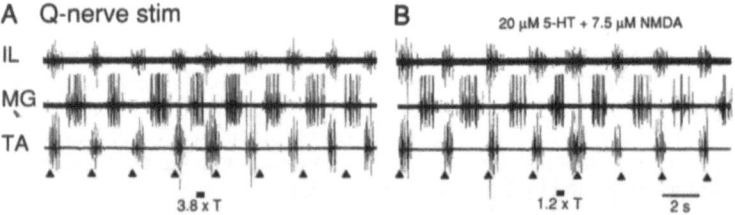

Figure 2. Afferent perturbation of the locomotor activity in the neonatal rat. A, alternating rhythmic flexor (IL and TA) and extensor (MG) activity induced by 5-HT (20μM) in combination with NMDA (7.5μM). The ongoing rhythm is reset to a new flexor phase by a brief train (3.8 or 1.2 x T, 10Hz, 5 impulses; pulse duration 0.2ms) of volleys in the quadriceps nerve (Q-nerve stim.).The triangles indicate the expected onset of the TA bursts had the rhythm not been perturbed. Data from a P0 rat. Previously unpublished data.

The main conclusion from these experiments is that afferents can perturb the centrally generated rhythm and have connections to the CPG at birth in the rat. The pattern of resetting from low threshold afferents appear to be different in the neonatal rat from the one found in the adult cat. Whether this is due to species or developmental differences is a the moment uncertain. No quantitative data on the degree of nerve myelination and conduction velocity for the main groups of afferents are available for 0-3 day old (or older) rats. This makes it impossible to assign certain groups of afferents to a given stimulus strength. The situation is further complicated by the fact that little is known about when peripheral sensory organs (Golgi tendon organs and muscle spindles) develop (see, however, Zelená and Soukup, 1977; Kudo and Yamada, 1985) and if for example vibration or muscle contraction can be used to differentiate between groups of muscle afferents. Experiments in rats aged 0-3 days and also in older rats are now in progress to determine if muscle contraction can reset the rhythm (Iizuka and Kudo). It might be that specific connections from Ib afferents to the CPG first develop later than 3 days after birth and that resetting from other afferents dominate in the very young animals.

Despite these uncertainties we suggest that our findings can provide an important tool for identifying neurones engaged in rhythm generation.

LOCALIZATION OF PUTATIVE LOCOMOTOR GENERATING NEURONES

One of the main problems when analyzing locomotor networks in the mammalian spinal cord is the large number of neurones. This makes localization of rhythm generating regions with conventional electrophysiological methods more difficult. To approach the problem, researchers have turned to techniques which can identify groups of neurones *en block*. Thus,

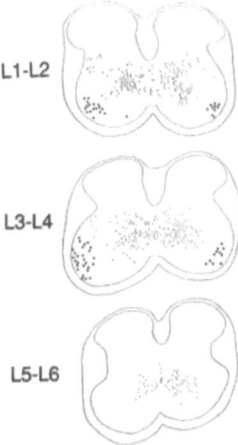

L1-L2

L3-L4

L5-L6

Figure 3. Distribution of sulforhodamine labelled cells in spinal segments L1-L6 in an isolated spinal cord preparation. Locomotor activity was maintained by a combination of NMDA (7.5 μM) and 5-HT (20 μM) for four hours in the presence of sulforhodamine. Each diagram includes all labelled cells in three representative sections from each double segment, and each dot represents one cell. Motoneurones are indicated with large dots. From Kjærulff, Barajon and Kiehn, 1994.

various activity-markers, like 2-deoxyglucose and c-*fos*, have been used to identify potential spinal locomotor neurones in the adult rabbit (Viala, Buisseret-Delmas and Portal, 1988) and cat (Dai, Douglas, Nagy, Noga and Jordan, 1990; Jordan, 1991). Recently sulforhodamine 101 has been introduced as an activity-marker. Sulforhodamine is a fluorescent probe which seems to be taken up in an activity-dependent manner in vertebrate central neurones (Keifer, Vyas and Houk, 1992), and has been used to label active neuronal circuits the turtle brain *in vitro* (Kriegstein, Avilla and Blanton, 1988; Keifer *et al.*, 1992). In an attempt to identify cells that belong to spinal locomotor networks in the neonatal rat, we have used sulforhodamine in combination with chemically induced locomotion in hindlimb-attached spinal cord preparations with intact or cut dorsal roots or completely isolated spinal cords (Kiehn, Kjærulff and Barajon, 1993; Kiehn, 1993; Kjærulff, Barajon and Kiehn, 1994). Spinal locomotor activity was maintained with NMDA in combination with 5-HT for 4-4.5 hours in the presence of low concentrations of sulforhodamine (0.0002-0.0005%). Previously sulforhodamine has been used in a concentration of 0.01 % (Lichtman, Wilkinson and Rich, 1985; Kriegstein *et al.*, 1988; Keifer *et al.*, 1992). We found, however, that this concentration disturbed the locomotor rhythm and we therefore reduced it 20-50 times. After experiments transverse sections of the lumbar spinal cord were cut on a cryostat and viewed in an epifluorescent microscope. In spinal cords with hindlimbs attached and intact dorsal roots labelled cells were distributed to laminae II-III in the dorsal horn, lamina VII and around the central canal (lamina X). Motoneurones were also labelled, although not consistently in all sections. It is likely that the labelling in the dorsal horn largely is due to afferent inflow, since after cutting the dorsal roots the number of labelled cells was reduced. A further reduction of labelling in the dorsal horn occurred after complete isolation of the cord (Fig. 3), mainly leaving labelled cells in a bilateral cluster close to the central canal and in lamina VII. In contrast to the distinctive distributions of labelled cells obtained in the different experimental situations few labelled cells were observed in non-locomoting control preparations which were exposed to sulforhodamine for a comparable time.

Based on the sulforhodamine labelling pattern we suggest that the cells located in the lamina VII and around the central canal are engaged in generating locomotor behaviour in the neonatal rat. In this context it is of interest that the above mentioned activity-markers label neurones in laminae VII and X in the rabbit (Viala, Buisseret-Delmas and Portal, 1988) and cat (Dai, Douglas, Nagy, Noga and Jordan, 1990). To further substantiate that these areas are of importance for rhythm generation in the neonatal rat spinal cord electrophysiological studies, perhaps in combination with ablation experiments (Ho and O'Donovan; 1993), are necessary. Our study might provide a framework for conducting such studies.

ACKNOWLEDGEMENTS. The research was supported by Carlsberg Fonden, Ib Henriksens Fond, Novos Fond, Fonden til Fremme af Eksperimentel Neurologisk Forskning, Wedell Wedellsborgs Fond, The Human Frontier Foundation and the Scandinavian-Japan Sasakawa Foundation. Ole Kiehn is a senior Research fellow supported by the Weimann Foundation.

REFERENCES

Andersson, O. and Grillner, S. (1983). Peripheral control of the cat's step cycle. II. Entrainment of the central pattern generators for locomotion by sinusoidal hip movements during 'fictive locomotion'. *Acta Physiologica Scandinavica* 118, 229-239.

Cazalets, J.R., Sqalli-Houssaini, Y. and Clarac, F. (1992), Activation of the central pattern generators for locomotion by serotonin and excitatory amino-acids in an *in vitro* new born rat spinal cord preparation. *Journal of Physiology* 455, 187-204.

Conway, B.A., Hultborn, H. and Kiehn, O. (1987). Proprioceptive input resets central locomotor rhythm in the spinal cat. *Experimental Brain Research* 68, 643-656.

Dai, X., Douglas, J.R., Nagy, J.I., Noga, B.R. and Jordan, L.M. (1990). Localization of spinal neurons activated during treadmill locomotion using the *c-fos* immunohistochemical method. *Society for Neuroscience Abstract* 16, 889.

Delcomyn, F. (1980). Neural basis of rhythmic behavior in animals. *Science* 210, 492-498.

Duysens, J. and Pearson, K. G. (1980). Inhibition of flexor burst generator by loading ankle extensor muscles in walking cats. *Brain Research* 187, 321-332.

Feldman, J.L., Smith, J. C. and Liu, G. (1991). Respiratory pattern in generation in mammals: *in vitro en bloc* analyses. *Current Opinion in Neurobiology* 1, 590-594.

Gossard, J.-P. and Hultborn, H. (1991). The organization of the spinal rhythm generator in locomotion. In *Plasticity of Motoneuronal Connections*, A. Wernig (ed), 385-404. Elsevier Science Publishers, Amsterdam.

Goudard, I., Orsal. D. and Cabelguen J.-M. (1992). An electromyographic study of the hindlimb locomotor movements in the acute thalamic cat. *European Journal of Neuroscience* 4, 1130-1139.

Grillner, S. (1981). Control of locomotion in bipeds, tetrapods and fish. In J.M. Brookhardt and V.B. Mountcastle (eds), *Handbook of Physiology*, section 2: The Nervous System, 1179-1236. American Physiological Society, Bethesda.

Grillner S. and Matsushima, T. (1991). The neural network underlying locomotion in lamprey. *Neuron* 7, 1-15.

Grunner, J. A. and Altman, J. (1980). Swimming in the rat: Analysis of locomotor performance in comparison to stepping. *Experimental Brain Research* 40, 374-382.

Harris-Warrick, R.M. (1988). Chemical modulation of central pattern generators. *In Neural Control of Rhythmic Movements in Vertebrates*, Cohen, A. H, Rossignol, S. and Grillner, S. (eds), 285-332. Willey, New York.

Harris-Warrick, R.M and Marder, E. (1991). Modulation of neural networks for behavior. *Annual Review of Neuroscience* 14, 39-57.

Ho, S. and O'Donovan,. M.J. (1993). Regionalization and intersegmantal coordination of rhythm-generating networks in the spinal cord of the chick embryo. *Journal of Neuroscience* 13, 1354-1371.

Jordan, L.M. (1991). Brainstem and spinal cord mechanisms for initiation of locomotion. In *Neurobiological Basis of Human Locomotion*, Shimamura, M., Grillner, S. and Edgerton, V.R. (eds), 3-20. Japan Scientific Societies Press, Tokyo.

Keifer J., Vyas, D. and Houk, J.C. (1992). Sulforhodamine labeling of neural circuits engaged in motor pattern generation in the *in vitro* turtle brainstem-cerebellum. *Journal of Neuroscience* 12, 3187-3199.

Kiehn, O. (1993). The *in vitro* neonatal rat spinal cord-hindlimb preparation for studying organization of locomotor networks and their afferent control. *IUPS abstract* XXXII, 31.11/O.

Kiehn, O., Iizuka, M. and Kudo, N. (1992). Resetting from low threshold afferents of N-methyl-D-aspartate-induced locomotor rhythm in the isolated spinal cord-hindlimb preparation from newborn rats. *Neuroscience Letters* 148, 43-46.

Kiehn, O., Kjærulff, O. and Barajon, I. (1993). Spinal locomotor network in the neonatal rat revealed by neuronal uptake of sulforhodamine. *Society for Neuroscience Abstract.* 19, 231.

Kjærulff, O. and Kiehn, O. (1993). Spatiotemporal characteristics of transmitter-induced locomotor activity in the neonatal rat. *IUPS abstract* XXXII, 280.2/P.

Kjærulff. O., Barajon, I. and Kiehn. O. (1994). Distribution of sulforhodamine labelled cells in the neonatal rat spinal cord following chemically induced locomotor activity in vitro. *Journal of Physiology* 478, 265-273.

Kriegstein, A.R., Avilla, J.G. and Blanton, M.G. (1988). Distribution of increased synaptic activity during focal and generalized epileptiform activity revealed by presynaptic uptake of fluorescent dyes. *Society for Neuroscience Abstract* 14, 471.

Kudo, N. and Yamada, T. (1985). Development of the monosynaptic stretch reflex in the rat: an *in vitro* study. *Journal of Physiology* **369**, 127-144.

Kudo, N. and Yamada, T. (1987). N-methyl-D,L-aspartate-induced locomotor activity in a spinal cord-hindlimb muscles preparation of the newborn rat studied *in vitro*, *Neuroscience Letters* **75**, 43-48.

Licthman, J.W, Wilkinson, R.S. and Rich, M.M. (1985). Multiple innervation of tonic endplates revealed by activity-dependent uptake of fluorescent probes. *Nature* **314**, 357-359.

Nicolopoulos-Stournas, S. and Iles, J. F. (1984). Hindlimb muscle activity during locomotion in the rat *(Rattus norvegicus)* (Rodentia: Muridae). *Jorurnal of Zoology* **203**, 427-440.

Pearson, K.G. (1993). Common principles of motor control in vertebrates and invertebrates. *Annual Review of Neuroscience* **16**, 265-97.

Roberts A., Soffe, S.R., and Dale, N. (1986) Spinal interneurones and swimming in frog embryos. In *Neurobiology of vertebrate locomotion* by Grillner, S., Stein, P.S.G., Stuart, D.G., Forssberg, H. and Herman R.M (eds), 279-306. Macmillan, London.

Smith, J. C. and Feldman. J. L. (1987). *In vitro* brainstem-spinal cord preparations for study of motor systems for mamalian respiration and locomotion. *Journal of Neuroscience Methods* **21**, 321-333.

Smith, J. C. Feldman, J. L. and Schmidt, B.J. (1988). Neuronal mechanisms generating locomotion studied in the mammalian brain stem-spinal cord *in vitro*. *FASEB Journal*.**2**, 2283-2288.

Viala, D., Buisseret-Delmas, C., and Portal, J.J. (1988). An attempt to localize the lumbar locomotor generator in the rabbit using 2-deoxy-[^{14}C] glucose autoradiography. *Neuroscience Letters* **86**, 139-143.

Zelená, I. and Soukup, T. (1976). The development of Golgi tendon organs. *Journal of Neurocytology* **6**, 171-194.

DUAL CONTROL OF CENTRAL PATTERN GENERATORS: NEONATAL RAT SPINAL CORD *IN VITRO*

J-R. CAZALETS

NBM, CNRS
31 chemin Joseph Aiguier BP71
13402 Marseille CEDEX 9

SUMMARY

Neuromodulatory control of the spinal network generating locomotor activity has been studied using an *in vitro* preparation of isolated brain stem/spinal cord. The activity produced in these conditions consists (locomotor-like activity) in alternating bursts of action potentials between the right and left side and between flexor and extensor which can be recorded in the ventral roots. Comparison with *in vivo* data shows that at birth the newborn rat can perform locomotor movements with a period and phase relationships comparable to that recorded *in vitro* (i.e. close to 1s). The locomotor-like activity can be induced by serotonin and excitatory amino-acids. In contrast GABA can suppress all rhythmic activity. It is suggested that the overall activity of the spinal network result from a fine interaction of these various transmitter systems.

INTRODUCTION

For producing the basic neuronal command involved in elementary behaviour, the central nervous system has developed specific neuronal structures called central pattern generators (CPGs). *In vitro* preparation have been successfully used both in invertebrates and lower vertebrates to demonstrate that these neuronal networks can generate an appropriate sequence of motor activity even in the absence of peripheral sensory feedback. The activity of these CPGs is however dependent on numerous extrinsic factors which will determine what the CPG will do at a given time. Of particular interest are the various neuromodulatory pathways which project onto the CPG and modify its activity. An *in vitro* preparation of isolated brain stem/ spinal cord of the new born rat has been used in order to investigate some neurotransmitter system properties involved in the control of mammalian spinal circuitry. It is only recently that this preparation has offered the opportunity to study the cellular mechanisms involved in locomotor processes without the limitations and constraints imposed by the use of acute animals. The first part of this paper will present some general characteristics of the isolated spinal cord and it will be shown that it can be considered as a suitable model for the analysis of locomotor behaviour. In a second part it will be shown how various neurotransmitters trigger and modulate the locomotor-like activity.

1. Properties and Characteristics of the Isolated Spinal Cord

An *in vitro* preparation of the isolated brain stem and spinal cord of the newborn rat (Fig. 1A) was initially developed by Konishi and Otsuka (1974). Since then it has been used for studies on respiratory processes (Suzue, 1984 ; Morin *et al.*, 1992), spinal reflexes, cellular properties of motoneurones (Fulton and Walton, 1986 ; Takahashi and Berger, 1990) and pharmacological studies (Connell and Wallis, 1989). Recently it appeared that when isolated the spinal cord generated motor patterns that were related to locomotor-like movements (Kudo and Yamada, 1987). The system can survives several hours (6 to 8h.) due to its small size and the absence of myelin, and one of its main advantage is the absence of a blood-brain barrier. Consequently, the effects of transmitters and of their agonists or antagonists can be directly studied by adding them in the superfusing saline. It should be noted, however, that this method involves some methodological pitfall. Even if it is likely that superfusion allows all drugs to gain ready access to the spinal tissue, experiments using tritiated serotonin revealed that it diffused through the whole spinal tissue but with a gradient decreasing from the periphery to the centre (Cazalets, unpublished results). This could explain that in most cases the range of concentration over which the various compounds are active is very narrow, and that dose-response curves are not log-linear.

To date, most of our data were obtained by recording the activity in the ventral roots (Fig. 1B), either during bath-application of neurochemicals (Fig. 1B1) or sensory stimulation of the coxygeal spinal cord (Fig. 1B2).

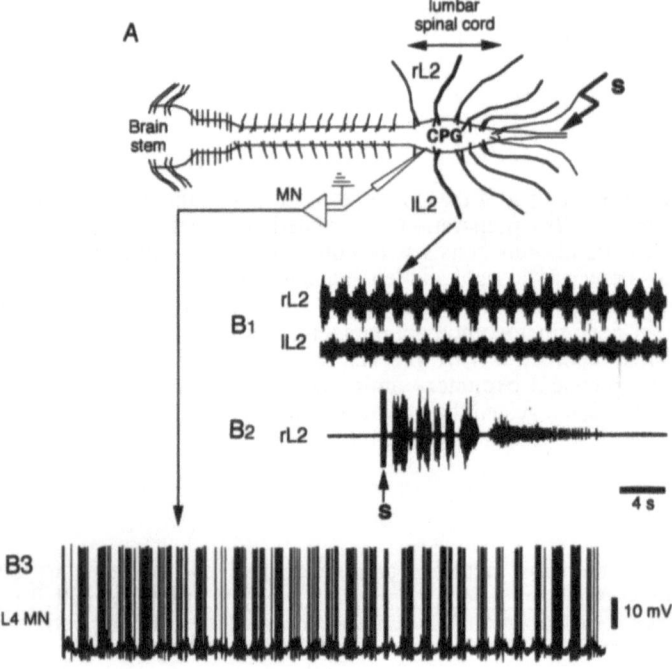

Figure 1. The isolated brain stem/spinal cord preparation. **A:** Drawing of the isolated preparation. Locomotor like activities are recorded from ventral roots at the lumbar level. **B:** Extracellular recording of the right and left alternating pattern evoked in response to chemical activation with an EAA (B1) or sensory stimulation of the coxygeal area (B2). Intracellular recording of a motoneurone located in the left fourth lumbar segment during EAA induced locomotion (B3). Abbreviations : MN, motoneurone ; IL and rL, left and right ventral roots; s, sensory stimulation.

Simultaneous intracellular recordings of motoneurones can also be obtained during chemically induced locomotor-like activity (Fig. 1B3). From a global analysis of extracellular activity one can study several features of the motor pattern expressed by the CPG in various conditions. The temporal characteristics of the locomotor-like activity i.e. the cycling period recorded in the ventral root issues from the CPG itself. According to this principle, a change in the period of the motor pattern will be interpreted as a change of the CPG timing itself. On the other hand a change in the amplitude can be due to an action on the motoneurone or on the synaptic drive to the motoneurone. Despite the fact that global extracellular recordings are made from the ventral roots, one can determine the flexion and extension phases during one cycle due to the intersegmental distribution of muscle units (Nicolopoulos-Stournaras and Iles, 1983). During a sequence of locomotor-like activity however, the global efferent discharge is such that the ipsilateral bursts of activity recorded in these two sets of ventral roots are phase-opposed (Cazalets *et al.*, 1992).

One main question when using *in vitro* preparations is whether or not the activity recorded can be reliably related to the behavioural activity observed in the intact animal. In rats, behavioural studies have shown that from birth to adulthood the complexity of the motor repertoire gradually increases (Altman and Sudarshan, 1975). Although at birth rat pups do not perform spontaneous locomotor movements, it has been shown that even at day 0, they perform efficient locomotor movements during swimming experiments (Bekoff and Trainer, 1979 ; Cazalets *et al.*, 1990 ; Menard *et al.*, 1991). The swimming pattern consists in an alternating gait (i.e. like stepping or trotting). The phase relationships between the right and left side were found to be the same in the intact animal as in the *in vitro* preparation, i.e. 0.5 which express strict alternation. In the intact swimming newborn rat the intraleg muscle relationships has not yet been studied. Electromyographic recordings performed in a semi-isolated preparations in which the legs were left attached, however, have shown that complex pattern of intra leg muscle relationships can be generated *in vitro* (Kjaerulff and Kiehn, 1993).

In intact animals, it has been shown that the period decreased from 1s at birth to 0.2s at adulthood (Menard *et al.*, 1991). From 0 to 3 days however, the period ranges from 1 to 0.5s which is similar to the minimal period recorded in vitro (Cazalets *et al.*, 1992 ; Sqalli *et al.*, 1993a). The dose response curves of the various excitatory amino-acids established in vitro (Sqalli *et al.*, 1993a) show that the range of period in which an organized rhythmic activity can be observed is quite wide (from 4 to 1s), while at the same developmental stage during swimming experiment the period was always around 1s. It can be thus concluded that in theses conditions *in vivo* the system is likely "operating at its maximum". Ii is interesting that the locomotor repertoire does not include gaits such as cantering and galloping. To date, the use of the *in vitro* preparation has not allowed us to elucidate whether this was due to some immaturity in spinal circuitry since some conflicting data have been collected on this subject. Smith *et al.*, (1988) reported that upon perfusing the brainstem with bicuculline they triggered right and left non-alternating activities with a 10 s period that they called galloping. Atsuta *et al.*, (1991) however, doing the same kind of experiment did not reproduce the same effect and rejected the previous conclusions. In our experiments it has never been possible to date to consistently evoked activity that could be related to gallop.

2. Mechanisms Affecting the Period of the Locomotor CPG

In this section it will be reviewed various mechanisms which contribute to the setting of the motor period and the generation of the motor pattern. Analysing the changes in the period of activity of the CPG for locomotion means that one will consider what makes the system works at a higher speed or in other words at a faster gait. It will be shown here how the neuromodulatory pathways which project onto the spinal CPG for locomotion interact with each others in order to elaborate the final motor pattern.

2. 1. Activating Systems

First of all, the locomotor period can be set *in vitro* by bath-applying an activatory transmitter at various concentrations. If one draw a parallel with the processes occurring in the intact animal, the various bath-applied concentrations would correspond to a more or less transmitter released or to a more or less intense discharge of one given activating pathway. We have investigated the action of an amine the serotonin (5-HT) and of various excitatory amino-acids (EAAs) which are able to trigger locomotor-like activity .

The 5-HT induced rhythm

It has long been known that the administration of L-DOPA in spinalized cats or rabbits can induce locomotor movements. To date these effects of L-DOPA have been attributed to an activation of noradrenergic receptors. Regarding the effects of serotonin, however, there has been some controversy as to whether 5-HT actually initiates the activity of the locomotor networks in mammals. In the curarised and decerebrated rabbit Viala and Buser

(1971) showed that the 5-HT precursor 5-HTP induced fictive locomotion, whereas in the spinal cat it has been reported that 5-HTP only enhanced muscle tone but failed to trigger locomotion (Grillner and Shik, 1975 ; Barbeau and Rossignol, 1990, 1991). Our results, collected on the isolated spinal cord, however, tend to favour the idea that the serotonergic system may trigger locomotion, acting on the premotoneuronal spinal network that organizes the alternating pattern. The superfusion of the isolated lumbar spinal cord with serotonin at concentrations ranging from 10^{-5}M to 10^{-4}M induced bursts of spikes alternating between the left and right sides of each lumbar segment. At threshold (10^{-5}M) serotonin occasionally elicited a very slow (with a 30s period) pattern. The period decreased from 2.5×10^{-5}M to 10^{-4}M, reaching at this concentration a plateauing period value (around 5s). A strict phase opposition between the two sides was maintained at all concentrations.

Our data on the pharmacological characterization of the receptors involved in this 5-HT action are in agreement with previous studies showing that the spinal receptors exhibit some of the 5-HT_1 and 5-HT_2 receptor features but with a somewhat different profile (Connell and Wallis, 1989). In order to identify the receptors affected by 5-HT, both antagonists and agonists were used. In the experiments in which antagonists were used, it was assumed that if 5-HT exerts a direct action on the locomotor CPGs, it was to be expected that the 5-HT antagonists would decrease the 5-HT induced rhythmicity. Antagonists of both 5-HT_1 ((-)-propanolol) and 5-HT_2 (ketanserin), progressively blocking the 5-HT induced activity at increasing concentrations (Cazalets et al., 1992). The 5-HT_3 antagonists tested, however, failed to block the 5-HT induced activity. Among the agonists tested, only 5-methoxytryptamine (a non selective 5-HT agonist) elicited locomotor-like activity, whereas selective 5-HT_1, 5-HT_2 and 5-HT_3 agonists were ineffective. The slowing down of the period of cycling provoked by 5-HT antagonists therefore strongly suggests that a blockade occurred at the CPG level itself.

The Excitatory Amino-Acids Induced Rhythm

Following the pioneering work performed in lower vertebrates (Brodin and Grillner, 1985; Dale and Roberts, 1984), it was recently found that excitatory amino-acids (EAAs) also play a crucial role in the activation of the locomotor CPGs in mammals. Kudo and Yamada (1987) and Smith et al., (1988) provided the initial evidence for the involvement of NMDA receptors in the genesis of locomotion in newborn rats. Recently, it was confirmed that NMDA receptors initiate locomotor-like activity but also that non-NMDA receptors, activated by kainate but not by AMPA were capable of triggering locomotor-like activity (Cazalets et al., 1992). In another study, using various endogenous EAAs, it was shown that in addition to glutamate and aspartate, the most "popular candidate" as endogenous ligands for EAA receptors, there exists other putative EAAs, the sulphur-containing amino-acids, which are able to generate locomotor-like activity (Sqalli et al., 1993a). Like glutamate and aspartate they activate the locomotor CPG via the same types of receptors since their action was blocked in a dose-dependent manner by phosphono-valeric acid (AP-5, a selective NMDA receptor blocker) and dinitroquinoxaline-2-3-dione (DNQX a non-NMDA receptor blocker). For all the EAAs tested it was shown that their effects were dose-dependent and that the antagonists progressively decreased the rhythm before complete cessation occurred. All these findings thus demonstrated that, like 5-HT, the EAAs act on the rhythmic component of the CPG. In this case, one cannot exclude an action also on the motoneurones directly, or on the presynaptic drive to the motoneurone as is suggested by the reduction in the burst amplitude when superfusing the antagonists.

2.2. Convergence of Activating Systems

In most experiments only one compound is used to trigger rhythmic activity. In fact it appears that the period can also be set by combining several activatory transmitters (5-HT and an EAA). In this condition, the final period will depend on the balance of each transmitter. These interactions have been studied for 5-HT and EAAs.

Fig. 2 presents evidence for a cumulative effects of these compounds when mixing them. In this experiment both 5-HT and NMA were bath-applied at subthreshold concentrations such that each individually failed to trigger a stable pattern. Their simultaneous bath-application however, induced a very regular alternating activity. Another type of synergic effects was provided by the fact that in about 25% of the experiments in which 5-HT and EAAs were bath-applied alone, they only initiated an irregular pattern that was rapidly followed by tonic activity (Cazalets *et al.*, 1992). In fact it appeared latter that in these apparently unsuccessful experiments, simultaneous bath application of 5-HT and NMA evoked a long lasting and stable activity (Sqalli *et al.*, 1993b).

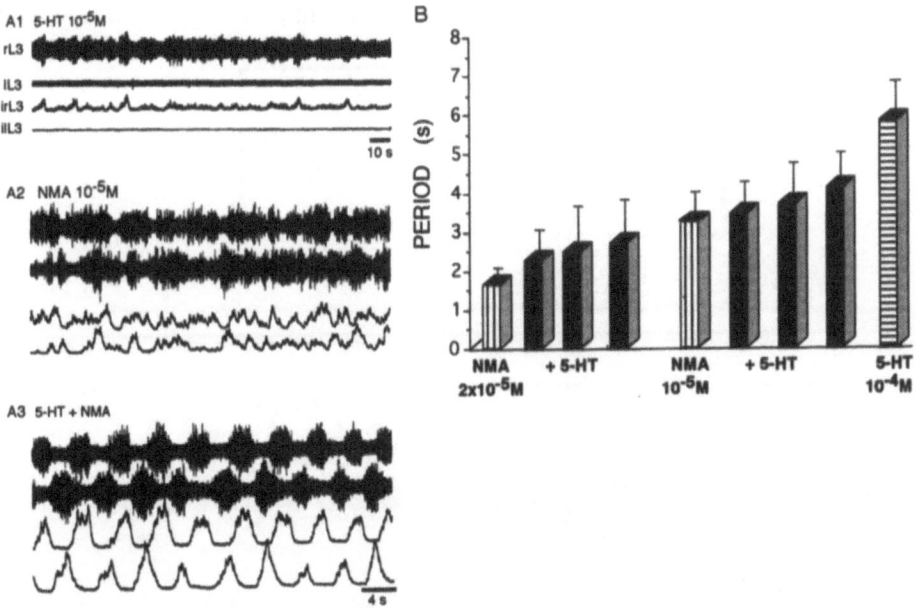

Figure 2. Interaction between activating systems. **A :** Raw (upper traces) and integrated (lower traces) recordings of lumbar ventral roots. At subthreshold concentrations neither 5-HT (A1) nor NMA (A2) triggered regular, rhythmic, alternating activity. In contrast, the combined bath-application of the two substances (A3) induced a very regular motor pattern. **B:** Interactions between 5-HT and EAAs modify the motor period. The histogram presents the period of the motor pattern recorded during systematic changes in the 5-HT and NMA concentration. Vertical striped bars are the period value for two different NMA concentrations bath -applied alone. For each of these NMA concentrations are plotted (black bars) the period obtained when adding 5-HT at increasing concentrations (10^{-5}M, 5×10^{-5}M, 10^{-4}M) to the NMA containing saline. The horizontal striped bar presents the period for 5-HT bath-applied alone. Values are mean ± SD.

It was then studied how 5-HT interacts with NMA to set the period value and the stability of the spinal locomotor CPGs. As these two substances were each individually activating the CPG it was expected that the period of the rhythm induced by combining them would be still shorter than when each compound was applied separately. In fact, it appeared that this was not the case and that the period was set at a value that was intermediate between that evoked by NMA alone and 5-HT alone (Fig. 2B; see also Sqalli *et al.*, 1993b).

In conclusion, although it is clear that from a qualitative point of view the two compounds have synergic effects, it appears that they have also a conflicting effect on the period and that it is possible to modulate the speed of the working CPG by varying their respective concentration.

2.3. Inactivating Systems

The third way to set the period is the presence of "inactivatory" pathways which strongly control the CPG for locomotion, and which when they are active, can stop all locomotor activity or slow down the rhythm. Our data supports the idea that a dual control is exerted on motor programmes, the activity of which may result from the equilibrium between antagonistic (i.e. inactivating and activating) systems.

It has been recently found that GABA modulate the oscillatory behaviour of the CPG for locomotion in mammals. There are several evidence for a physiological role played by GABA. First, it was demonstrated that GABAergic inputs to the locomotor CPGs exist and that endogenously released GABA modifies the period of locomotor-like activity. This is shown in Figure 3. In these experiments, the descending supraspinal pathways (brain stem and thoracic spinal cord) were chemically stimulated by bath-applying an excitatory amino acid (Fig. 3A1). This bath-application was unable by itself to trigger rhythmic activity (Fig. 3A2). However the simultaneous bath-application of the GAB_{AA} antagonist bicuculline, or the GAB_{AB} antagonist phaclofen, to the lumbar spinal cord only, triggered a very stable

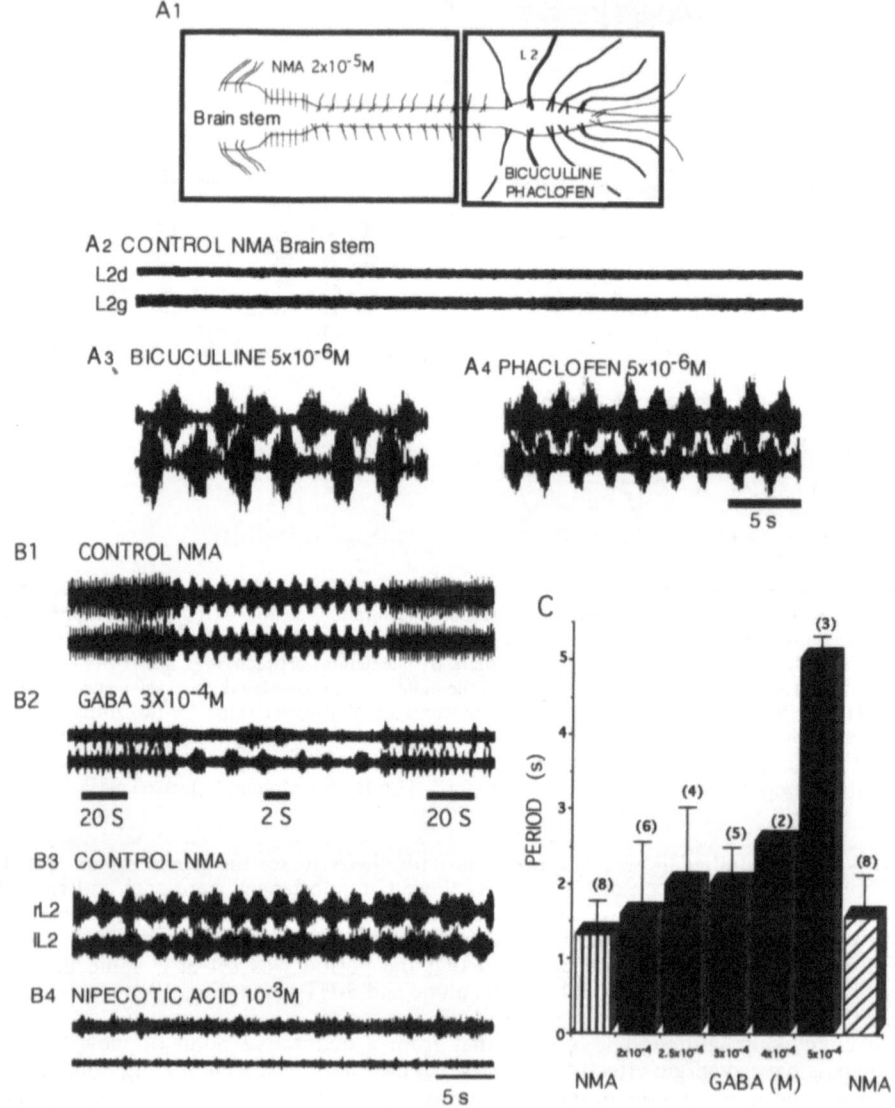

Figure 3. GABAergic control of the CPG for locomotion. **A:** Unmasking of rhythmic activity by GABA antagonists. Diagram (A1) of the experimental preparation. NMA (an NMDA receptor agonist) was only bath-applied at the brain stem and thoracic level, whereas bicuculline or phaclofen were bath applied only at the lumbar level. Under control conditions (A2), the NMA bath application did not induce any activity in the ventral roots. In contrast the simultaneous bath-application of bicuculline (GAB_{AA} antagonist ; A3) or phaclofen (GAB_{AB} antagonist ; A4) triggered regular alternating activity. **B:** Effects of GABA on locomotor pattern. During NMA induced locomotion (B1), bath application of GABA slowed down the rhythm (B2). Increasing the level of endogenous released GABA using the GABA uptake inhibitor nipecotic acid stopped the rhythm (B4). C: Dose dependent effects of GABA on the motor pattern. The vertical-striped bar shows the period for NMA alone at the beginning of the experiment, while the oblique-striped bar shows the period for NMA following GABA application. Values are means ± SD.

192

and long-lasting rhythmic activity (up to 20 minutes) in a dose dependent manner (Fig. 3A3 and 3A4, see also Cazalets *et al.*, 1994). These experiments thus demonstrated that the expression of the locomotor CPG was blocked by the action of endogenously released GABA. The direct bath-application of GABA itself during NMA induced locomotor-like activity decreased or stopped the rhythm (Fig. 3B1-B2). The effects of GABA were dose-dependent (Fig. 3C). A slowing down in the cycling period or a complete cessation of rhythmic motor activity also occurred in response to the bath-application of $GABA_A$ (muscimol) or $GABA_{AB}$ (baclofen) agonists.

Further evidence was provided by the use of an inhibitor of the uptake system for GABA, nipecotic acid. Bath-application of this substance during NMA induced locomotor-like activity, decreased or suppressed the CPG activity (Fig. 3B3-B4).

These data thus supports the idea that a dual control is exerted on motor programmes, the activity of which may result from the equilibrium between antagonistic (i.e. activating and inactivating) systems.

CONCLUSION.

In conclusion several points can be highlighted. The extracellular analysis of the motor pattern generated by the isolated spinal allows to study the effects of transmitters on the locomotor spinal networks. Our study demonstrates that in order to regulate the speed of locomotion, the central nervous system can use various strategies. The existence of a dual control exerted by activating and inactivating systems acting in parallel is probably of major importance for motor behaviour. An inactivating system might play multiple role in the overall control of locomotion. First, by acting in an all-or-none manner, the GABAergic system can be used as a "brake" to cause a complete cessation of all locomotor activity. In this case a sudden arrest while walking would not only be attributable to the cessation of excitatory neuronal firing, but also to the involvement of the GABAergic inputs. In this framework, the supra-spinal structures might be equipped with an accelerator (the aminergic descending pathways and the excitatory amino acidergic inputs) as well as a brake (the GABAergic system) serving to start and stop the system, respectively. Secondly, the GABAergic system may operate in a dynamic equilibrium with activatory influences, so as to finely adjust the level of activity of spinal neurones, thus setting the motor outflow.

Another point that may be emphasised is the validity of the isolated spinal cord of the neonatal rats as a model for the study of mammalian locomotor processes. Despite its immaturity it is undoubtedly the case that some general properties of vertebrate oscillatory systems can be studied with this preparation. Its main advantage is that one can use similar methodological approaches as in isolated nervous systems preparation of invertebrates or lower vertebrates (Harris-Warrick, 1988). This preparation is still largely unexplored. Once intracellular recordings can routinely be performed, it may be expected that the isolated spinal cord will permit the investigation of general networks properties in mammals.

REFERENCES

Altman, J. and Sudarshan, K. (1975). Postnatal development of locomotion in the laboratory rat. *Animal Behavior* **23**, 896-920.

Atsuta, Y., Abraham, P., Iwahara, T., Garcia-Rill, E. and Skinner, R.D. (1991). Control of locomotion in vitro: II. Chemical stimulation. *Somatosensory and Motor Research* **8**, 55-63.

Barbeau, H. and Rossignol, S. (1990). The effects of serotonergic drugs on the locomotor pattern and on cutaneous reflexes of the adult chronic spinal cat. *Brain Research* **514**, 55-67.

Barbeau, H. and Rossignol, S. (1991). Initiation and modulation of the locomotor pattern in the adult chronic spinal cat by noradrenergic, serotonergic and dopaminergic drugs. *Brain Research* **546**, 250-260.

Bekoff, A. and Trainer, W. (1979). The development of interlimb coordination during swimming in postnatal rats. *Journal of Experimental Biology* **83**, 1-11.

Brodin, L., Grillner, S. and Rovainen, C.M. (1985). NMDA, kainate and quisqualate receptors and the generation of fictive locomotion in the lamprey spinal cord. *Brain Research* **325**, 302-306.

Cazalets, J.R., Menard, I., Cremieux, J. and Clarac, F. (1990). Variability as a characteristic of immature motor systems: an electromyographic study of swimming in the newborn rat. *Behavioural Brain Research* **40**, 215-225.

Cazalets, J.R., Sqalli-Houssaini, Y. and Clarac, F. (1992). Activation of the central pattern generators for locomotion by serotonin and excitatory amino acids in neonatal rat. *Journal of Physiology* **455**, 187-204.

Cazalets, J.R., Sqalli-Houssaini, Y. and Clarac, F. (1994). GABAergic inactivation of the central pattern generators for locomotion in isolated neonatal rat spinal cord. *Journal of Physiology* **474**, 173-181.

Connell, L.A. and Wallis, D.I. (1989). 5-Hydroxytryptamine depolarizes neonatal rat motoneurones through a receptor unrelated to an identified site. *Neuropharmacology* **28**, 625-634.

Dale, N. and Roberts, A. (1984). Excitatory amino-acids receptors in Xenopus embryo spinal cord and their role in the activation of swimming. *Journal of Physiology* **348**, 527-543.

Fulton, B.P. and Walton, K. (1986). Electrophysiological properties of neonatal rat motoneurones studied in vitro. *Journal of Physiology* **370**, 651-678.

Grillner, S. and Shik, M.L. (1975). On the descending control of the lumbosacral spinal cord from the "mesencephalic locomotor region". *Acta Physiologica Scandinavica* **87**, 320-333.

Harris-Warrick, R.M. (1988). Chemical modulation of central pattern generators. In : *Neural control of rhythmic movements in vertebrates*, eds. Cohen, A.H., Rossignol, S. and Grillner, S., pp. 285-331. John Wiley and Son, New York.

Kjaerulff, O and Kiehn, O. Spatiotemporal characteristics of transmitter-induced locomotor activity in the neonatal rat. 280.2P XXXII congress IUPS, Glasgow 1993.

Konishi, S. and Otsuka, M (1974). Electrophysiology of mammals spinal cord *in vitro*. *Nature* **252**, 733-735.

Kudo, N. and Yamada, T. (1987). N-Methyl-D,L-aspartate-induced locomotor activity in a spinal cord-hindlimb muscles preparation of the newborn rat studied in vitro. *Neuroscience Letters* **75**, 43-48.

Menard, I., Cremieux, , J. and Cazalets, J.R. (1991). Evolution non linéaire de la fréquence des mouvements de nage pendant l'ontogenèse chez le rat. *C.R. Académie des Sciences Paris* **312**, 233-240.

Morin, D., Monteau, R. and Hilaire, G. (1992). Compared effects of serotonin on cervical and hypoglossal inspiratory activities : an *in vitro* study. *Journal of Physiology* **451**, 605-629.

Nicolopoulos-Stournaras, S. and Iles J.F. (1983). Motor neuron columns in the lumbar spinal cord of the rat. *Journal of Comparative Neurology* **217**, 75-85.

Smith, J.C., Feldman, J.L. and Schmidt, B.J. (1988). Neural mechanisms generating locomotion studied in mammalian brain stem-spinal cord in vitro. *FASEB Journal* **2**, 2283-2288.

Sqalli-Houssaini,, Y., Cazalets J.R., Martini, F. and Clarac, F. (1993a). Induction of fictive locomotion by sulphur-containing amino acids in an in vitro newborn rat preparation. *European Journal of Neuroscience* **5**, 1266-1232.

Sqalli-Houssaini, Y., Cazalets, J.R. and Clarac, F. (1993b). Oscillatory properties of the central pattern generator for locomotion in neonatal rats. *Journal of Neurophysiology* **70**, 803-813.

Suzue, T. (1984). Respiratory rhythm generation in the in vitro brainstem-spinal cord of the neonate rat. *Journal of Physiology* **354**, 173-183.

Takahashi, T. and Berger, A.J. (1990). Direct excitation of rat spinal motoneurones by serotonin. *Journal of Physiology* **423**, 63-76,.

Viala, D. and Buser, P. (1971). Modalités d'obtention de rythmes locomoteurs chez le lapin par traitement pharmacologique (DOPA, 5-HTP, D-amphétamine). *Brain Research* **35**, 151-165.

RHYTHMIC ACTIVITY PATTERNS OF MOTONEURONES AND INTERNEURONES IN THE EMBRYONIC CHICK SPINAL CORD.

M. O'DONOVAN AND A. RITTER

Section on Developmental Neurobiology
Laboratory of Neural Control
NINDS, NIH
Bethesda, MD 20817

SUMMARY

We have studied the organization of patterned motor activity in the developing spinal cord of the chick embryo using optical and electrophysiological methods. Optical imaging of motoneurones filled with calcium dyes revealed the presence of fluorescence transients that were synchronized with the electrical activity recorded from hindlimb muscle nerves. Optical imaging was also used to demonstrate the rhythmic activity of a population of interneurones that is believed to provide some of the excitatory drive to motoneurones. Calcium transients were synchronized between motoneurones and these interneurones. Whole cell recordings from interneurones around the lateral motor column confirmed the optical findings and showed that many interneurones receive rhythmic synaptic drive in phase with that of motoneurones. Dye injections into individual interneurones failed to demonstrate the existence of widespread dye-coupling implying that synaptic mechanisms are responsible for the widespread synchrony in the activity of embryonic spinal neurones.

INTRODUCTION

An understanding of the networks controlling rhythmic motor activity requires characterization and identification of the constituent neurones as well as detailed information about their activity, properties and connectivity. Most progress in accomplishing these goals has been made in the lamprey (for review see Grillner and Matsushima 1991) and in the *Xenopus* embryo (Roberts *et al.*, 1986), both of which have comparatively simple nervous systems. The increased complexity of the nervous system in higher vertebrates and the attendant difficulties of identifying the relevant interneurones has resulted in slower progress in understanding the locomotor networks of birds and mammals. To overcome some of these problems we have been studying the genesis of rhythmic motor activity in the isolated spinal cord of the chick embryo. This preparation offers a number of advantages for the analysis of rhythmically active networks. These include relative anatomical simplicity (compared to the adult), spontaneously active spinal networks in highly reduced preparations in vitro (Ho and O'Donovan 1993) and the ability to employ whole cell recording methods for single cell analysis and optical recordings for population activity (O'Donovan *et al.*, 1992; O'Donovan et. al.,1993). In this chapter we review our work on identifying the activity patterns of both motoneurones and interneurones and discuss the implications of these patterns for the organization and function of networks in the developing spinal cord.

Our experiments are performed on using an isolated preparation of the spinal cord that has been described in detail previously (O'Donovan and Landmesser 1987, O'Donovan 1989). In our most recent experiments the dissection is made in a Cl⁻ free medium in which sucrose replaces NaCl (Aghajanian and Rasmussen 1989). This procedure appears to improve the viability of the preparations. Standard methods are used to stimulate and record electrical activity from muscle nerves or interneuronal axons in the ventrolateral white matter. Electrical recordings are often made with a wide bandwidth (DC-2kHz) to resolve slow electrotonic potentials which provide a good monitor of intracellular membrane changes (O'Donovan 1989). In optical experiments the electrical signals are also recorded on one of the audio tracks of a video tape recorder to allow synchronization with the video data. Whole cell current and voltage clamp recordings are obtained from neurones using the 'blind patch' technique (Blanton, La Turco and Kriegstein 1989) whose application to the embryonic chick cord has been discussed elsewhere (Sernagor and O'Donovan 1991).

We have also developed optical methods to allow us to image the activity of populations of motoneurones and interneurones using real-time calcium imaging (O'Donovan et al., 1992, O'Donovan et al., 1993, O'Donovan et al., 1994). For this purpose spinal neurones are loaded with calcium sensitive dyes. This is accomplished either by using a membrane permeant form of the dye (fura2-am, Grynkiewicz et al., 1985) that is used to label neurones in the cut, transverse face of the spinal cord (O'Donovan et al., 1992 ,1994) or alternatively by retrograde transport of dextran-conjugated dyes (O'Donovan et al., 1993). Activity-dependent changes in fluorescence are detected using an intensified videomicroscopy and stored on video tape.

OPTICAL IMAGING OF SPINAL NEURONES DURING RHYTHMIC MOTOR ACTIVITY

A. Recordings from Motoneurones

Dye loaded motoneurones can be identified by their location within the lateral motor column and following antidromic stimulation of motor nerves. Fig. 1 illustrates a preparation in which motoneurones were retrogradely filled with calcium green dextran applied to the spinal nerves of the crural plexus. The filled motoneurones (outlined) extended in a rostrocaudal column and were clearly visible through the ventral white matter. The ventral roots were also well filled and sometimes cells on the ventral surface of the cord were also partially labelled, presumably due to dye leakage from the site of the fill. Antidromic stimulation of the ventral roots resulted in a large frequency-dependent change in the fluorescence of the filled column of cells as described in earlier reports in which motoneurones were visualized in the cut, transverse face of the cord (O'Donovan et al., 1993). The active cells were visualized by subtracting an image acquired during the stimulation from a control image obtained just before the stimulus. This is illustrated in fig.1B which shows that the stimulus-evoked fluorescence change is restricted to the motor column and the ventral roots. We consistently observed changes in the fluorescence of the ventral roots following antidromic stimulation which suggests that calcium entry accompanies action potentials in axons.

We determined that the antidromically induced fluorescence changes are nearly abolished in the absence of extracellular calcium (O'Donovan et al., 1993) and can be depressed by calcium channel blockers (O'Donovan and Ho, 1992). These findings suggest that the somatic fluorescence changes accompanying action potentials are due to increased intracellular free calcium entering through voltage-gated channels.

We also measured the fluorescence changes in motoneurones during episodes of spontaneous motor activity. Studies with membrane permeant fura2-am showed that motoneurones exhibit rhythmic fluorescence transients that occur in phase with discharge and synaptic drive potentials recorded from the ventral roots or muscle nerves. The rhythmic fluorescence changes were synchronized in different motoneurones and occurred in the perinuclear region as well as in the cytoplasm (O'Donovan et al., 1992, O'Donovan et al., 1994). The observation that calcium transients in motoneurones are synchronized is perhaps surprising because flexor and extensor motoneurones are known to fire out of phase (Landmesser and O'Donovan 1984, O'Donovan and Landmesser 1987, O'Donovan 1989). This discrepancy probably arises for several reasons. These include the existence of calcium entry during the rhythmic drive potentials that are synchronized in flexor and extensor

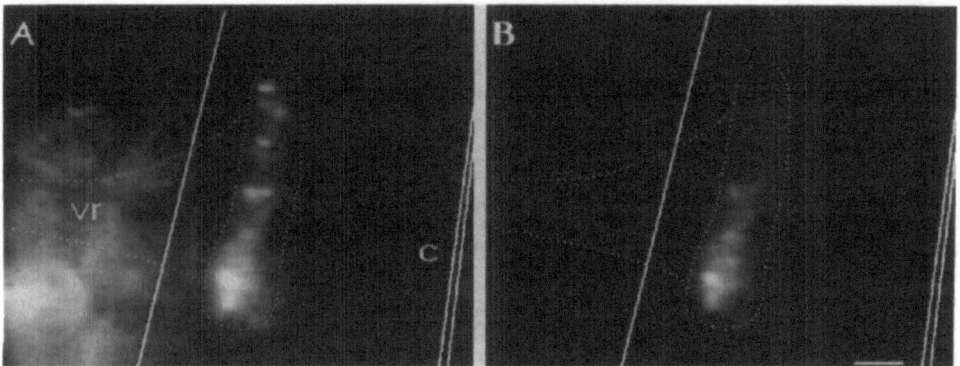

Figure 1. a) Video micrograph of the ventral surface of an E11 lumbosacral spinal cord in which motoneurones had been retrogradely loaded with calcium green dextran. The labelled column of motoneurones has been outlined and the ventral roots (vr), which exit to the left of the column, are identified. The central canal is also labelled (c). The image in (B) was obtained during antidromic stimulation of the ventral roots at 20Hz. This difference image was generated by subtracting an image averaged (30 frames) during antidromic stimulation a control image averaged from 30 frames taken 1 second before the onset of stimulation from the calibration bar in B is 100μM

motoneurones and the slow time course of the calcium signal. Such observations emphasize that caution must be exercised when inferring firing patterns from the calcium signals alone.

Synchronized rhythmic activity was also observed when the motoneurones were retrogradely loaded with calcium sensitive dyes. An example of this is illustrated in Fig. 2A. which compares the optical responses from a cluster of labelled lumbosacral motoneurones in LS2-3 (from the preparation illustrated in fig.1), with the electrical activity recorded from the femorotibialis muscle nerve. The optical signals were well correlated with neural discharge. This type of observation taken together with the results of antidromic stimulation provide good evidence that action potentials are a major determinant of the somatic calcium changes in rhythmically active neurones.

The amplitude of the rhythmic fluorescent changes illustrated in Fig. 2A was between 50-60% and was similar to the change following antidromic stimulation at 20Hz. In earlier studies, using fura-2am, cells were imaged on the cut transverse face of the spinal cord. Under these conditions, the fluorescent change during rhythmic activity was closer to that generated by 5-10Hz antidromic stimulation (O'Donovan et al., 1994). One possible reason for this difference is that neurones in the cut face may be partially de-afferented, resulting in weaker excitation of the surface cells.

B. Recordings from Interneurones

We also obtained recordings from rhythmically active interneurones lying outside the lateral motor column. Studies with bath-applied fura2-am (O'Donovan et al., 1992, 1994) revealed a widespread distribution of the active cells with a particularly high concentration in the region dorsomedial to the lateral motor column. Some of the cells behaved like motoneurones exhibiting rhythmic oscillations in phase with ventral root or muscle nerve activity whereas others displayed tonic signals that persisted for the duration of the episode. This variability could reflect differences in the firing behavior of the interneurones, although differences in the processing of intracellular calcium cannot be excluded.

One difficulty with experiments employing bath-applied dyes is the interneurones are not identified. To address this problem we retrogradely labelled interneurones by applying calcium dyes to their axons traveling in the ventrolateral white matter. We had already established, in earlier experiments, using biocytin as a retrograde tracer (Ho and O'Donovan 1991) that the cell bodies of VLT axons are located in two columns, one on each side of the cord. On the ipsilateral side, cells are distributed in the lateral intermediate region, dorsal and medial to the lateral motor column. The contralateral cell group is located in the medial part of the cord, near the central canal.

We found that both the ipsi- and contralateral cells groups were rhythmically active in phase with each other and with motoneurones (Ho and O'Donovan 1992). As illustrated in

Fig. 2B cells on the ipsilateral side of the cord exhibit rhythmic oscillations of fluorescence in phase with motor nerve discharge. A detailed analysis of the activity of single cells was precluded because it was often difficult to isolate individual cells from the background fluorescence arising from other labelled cells in the group. Nevertheless, we did determine that many neurones in both cell groups do not exhibit optical signals during rhythmic activity. This situation probably arises for two reasons. First we may not be able to detect somatic fluorescence changes during subthreshold synaptic activity, either because the somatic calcium changes are too small or because the sensitivity of the present system is too low to detect them. The existence of subthreshold, rhythmic synaptic inputs to interneurones in the region of the VLT cells has been confirmed by whole cell recording. The second reason is that the labelled neurones projecting into the ventrolateral tract are a heterogeneous group including propriospinal and ascending cells. It seems reasonable to assume that not all of these will be rhythmically active, although whole cell recording from interneurones in the region has revealed that most cells do receive some level of rhythmic synaptic input.

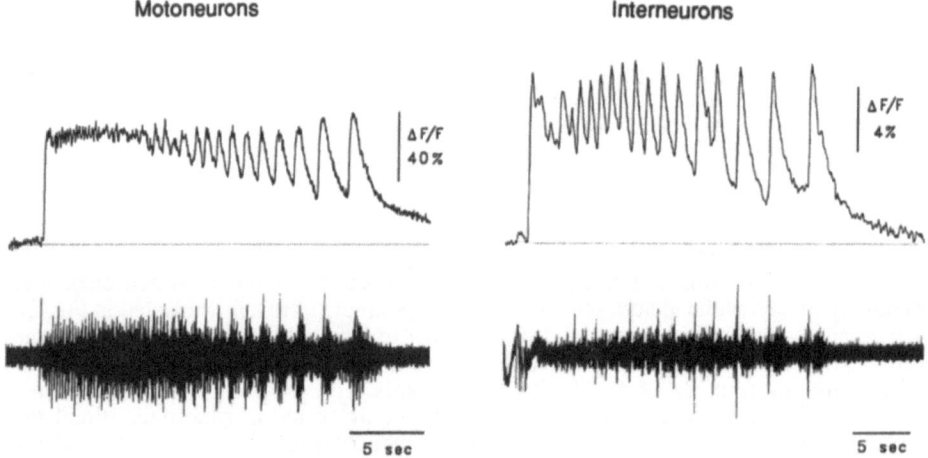

Figure 2. Video-rate measurements of the fluorescence change in lumbosacral motoneurones and lateral tract interneurones during episodes of rhythmic motor activity. In both cases the labeled cells were imaged through the ventral surface of the lumbosacral cord.

WHOLE CELL RECORDING FROM SPINAL INTERNEURONES DURING RHYTHMIC ACTIVITY

The fact that the propriospinal neurones are rhythmically active and supply some of the excitation to motoneurones (Ho and O'Donovan 1993) makes them candidates for the origin of the rhythmicity and the synchronization of network activity. Therefore, to investigate the properties and activity of these cells in more detail we obtained whole cell patch recordings from interneurones located in the region of the two VLT cell groups (Ritter and O'Donovan 1993). We found that the majority of interneurones were active during episodes of rhythmic activity. The active neurones received rhythmic depolarizing synaptic drive which was synchronized with muscle nerve activity and with rhythmic, electrotonic potentials recorded from the ventrolateral tract (Fig. 3). In some neurones current injection revealed the presence of depolarizing IPSPs in addition to the excitatory synaptic drive. Some of the interneurones fired like flexor motoneurones with a pause in their firing during each cycle, whereas others fired throughout the cycle like extensor motoneurones.

Some of the cells were injected with biocytin (Horikawa and Armstrong 1988) or neurobiotin (Kita and Armstrong 1991) to reveal their morphology and location and to establish the extent of dye-coupling. We found evidence for dye coupling in only one cell suggesting that electrical coupling of cells is probably not widespread. As a result electrical coupling is probably not responsible for the synchrony among interneurones and motoneurones.

We also examined the synaptic projections of the VLT axons onto interneurones. We found that most rhythmically active cells receive synaptic input from the VLT. VLT inputs onto interneurones appeared to be composed of mono- and polysynaptic components on the basis of their frequency dependence. Axons in the VLT also project strongly onto motoneurones (Ho and O'Donovan 1992). The fact that projections from tract cells are so widespread suggests that they may be involved in the synchronization of network activity.

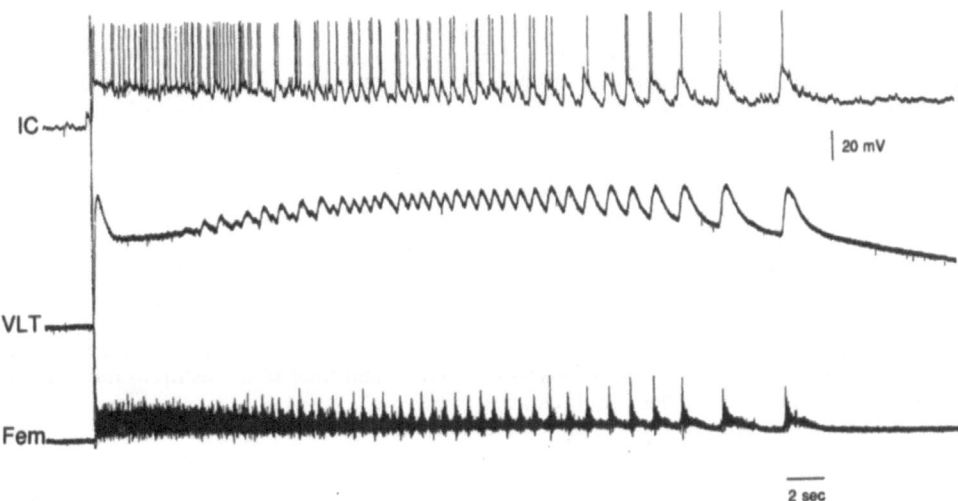

Figure 3. Whole cell recording from an interneurone (IC) in the intermediate region of the lumbosacral cord where many of the ventrolateral tract interneurones are found. Also shown are DC recordings of the simultaneous activity in the axons of the ventrolateral tract (VLT) which have been dissected from the side of the cord and from the femorotibialis muscle nerve (Fem). The voltage calibration applies to the intracellular record

CONCLUDING REMARKS

The imaging and single cells studies have revealed that each cycle of rhythmic synaptic input is remarkably synchronous in both interneurones and motoneurones. Despite their synchronized synaptic input motoneurones are known to discharge out of phase with each other (Landmesser and O'Donovan 1984, O'Donovan and Landmesser 1987). This situation arises because some motoneurones (extensors) receive a predominantly excitatory synaptic drive whereas in others (flexors) inhibitory synaptic inputs are synchronized with the excitatory inputs. The inhibitory inputs interrupt flexor firing at the time of peak extensor activity (O'Donovan 1989, Sernagor and O'Donovan 1991). In interneurones it seems likely that a similar mechanism operates because we have found cells with firing patterns like both classes of motoneurone and some cells that receive inhibitory and excitatory synaptic inputs.

An important question that we cannot answer fully yet concerns the mechanism and functional significance of the synchronized activity. Our failure to detect significant dye-coupling among interneurones suggests, but does not prove, that electrical coupling is probably not the mechanism underlying the synchrony. We favor instead the idea that the synchrony is mediated by strong recurrent excitatory interconnections amongst the interneurones and between the interneurones and motoneurones. In favor of this idea are several observations. Almost all of the rhythmically active interneurones and motoneurones receive synaptic input from the ventrolateral tract where their axons project. In the case of motoneurones these inputs are unusually strong and capable of sustaining stimulation frequencies up to 30Hz (Ho and O'Donovan unpublished observations). If we assume such axons are responsible for the VLT-evoked synaptic potentials in interneurones and in motoneurones then this implies that the population of VLT cells is synaptically interconnected and that there are powerful projections, probably monosynaptic, onto motoneurones. This appears likely because cutting the axons traveling in the VLT reduces the amplitude of rhythmic synaptic potentials in motoneurones caudal to the cut. However, stronger support for this pattern of connectivity requires us to eliminate the possibility that the VLT-evoked synaptic potentials in motoneurones and interneurones are mediated by the descending axons present in the in the VLT.

The functional role of the synchronous activation is currently unknown but is likely to be related to the role of spontaneous activity in the developing nervous system. Spontaneous activity has been implicated in the development and differentiation of several regions of the nervous system. In the spinal cord activity is thought to be involved in the formation of synaptic connections (Fields *et al.*, 1991) and in the regulation of gene expression, particularly neurotransmitter synthesis and expression (Garner *et al.*, 1992; Mendelson, 1992, Eiden *et al.*, 1988; Foster *et al.*, 1989). These effects may be mediated by the calcium elevations that accompany rhythmic activity because Agoston *et al.*, (1991) have shown calcium entry through L-type calcium channels is necessary for activity-dependent regulation of certain neurotransmitters in spinal cord cell cultures.

An important direction of future research is to establish how these embryonic circuits become transformed into the locomotor machinery of the adult, and to what extent the circuitry present in the developing cord in preserved in the adult. To understand this requires more detailed analysis of the spinal networks generating rhythmic activity. Outstanding questions include the mechanism of rhythmicity in the developing spinal cord, and the extent to which this is a network phenomenon or one based on the endogenous properties of individual neurones. In addition, further progress must be made in identifying specific classes of interneurone, their role in rhythmic activity and their relationship to the known classes of interneurone in other vertebrates.

REFERENCES

Aghajanian, G.K. and Rasmussen , K. (1989) Intracellular studies in the facial nucleus illustrating a simple new method for obtaining viable motoneurones in adult rat brain slices. *Synapse*. **3**: 331-338

Agoston, D.V., Eiden, L.E. & Brenneman, D.E. (1991) Calcium-dependent regulation of the Enkephalin Phenotype by Neuroneal Activity during early ontogeny. *Journal of Neuroscience Research*. **28**:140-148.

Blanton, M. G., LoTurco, J.J. and Kriegstein, A.R. (1989) Whole cell recording from neurones in slices of reptilian and mammalian cerebral cortex. *Journal of Neuroscience Methods*. **30**: 203-210.

Eiden, L.E., Siegel, R.E., Giraud, P. and Brenneman, D.E. (1988) Ontogeny of enkephalin- and VIP-containing neurones in dissociated cultures of embryonic mouse spinal cord and dorsal root ganglia. *Developmental Brain Research*. **44**: 141-150

Fields, R.D., Yu, C. and Nelson, P.G. (1991) Calcium, network activity and the role of NMDA channels in synaptic plasticity in vitro. *Journal of Neuroscience*. **11**:134-146.

Foster, G.A., Eiden, L.E., and Brenneman, D.E. (1989) Regulation of a discrete subpopulation of transmitter-identified neurones after inhibition of electrical activity in cultures of mouse spinal cord. *Cell and Tissue Research*. **256**: 543-552

Garner, L.K., Mendelson, B.M., Albers, K.M., Kindy, M. and B. M. Davis (1992) Effect of activity on Enkephalin and Substance P mRNA in the developing chick spinal cord. *Society for Neuroscience Abstracts*. **18**: 420

Grillner, S., and Matsushima T. (1991) The neural network underlying locomotion in lamprey-Synaptic and cellular mechanisms. *Neurone*. **7**: 1-15.

Grynkiewicz, G., Poenie, M. and Tsien, R.. (1985) A new generation of Ca2+ indicators withgreatly improved fluorescence properties. *Journal of Biological Chemistry*. **260**: 3440-3450

Ho, S. and O'Donovan, M. J. (1991) Properties of propriospinal neurones involved in the rhythmic excitation of motor pools in the isolated embryonic chick spinal cord. *Society for Neuroscience Abstracts*. **17**: 120

Ho, S., and O'Donovan, M.J. (1992) Optical and pharmacological studies of propriospinal neurones involved in rhythmic motor activity in the embryonic chick spinal cord. *Society for Neuroscience Abtstracts*. **18**: 1057

Ho, S. and O'Donovan M.J. (1993) Regionalization and inter-segmental coordination of rhythm generation networks in the spinal cord of the chick embryo. *Journal of Neuroscience*. **13**: 1354-1371

Horikawa, K. and Armstrong, W.E. (1988) A versatile means of intracellular labeling:injection of biocytin and its detection with avidin conjugates. *Journal of Neuroscience Methods*. **25**:1-11

Kita, H. and Armstrong, W. (1991) A biotin-containing compound N-(2-aminoethyl) biotinamide for intracellular labeling and neuroneal tracing studies:comparison with biocytin. *Journal of Neuroscience Methods*. **37**: 141-150

Landmesser, L.T. and O'Donovan M.J. (1984) Activation patterns of embryonic chick hind limb muscles recorded in ovo and in an isolated spinal cord preparation. *Journal of Physiology*. **347**:189-204.

Mendelson, B. (1992) Activity dependent alterations in substance P and CGRP immunoreactivity in neurones and fibers in the embryonic chick spinal cord. *Society for Neuroscience Abstracts*. **18**: 420

O'Donovan M.J. (1989) Motor activity in the isolated spinal cord of the chick embryo: Synaptic drive and firing pattern of single motoneurones. *Journal of Neuroscience.* **9:** 943-958.

O'Donovan, M.J. and Ho, S. (1992) The role of extracellular calcium and calcium channels in activity dependent intracellular calcium changes in embryonic chick motoneurones. *Society for Neuroscience Abstracts.* **18:** 1303

O'Donovan, M.J. and Landmesser L.T. (1987) The development of hindlimb motor activity studied in an isolated preparation of the chick spinal cord. *Journal of Neuroscience.* **7:**3256-3264.

O'Donovan, M., Sernagor, E.. Sholomenko, G., Ho S., Antal, M., and Yee, W. (1992) Development of spinal motor networks in the chick embryo. *Journal of Experimental Zoology.* **261:**261-273.

O'Donovan, M.J., Ho, S., Sholomenko, G. and Yee, W. (1993) Real-time imaging of neurones retrogradely and anterogradely loaded with calcium sensitive dyes. *Journal of Neuroscience Methods.* **46:**91-106

O'Donovan, M.J., Ho., S and Yee, W. (1994) Calcium imaging of rhythmic network activity in the developing spinal cord of the chick embryo. *In Press.* J.Neuroscience

Ritter, A. and O'Donovan, M.J. (1993) Firing patterns and membrane properties of rhythmically active interneurones in the embryonic chick spinal cord. *Society for Neuroscience Abstracts.* **19:**557.

Roberts, A., Soffe, S.R. and Dale, N. (1986) Spinal interneurones and swimming in frog embryos. In *"Neurobiology of Vertebrate Locomotion"* Ed. S. Grillner, P.S.G. Stein, D.G. Stuart, H.Forssberg and R.M. Herman pp. 279-306, Macmillan.

Sernagor, E. and O'Donovan, M. J. (1991) Whole cell patch clamp of rhythmically active motoneurones in the isolated spinal cord of the chick embryo. *Neuroscience Letters.* **128:**211-216.

201

... ...
... ...

POSTEMBRYONIC MATURATION OF A SPINAL CIRCUIT CONTROLLING AMPHIBIAN SWIMMING BEHAVIOUR

K.T. SILLAR, J.F.S. WEDDERBURN and A.J. SIMMERS*

School of Biological and Medical Sciences
Gatty Marine Laboratory
University of St. Andrews
St. Andrews, Fife KY16 8LB, Scotland, U.K.

*CNRS et Université de Bordeaux I
Laboratoire de Neurobiologie et Physiologie Comparées
Place du Docteur Peyneau
33120 Arcachon, FRANCE

SUMMARY

In most vertebrates, including humans, the basic neural circuitry responsible for generating rhythmic locomotor movements appears to be resident in the spinal cord at very early stages in development, often at times when locomotion is not even possible. After hatching or birth, this naive circuitry then matures, often over a prolonged period, until the complex and sophisticated repertoire of adult locomotion is attained. To address the developmental mechanisms responsible for locomotor circuit maturation we are studying the ontogeny of locomotion in a simple model system - swimming activity in postembryonic amphibian (*Xenopus laevis*) tadpoles. The neuroanatomical simplicity of these organisms, together with their rapid development has enabled detailed description of: 1) the timecourse of developmental changes in rhythmic swimming activity of immobilized animals; 2) the axial progression of these changes; 3) the modulation of spinal neuronal properties as the swimming system develops; and 4) the central mechanisms underlying swimming rhythm maturation. We find that as the swimming circuit matures, a bursty ventral root pattern becomes established during the first 24 hours after hatching from the egg membranes. This occurs in association with a transition from single to multiple impulses per cycle in myotomal motoneurones. As we describe here this developmental change imparts flexibility on a simple stereotyped embryonic system at the precise time in ontogeny when behavioural manoeuvrability is likely to enhance survival. The acquisition of ventral root bursts follows a rostrocaudal path during larval development indicating the possible involvement of a descending neuronal process and evidence suggests that rhythm maturation is controlled by serotonergic raphespinal interneurones.

INTRODUCTION

The immense complexity of the adult vertebrate central nervous system masks a further dimension of intricacy; namely the way in which the neural networks that control behaviour are assembled during development. This is not a simple task even from an engineering viewpoint - most man-made machines are constructed solely to perform 'adult' tasks and do not need to function during their assembly. The biological machinery which generates

Neural Control of Movement, Edited by W.R. Ferrell
and U. Proske, Plenum Press, New York, 1995

patterns of behaviour, like locomotion for example, must also serve the functional requirements of successive stages in the development of the adult organism. Consequently, any modifications to developing central circuitry must suit the behavioural repertoire of a given stage in development, yet at the same time lay the appropriate foundations for the eventual control of adult behaviour. The problems of neural ontogeny are further compounded, because the biomechanical constraints which limit behaviour are also in a state of flux during development and the behavioural requirements can differ markedly at different developmental stages.

Little is currently known about how neural circuits underlying locomotion mature but one emerging view is that the basic rhythm generating networks of the spinal cord are constructed first, *in ovo* or *in utero* and their output is then modulated by descending control systems which themselves continue to develop for some period after hatching or birth. We have chosen to study spinal circuit maturation in a rapidly developing model vertebrate system, that of locomotion in postembryonic amphibian tadpoles. Our work focusses on the early post-hatching stages of the South African clawed frog, *Xenopus laevis*. This preparation is particularly well suited to studies on neural circuit development for the following reasons. **Firstly**, much is already known about the neural circuitry driving swimming behaviour in *Xenopus* embryos near the time of hatching (stage 37/38; Nieuwkoop and Faber, 1956; for recent review, see Roberts, 1990). Of the 8 distinct classes of differentiated spinal neurone (Roberts and Clarke, 1982), a functional role for 6 of these has already been established. Three of them are important for swimming: excitatory (1) and inhibitory (2) interneurones (Dale and Roberts, 1985; Dale, 1985) drive the motoneurones (3) in a characteristic pattern of discharge to produce the swimming rhythm (Roberts, 1990). Thus, swimming is generated by an anatomically restricted and well described neural circuit, so providing an extensive database from which the subsequent development of this circuit can be investigated. **Secondly**, the spinal cord is alone sufficient to generate a basic pattern of rhythmic ventral root activity appropriate for swimming behaviour (Kahn and Roberts, 1982). The intact and spinal patterns of immobilized embryos are essentially indistinguishable, indicating that neither peripheral feedback nor descending control systems play an important role in rhythm generation or its modulation. The immature spinal circuit, therefore, has yet to be exposed to extrinsic inputs which might be expected to become established during larval development. **Thirdly**, the motor output for swimming is remarkably simple, stereotyped and well coordinated; very brief ventral root discharge on each cycle alternates on opposite sides of the body (Figure 1Aii) and a wave of activity passes from front to back with a brief rostrocaudal phase delay. Within an episode of swimming, each myotomal motoneurone fires just one impulse in each cycle (Figure 1Aiii), while rhythm frequency starts high (ca. 20Hz) and gradually declines to about 10Hz before activity ceases. Thus, the embryonic locomotor pattern is simple and clearly defined in its basic coordination pattern, making developmental changes relatively easy to identify. **Fourthly**, postembryonic development and growth of the organism as a whole is a relatively rapid process, including changes both in the central nervous (e.g. the descent of brainstem neurones into the spinal cord) and neuromuscular systems. The existing knowledge of the embryonic system, accrued over the last decade or so, therefore presents a unique opportunity to observe, describe and understand how a very simple and stereotyped spinal locomotor rhythm matures.

The *Xenopus* embryo swimming system can be viewed as a basic, though very effective "starter kit" for locomotion. It differs from the locomotor rhythms of more mature vertebrates in its inherent lack of flexibility. Since myotomal motoneurones fire no more than once on each cycle of swimming, and appear to do so every cycle of an episode (Sillar and Roberts, 1993), neither burst duration nor intensity can normally be altered on a cycle by cycle basis (see, however, Sillar and Roberts, 1988). In this article we will describe the way in which the starter kit for swimming develops over a brief (ca. 24 hour) period after hatching and focus on the increased flexibility in the motor output which these changes impart. Finally, we review some of the evidence that a descending brainstem control system is largely responsible for the development of the swimming circuit.

DEVELOPMENT OF A BURSTY LARVAL SWIMMING PATTERN

About 24 hours after hatching from their egg membranes (from developmental stage 37/38; Figure 1Ai to stage 42, Figure 1Bi; Nieuwkoop and Faber, 1956), *Xenopus* larvae have consumed most of their yolk sac and grown in length by about 40% (to 7mm). Existing myotomes have enlarged and more have been added caudally. Using extracellular recordings from ventral roots located in intermyotome clefts of -bungarotoxin-immobilized animals, we discovered a very dramatic change to the swimming rhythm at stage 42 compared with the embryo (Figure 1Aii, cf. Bii). Thus, the very brief biphasic discharge on each cycle seen in embryo swimming (Figure 1Aii) is replaced by a much longer burst of activity on each cycle at stage 42 (Figure 1Bii). The normal phasing of activity does not change radically, however, with strict left/right alternation and a rostrocaudal phase-delay down the body (see also below) evident at both stages in development.

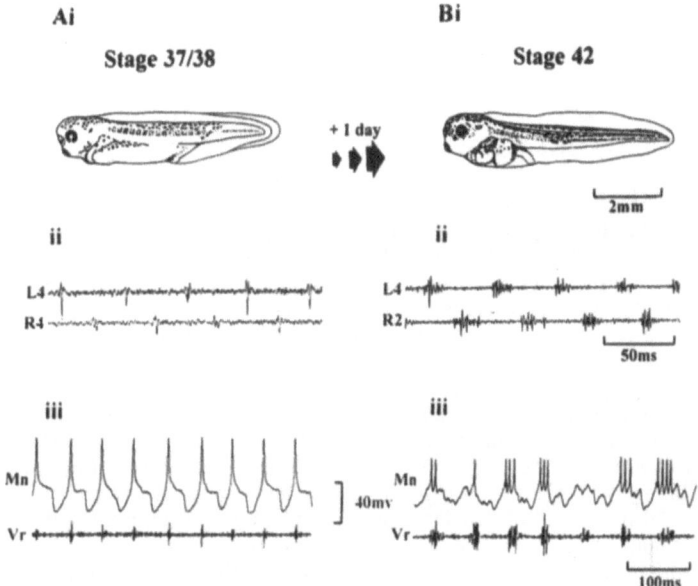

Figure 1. Development of motor bursts for swimming in postembryonic *Xenopus* tadpoles. **A:** At the time of hatching (Ai; stage 37/38), the embryonic motor pattern involves brief (ca. 7ms) ventral root discharge per cycle, alternating on opposite sides (Aii). Intracellularly recorded motoneurones (Aiii, Mn, top trace) fire one impulse per cycle in phase with ipsilateral ventral root discharge (Aiii, lower trace) **B:** At larval stage 42 (Bi), the motor ouput now involves longer bursts of ventral root discharge on each cycle, which still alternates on opposite sides of the body (Bii). Motoneurones can now fire more than once on each cycle (Biii). A, B are adapted from Sillar *et al.,* 1991 and Sillar *et al.,* 1992a. Recordings are denoted according to side (left/right) and position on the body relative to the otic capsule with intermyotome cleft 1 the most rostral.

The development of bursts of discharge in the ventral roots during larval swimming activity involves a transition in the firing properties of spinal neurones during postembryonic development, so that by stage 42 they can now fire more than once on each cycle (Figures 1Biii, 3; Sillar *et al.,* 1992a). The single spike capability of embryonic motoneurones is an intrinsic cellular property apparently imposed by the clamping effect of a slowly inactivating voltage-dependent K^+ conductance (Soffe, 1990). It seems reasonable to assume that to enable larval neurones to discharge multiple impulses per cycle, the relative potency of this conductance must somehow be reduced during development, although other contributing factors cannot be discounted. Analysis of the synaptic drive underlying swimming also indicates a developmental change in the firing pattern of premotor rhythm generating interneurones. In embryos, three types of synaptic potential are detectable in rhythmically active neurones. Throughout an episode, activity is superimposed upon a tonic level of depolarization mediated by NMDA receptor activation. Impulses are triggered off a phasic compound EPSP mediated at non-NMDA receptors and a large compound glycinergic IPSP occurs midcyle when neurones on the opposite side are active (Figure 1Aiii). In rhythmic stage 42 neurones the tonic excitation is still present (Figure 3A), but now impulses are triggered off a train of phasic summating EPSPs and midcyle a barrage of inhibitory potentials occurs (Figure 1Biii, 3B; Sillar *et al.,* 1992a). The simplest explanation for this observation is that premotor rhythm generating interneurones, like the motoneurones that they drive, now also discharge multiple impulses on each cycle of rhythm. However, in the absence of direct evidence from intracellular recordings of the interneurones it remains feasible that they continue to fire once per cycle but that significant desynchronization of activity in the premotor pool exciting or inhibiting a given motoneurone has occurred.

Insight into the temporal and spatial characteristics of rhythm development was obtained by recording swimming activity at an intermediate stage (stage 40), some 12 hours after hatching (Sillar *et al.*, 1991). At this stage, bursts of activity are recorded from rostral ventral roots but activity in more caudal segments consists of brief embryonic-like activity. This was the first indication that the development of ventral root bursts followed a rostrocaudal sequence occurring first at the front of the animal at stage 40 and 12 hours later, at stage 42, progressing to caudal ventral roots. This was consistent with rhythm development being dependent upon a neuronal influence which rapidly descends the spinal cord early in larval life.

DEVELOPMENT OF ROSTROCAUDAL PHASE-DELAY

In mature vertebrate locomotor systems, like lampreys and teleost fish, the magnitude of the phase-delay down the body correlates with swimming cycle period (Wallén and Williams, 1984). This ensures that during swimming the body oscillates at a constant wavelength (~1), irrespective of swimming frequency. We have examined this relationship in *Xenopus* during development. The embryonic swimming rhythm differs from that of more mature vertebrates in the lack of phase-constancy, but once the bursty pattern is established, a clear relationship between delay and cycle period emerges (Tunstall and Sillar, 1993). Thus, the swimming rhythm in *Xenopus* tadpoles has matured in a brief 24 hour period and now more closely resembles the activity generated by adult locomotor rhythm generators in two important respects - i) bursts of motor discharge occur on each cycle and ii) rostrocaudal phase delay and cycle period are positively correlated. Is there a causal link between these two phenomena? This seems a plausible suggestion since intuitively one would expect the delay to spiking in a caudal neurone to be at least partly dependent on the temporal summation of descending presynaptic excitatory inputs. At short cycle periods, when synaptic drive is intense, presynaptic neurones are likely to be firing at higher frequency so postsynaptic EPSPs are likely to summate quickly. In contrast when cycle periods are longer, and the drive less intense, the frequency of firing of presynaptic neurones will be lower and temporal summation will be correspondingly slower. It should therefore take longer for postsynaptic neurones to cross threshold for firing. If, in addition, a proportion of premotor interneurones cease firing at long cycle periods, as has been shown in the embryo (Sillar and Roberts, 1993), fewer EPSPs will be generated in a given postsynaptic neurone. However, this hypothesis awaits direct experimental testing.

FLEXIBILITY OF THE BURSTY PATTERN

An important consequence of the development of the bursty larval pattern is that the motor control system for swimming is more flexible. This is because the duration and intensity of ventral root discharge can now be varied on a cycle by cycle basis. Assuming that this translates into variability in the strength and duration of myotomal muscle contractions, then swimming behaviour itself will be under a greater degree of central control. One important feature of this enhanced variability is that the bursts can be modulated at different locations in the spinal cord. For example, bursts can be enhanced bilaterally (Figure 2A) or on one side relative to the other (Figure 2B), so that centrally generated changes in burst intensity that could signal an acceleration or fictive turning manoeuvre are often seen in larval recordings, but have not been reported in the embryo. Such changes occur apparently spontaneously, and could be triggered by descending commands. The flexibility of the larval rhythm is also witnessed in two related phenomena that differ from embryo preparations. Firstly, the larval swimming system spans a wider range of cycle periods than the embryo (Sillar *et al.*, 1991), so that frequencies within an episode can range from 7 to 35Hz (cf. 10 to 20Hz at stage 37/38). Secondly, rapid changes in frequency within an episode are a common event in larval preparations. An example is shown in Figure 2A, where cycle period had fallen from about

30Hz near the start of an episode to about 12Hz, but then suddenly accelerated to 25Hz in the space of a single cycle. In embryos, the frequency of swimming within an episode characteristically starts high (ca. 20Hz), drops to a plateau and then gradually falls to 10 or 12 Hz before swimming finally ceases.

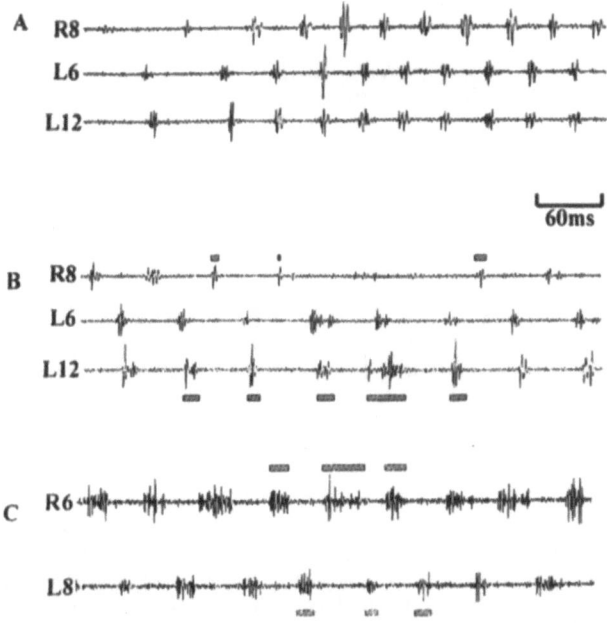

Figure 2. Intrinsic flexibility of motor output in the larval (stage 42) swimming rhythm. **A:** Within an episode, spontaneous accelerations are often recorded (in contrast to the embyro); in this case rhythm frequency approximately doubled in the space of two cycles. Note correlated increase in burst intensity. **B and C:** Burst duration can be spontaneously increased on one side relative to the other. In B, ventral root activity is enhanced down the left side of the animal (L6 and L12) and simultaneously suppressed on the right (R8). By contrast in C, burst durations are increased on the right (R6) and suppressed on the left (L8). Bars above and below recordings represent approximate durations of vr bursts. Note in B, vr activity on right side almost disappears when left vr bursting increases.

The mechanisms responsible for some of these centrally generated modifications in the swimming pattern are not yet known, but evidence for neuronal recruitment as one contributory factor has been obtained. In the embryo, for example, it has been shown that the probability of firing in premotor interneurones is reduced as cycle period lengthens and this may be responsible for the gradual reduction in frequency within an episode (Sillar and Roberts, 1993). If the same occurs in *Xenopus* larvae then rapid speed changes may result from the recruitment of inactive interneurones into the premotor pool. In contrast, embryo motoneurones, unlike presynaptic interneurones (Sillar and Roberts, 1993), seem to be active throughout an episode, irrespective of cycle periods. Thus, motor recruitment as a means of grading muscle force appears not to operate in the embryonic locomotor system. It would seem strange if this were also true in the larval swimming system where burst duration and intensity can vary widely within an episode. Changes in motor bursts structure could in principle occur in two ways: either the firing frequency of motoneurones decreases, and/or some motoneurones cease firing at long cycle periods, when activity is generally less intense. Our evidence from intracellular recordings of ventrally-located neurones, presumed to be motoneurones, supports both possibilities. For example, the recordings illustrated in Figure 3 show that motoneurones at stage 42 do not always fire on every cycle of an episode of fictive swimming and often drop out towards the end of an episode when cycle period lengthens and synaptic drive weakens. Examination of such records shows that the number of impulses per cycle is usually related to the duration and intensity of the ventral root burst in neighbouring intermyotome clefts. (However, some neurones must fire on every cycle for ventral root activity to be recorded at all!). Once motoneurones have ceased firing within an episode they continue to receive rhythmic synaptic drive and they can then be recruited back into the motor pool, for example during spontaneous accelerations in rhythm frequency or when a switch to an alternative motor programme, (that underlying struggling), occurs (Figure 3B; Kahn and Roberts, 1982).

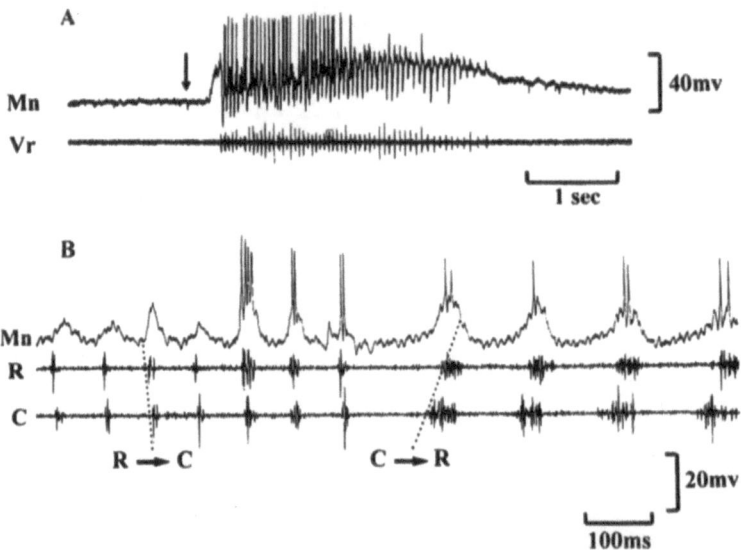

Figure 3. Mutiple firing and motor recruitment in stage 42 *Xenopus* larvae. **A:** a motoneurone (Mn) ceased firing before end of episode of swimming evoked by dimming the illumination (at arrow), but would fire up to 3 impulses per cycle near the start. **B:** In a different preparation a Mn dropped out of the rhythm shortly after start of episode (not illustrated), but continued to receive rhythmic synaptic drive during swimming (first four cycles of illustrated activity). In correlation with spontaneous increase in burst intensity (vr), synaptic drive to Mn increases in amplitude and spiking was resumed. After a spontaneous switch in motor output Mn continued to fire in the struggling motor pattern (Kahn and Roberts, 1982). Dotted lines in B show rostrocaudal phase-delay in swimming (R→C) and reversed caudorostral (C→R) delay in struggling.

DESCENDING CONTROL OF RHYTHM DEVELOPMENT

The rostrocaudal progression of burst development during larval swimming activity indicated the possible involvement of a descending neuronal control system. A range of neural systems invade the spinal cord at these early stages in development (Nordlander, 1984; van Mier and ten Donkelaar, 1989), but we chose to examine first the possible role and influence of raphespinal interneurones of the rostral ventral medulla (van Mier *et al.*, 1986). The descending axons of raphespinal neurones reach the cord relatively early in development, and project to ventral (motor) regions over the first few days of larval life. In addition to this temporal coincidence between rhythm development and the descent of axons into the ventral cord from the raphe nucleus, our interest in this population of neurones stemmed largely from their transmitter phenotype. In *Xenopus*, as in all other vertebrates so far studied, the vast majority of raphe neurones are immunopositive for antibodies raised against the neuromodulatory amine, 5HT. A major influence of 5HT on the function of vertebrate motoneurones is to enhance firing properties, notably during locomotor rhythm generation (lamprey - Harris-Warwick and Cohen, 1985; cat - Barbeau and Rossignol; 1990; rabbit - Viala and Buser, 1969), often through an action on K^+ conductances (Wallen *et al.*, 1989; Grillner *et al.*, 1991). Since the development of the swimming rhythm involves a transition from single impulses (evidently imposed by a K^+ conductance; Soffe, 1990) to multiple impulses on each cycle it seemed conceivable that this transition is causally related to the release of 5HT from developing raphespinal interneurones.

In an initial series of experiments we examined the modulatory influences of exogenous 5HT applied to preparations at various stages in development. We found that 5HT (at 1 - 5μM) mimicked the normal development of the swimming system, but preceded it by about

12 hours. Thus, when 5HT was applied to the hatchling embryo, bursts of activity were recorded rostrally, but there was no effect on the rhythm recorded caudally. This effectively transformed the embryo rhythm into one typical of a stage 40 larva. At stage 40, where normally bursts are recorded rostrally, but embryo-like single spikes recorded caudally, 5HT enhanced the duration of rostral bursts, and converted caudal single spikes into brief bursts, again advancing the activity to that typical of an animal 12 hours older, at stage 42. At this stage 5HT affected rhythm generation in much the same way as seen in other vertebrate systems, enhancing burst durations at all axial levels in the cord. These results showed that, at least in principle, 5HT release could orchestrate rhythm development and in addition implied a dynamic rostrocaudal sequence in the expression of 5HT receptors. This latter notion was further supported by the observation that at prehatchling embryonic stages, 5HT did not affect ventral root activity at any level in the cord (Sillar *et al.*, 1992b). And by preloading serotonergic projections with 5HT (using the metabolic precursor, 5HTP) evidence for the former suggestion was obtained; enhanced endogenous 5HT release at the three developmental stages could also mimic the normal rostrocaudal acquisition of motor bursts. Our evidence therefore strongly favours the conclusion that the developmental incorporation of raphespinal axons into the spinal cord network for swimming and the release of 5HT from presynaptic terminals is responsible for the postembryonic development of the more flexible motor system that we describe in *Xenopus* larvae.

DISCUSSION

An emerging concept in the developing field of locomotor ontogeny is that the basic neural circuits responsible for rhythmic body or limb movements are assembled very early in development, often at times when locomotion is limited or not even possible. In their immature form these circuits are only capable of primitive and relatively unrefined locomotor movements, but over the course of early larval or postnatal development, the same pre-existing circuits acquire precision and flexibility through modulation by descending control systems that continue to develop for some time after birth or hatching. The results of our recent work suggest that a very similar strategy governs the development of swimming in a lower vertebrate, the amphibian, *Xenopus laevis*. At the time of hatching from their egg membranes, the swimming activity of *Xenopus* embryos is remarkably stereotyped and inflexible, and it is probable that, at this stage, neither peripheral feedback nor descending modulatory inputs are fundamental to the operation of the spinal circuit. This makes the subsequent larval development of the swimming system attractive for study because extrinsic control systems can be examined sequentially as they become incorporated into networks of the spinal cord.

In a brief 24 hour period after hatching, the motor programme in *Xenopus* adopts a radically different form; neurones acquire a multiple spike capability and ventral root bursts develop (Sillar *et al.*, 1991; 1992a). Because of this simple transition in the properties of component neurones in the circuit the swimming system, the flexibility of the swimming rhythm increases markedly. Burst duration can now be varied on a cycle by cycle basis, so presumably the strength of muscle contractions can be graded under different circumstances. Even within a single cycle, bursts can now be enhanced on one side and reduced on the opposite side so the inherent ability to execute turning manoeuvres becomes feasible. This increased variability in burst structure is accompanied by an increase in the range of cycle periods that can be sustained within an episode and the rhythm is now capable of rapid accelerations. Such alterations in the output of a locomotor circuit presumably meet the survival requirements of free swimming larval life, to which the organism is very suddenly exposed after hatching.

A growing body of circumstantial evidence supports the conclusion that the postembryonic development of the swimming system is causally linked to the anatomical ingrowth of descending raphespinal interneurones to the ventral regions of the spinal cord and their release of the neuromodulator, 5HT (Sillar *et al.*, 1993). Both exogenously applied and endogenously released 5HT parody the normal rostrocaudal sequence of rhythmic burst development. In more recent experiments direct evidence for a causal role of raphespinal interneurones has been obtained: neurotoxic ablation of serotonergic spinal projections (with a selective neurotoxin; 5,7 DHT) prevents the development of ventral root bursts during larval swimming (Sillar *et al.*, in preparation). In this species it appears that the serotonergic system plays a dual role in motor control, firstly adapting a pre-existing spinal network into a modulable form upon which it can act at later stages in development.

To what extent might our findings begin to unveil general principles governing the development of vertebrate locomotor circuitry? This is clearly open to speculation but some parallels already exist in the role of 5HT in adult preparations. In the lamprey, the cat and the rabbit, for example, 5HT enhances the intensity and duration of rhythmic locomotor bursts. In most vertebrates, moreover, there is generally a massive proliferation of 5HT projections to the spinal cord after birth. It is plausible, therefore, that in other systems the ensuing release of 5HT during development will modulate locomotor output in a way which parallels the effects of the amine that we find in *Xenopus* tadpoles. Some supportive evidence for this possible generality has arisen from a comparative study of swimming in hatchling *Rana temporaria* embryos. Here the embryonic motor pattern differs from *Xenopus* in that at the time of hatching it already involves bursts of ventral root activity on each cycle and motoneurones are capable of firing multiple impulses (Soffe and Sillar, 1991). In contrast to *Xenopus* embryos, but in common with *Xenopus* larvae the descending 5HT system is by now well established in the ventral aspect of the spinal cord (Sillar *et al.*, 1992c, and in preparation). It is already well known that 5HT exerts profound influences on the physiological properties of a wide range of central neurones, but can also act as a developmental signal, promoting cell division, differentiation and synaptogenesis (Lauder *et al.*, 1981; Lipton and Kater, 1989). Perhaps in other vertebrates too 5HT released from developing raphespinal projections sculpts target circuitry in the spinal cord into a form that is then susceptible to modulation.

ACKNOWLEDGEMENTS. This work was supported by the Royal Society of London, the SERC (UK), the Wellcome Trust, and in part by the European Science Foundation, to whom we are grateful. K.T.S is a Royal Society 1983 University Research Fellow, J.F.S.W. was a SERC research student and A.J.S. was a Royal Society/CNRS Exchange Fellow.

REFERENCES

Barbeau, H. and Rossignol, S. (1990) The effects of serotonergic drugs on the locomotor pattern and on cutaneous reflexes of the adult chronic spinal cat. *Brain Research* **514,** 55-67.

Dale, N. (1985) Reciprocal inhibitory interneurones in the spinal cord of *Xenopus laevis*. *Journal of Physiology* **363,** 527-543.

Dale N. and Roberts A. (1985) Dual component amino acid-mediated synaptic potentials: excitatory drive for swimming in *Xenopus* embryos. *Journal of Physiology* **363,** 35-59.

Grillner, S., Wallén, P., Brodin, L. and Lansner, A. (1991) Neuronal network generating locomotor behavior in lamprey. *Annual Review of Neuroscience* **14,** 169-199.

Harris-Warwick, R.M. and Cohen, A.H. (1985) Serotonin modulates the central pattern generator for locomotion in the isolated spinal cord of the lamprey. *Journal of Experimental Biology* **116,** 27-46.

Kahn, J.A. and Roberts, A. (1982) The neuromuscular basis of rhythmic struggling movements in embryos of *Xenopus laevis*. *Journal of Experimental Biology* **99,**197-205.

Lauder, J.M., Wallace, K.A. and Krebs (1981) Roles for serotonin in neuroembryogenesis. *Advances in Experimental and Medical Biology* **133,** 477-506.

Lipton, S.A. and Kater, S.B. (1989) Neurotransmitter regulation of neuronal outgrowth, plasticity and survival. *Trends in Neuroscience* **12,** 265-270.

Nieuwkoop, P.D. and Faber, J. (1956) Normal tables for *Xenopus laevis* (Daudin). Amsterdam, North Holland.

Nordlander, R. (1984) Developing descending neurons in the early *Xenopus* tail spinal cord. *Journal of Comparative Neurology* **228,** 117-128.

Roberts, A. (1990) How does a nervous system produce behaviour: a case study in neuroethology. *Science Progress* **74,** 31-51.

Roberts, A. and Clarke, J.D.W. (1982) The neuroanatomy of an amphibian embryo spinal cord. *Philosophical Transactions of the Royal Society Series B.* **296,** 195-212.

Sillar, K.T.and Roberts, A. (1988) Unmyelinated cutaneous afferent neurones activate two types of excitatory amino acid receptor in the spinal cord of *Xenopus laevis* embryos. *Journal of Neuroscience* **8,** 1350-1360.

Sillar, K.T. and Roberts, A. (1993) Control of frequency during swimming in *Xenopus* embryos: a study on interneuronal recruitment in a spinal rhythm generator. *Journal of Physiology* **472,** 557-572.

Sillar, K.T., Wedderburn, J.F.S. and Simmers, A.J. (1991) The development of swimming rhythmicity in post-embryonic *Xenopus laevis*. *Proceedings of the Royal Society Series B.* **246,** 147-153.

Sillar, K.T., Simmers, A.J. and Wedderburn, J.F.S. (1992a) The post-embryonic development of cell properties and synaptic drive underlying locomotor rhythm generation in *Xenopus* larvae. *Proceedings of the Royal Society Series B.* **249,** 65-70.

Sillar, K.T., Wedderburn, J.F.S. and Simmers, A.J. (1992b) Modulation of swimming rhythmicity by 5-hydroxytryptamine during post-embryonic development in *Xenopus laevis. Proceedings of the Royal Society Series B.* **250,** 107-114.

Sillar, K.T., Woolston, A-M. and Wedderburn, J.F.S. (1992c) Development and role of serotonergic innervation to the spinal cord of hatchling *Rana temporaria* and *Xenopus laevis* tadpoles. *Journal of Physiology* **446,** 323P.

Sillar, K.T., Wedderburn, J.F.S., Woolston, A-M. and Simmers, A.J. (1993) Control of locomotor movements during vertebrate development. *News in Physiological Sciences* **8,**107-111.

Soffe, S.R. (1990) Active and passive membrane properties of spinal cord neurones during fictive swimming in frog embryos. *European Journal of Neuroscience* **2,** 1-10.

Soffe, S.R. and Sillar, K.T. (1991) Patterns of synaptic drive to ventrally located spinal neurones in *Rana temporaria* embryos during rhythmic and non-rhythmic motor responses. *Journal of Experimental Biology* **156,** 101-118.

Tunstall, M.J. and Sillar, K.T. (1993) Physiological and developmental aspects of intersegmental coordination in *Xenopus* embryos and tadpoles. *Seminars in the Neurosciences* **5,** 29-40.

van Mier, P. and ten Donkelaar, H.J. (1989) Structural and functional properties of reticulospinal neurones in the early swimming stage *Xenopus* embryo. *Journal of Neuroscience* **9,** 25-37.

van Mier, P., Joosten, H.J.W., van Rheden, R. and ten Donkelaar, H.J. (1986) The development of serotonergic raphe spinal projections in *Xenopus laevis. International Journal of Developmental Neuroscience* **4,** 465-476.

Viala, D. and Buser, P. (1969) The effects of DOPA and 5HTP on rhythmic efferent discharges in hindlimb nerves in the rabbit. *Brain Research* . **12,** 437-443.

Wallén, P. and Williams, T.L. (1984) Fictive locomotion in the lamprey spinal cord *in vitro* compared with swimming in the intact and spinal animal. *Journal of Physiology* **347,** 225-239.

Wallén, P., Buchanan, J.T., Grillner, S., Christenson, J. and Hokfelt, T. (1989) The effects of 5-hydroxytryptamine on the afterhyperpolarization, spike frequency regulation and oscillatory membrane properties in lamprey spinal neurons. *Journal of Neurophysiology* **61,** 759-768.

CEREBELLAR MECHANISMS

THE CEREBELLUM AS A PREDICTIVE MODEL OF THE MOTOR SYSTEM: A SMITH PREDICTOR HYPOTHESIS

R.C. MIALL and D.M. WOLPERT*

University Laboratory of Physiology
Oxford OX1 3PT, U.K.

*Department of Brain and Cognitive Sciences
M.I.T., Cambridge, MA 02139, U.S.A.

SUMMARY

The performance of motor systems with large feedback delays can be significantly enhanced by the use of internal predictive representations of the motor apparatus. The cerebellum is a likely site for these internal models, and we show that ataxic patients appear to have reduced awareness of the hand position during movement, suggesting that a sensory predictor within the cerebellum is impaired. We recently suggested that the cerebellum holds two types of neural model which together form a 'Smith Predictor'. One is a model of the motor apparatus (limbs and muscles) which provides a rapid prediction of the sensory consequences of each movement. The other model is of the time delays in the feedback control loop (conductance delays, muscle latencies, sensory processing). This delays a copy of the rapid prediction, so that it can be compared with actual sensory feedback; any errors are then used both to correct the movement and to update the internal representations of the motor apparatus. We propose mechanisms by which both parts of the Smith Predictor could be formed within the cerebellum, and present a neural network simulation based on these ideas.

INTRODUCTION

We have recently suggested that the cerebellum acts as a 'Smith Predictor" (Miall et al., 1993a). This is a type of controller originally devised for engineering control systems that suffer long feedback delays -- for example catalytic crackers in steel mills (Smith, 1959). Its principles are however equally well applied to the control of human movement and we propose the scheme for the cerebellar control of visually guided movements. Here the unavoidable delays are due mainly to visual processing, with contributions from visuo-motor integrative processes, axonal conduction delays and muscular latencies. Together these delays may add up to a feedback loop time of 150-200msec, which is long with respect to the duration of many visually guided movements.

The Smith Predictor controller is based on the idea of internal models or neural representations: it holds a predictive model of the motor system, and its output is a prediction of the results of movements (Figure 1). Thus, we propose, the cerebellum receives a copy of a motor command being generated by 'upstream' motor regions (posterior parietal cortex or primary motor cortex), and uses its knowledge of the 'downstream' motor system (joints, muscles etc.) to estimate what the outcome of the movement would be, given the current state of the body. Since this output is a sensory prediction, it is not used to directly control the movement. However, this sensory prediction avoids all the delays within the feedback system. It can be rapidly compared with the intended outcome of the movement, and the difference between the two forms an error signal used to correct the final stages of the movement (Comparator 2 in Figure 1). So by using a predictive model within an internal negative feedback loop, rapid control can be achieved even in the face of long sensory and motor delays.

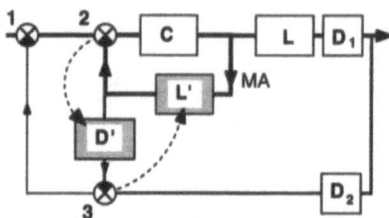

Figure 1. *A schematic diagram of the Smith Predictor:* The Smith Predictor (shaded boxes) lies within a negative feedback loop in which errors (sensed by Comparator 1) are converted by a PD controller (box labelled C) into torques sent to the limb (L). The feedback loop contains delays on the forward and backward paths (D_1 and D_2). The dynamic model of the limb (L') lies on a fast internal feedback loop receiving a moving average of the torques sent to the muscles (MA), and sending its output to Comparator 2. A copy of its output is also passed through a model delay ($D'=D_1+D_2$) before comparison with the delayed feedback (Comparator 3). The dashed lines indicate training signals used to modify the two models.

Two additional properties of this control scheme should be mentioned. First, the prediction is used instead of the actual feedback within the negative feedback loop, so that it cancels the outstanding sensory errors (which remain present in the feedback system until the true sensory feedback returns some 150msec later) and further, unneeded, movements are not generated. It is these inappropriate responses to out-of-date signals that cause instability in feedback systems. Second, the internal model can be used to plan movements independent of their execution -- the brain can ask "what if" questions and rehearse, modify and learn movements on the basis of the internal predictions of their sensory consequences.

Evidence for these ideas is provided by cerebellar patients and monkeys with cerebellar inactivating lesions, whose movements are poorly co-ordinated and ataxic, as might be expected if their internal predictions of their movements were impaired. We have recently examined a group of ataxic patients, shown by MR imaging to have significant damage to the superior cerebellar peduncles, the output tract of the lateral cerebellum (Haggard *et al.*, 1994). These patients performed a visually guided tracking task but showed considerable intermittency in their responses (Figure 2A), indicating that they were especially reliant on visual inputs to perform the task (see Miall *et al.*, 1993b). If the visual target that they were following was briefly blanked from the screen, they could continue to tracking with only slightly increased errors. This implies that they had no difficulty in predicting the target's movement. However, if the joystick-controlled cursor was blanked off briefly, their movements became significantly less accurate (Figure 2A); this was not the case for control subjects (Miall *et al.*, 1993c). This therefore suggests that the cerebellar damage had interrupted an internal prediction of their arm movement, and they were only able to track successfully when provided with visual feedback of their arm position. We have also demonstrated the same effect in a monkey whose dentate nucleus was temporarily inactivated by an infusion of the local anaesthetic lignocaine (Figure 2B,C).

The Smith Predictor includes a second important component that differentiates it from other schemes based on internal predictive models. This second part is an internal model of the very feedback delays that we need to avoid. This time delay model receives the output of the sensory predictor and delays it by an amount equal to the delay suffered by the actual sensory feedback signals (Figure 1). Thus its output is a copy of the prediction delayed to be in synchrony with real feedback. Differences between these two signals (Comparator 3 in Figure 1) indicate a failure of the predictions, and these differences can be treated as sensory errors that must be corrected. Moreover, they can also be used to improve future predictions of the model -- they form an appropriate training signal for the brain (and we suggest specifically the cerebellum) to learn accurate predictive models.

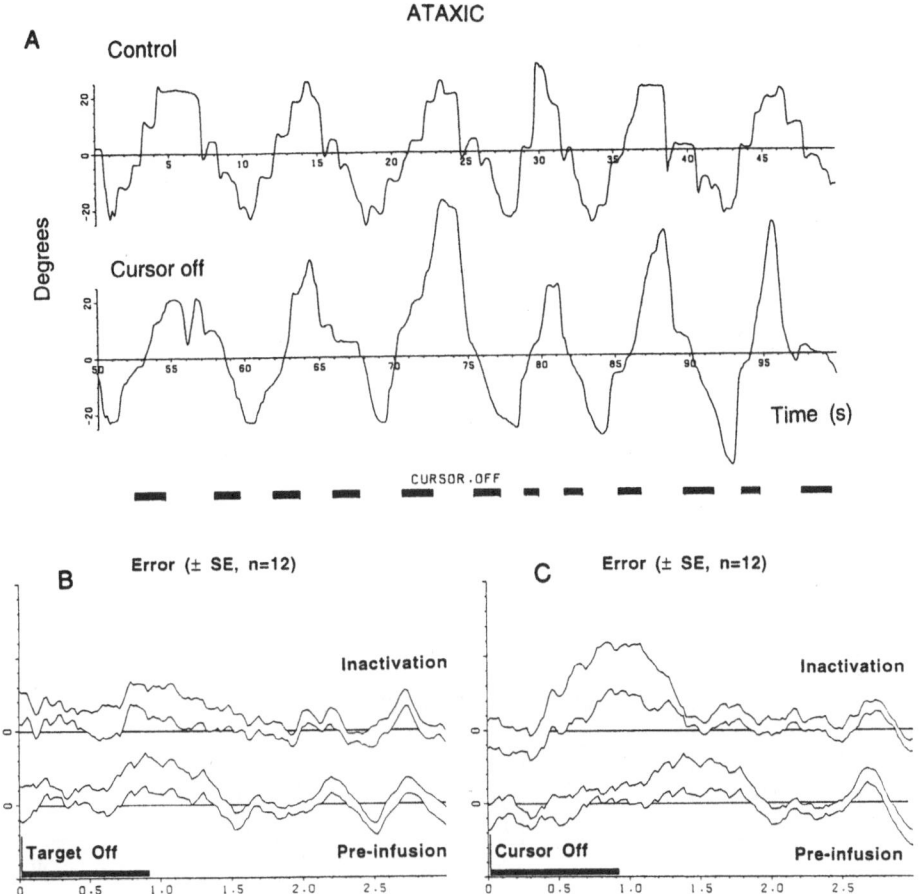

Figure 2. *Visually guided tracking during cerebellar dysfunction:* **A:** tracking of a ramp waveform by an ataxic patient. The upper record shows his normal tracking movements, which are much less smooth than those of normal subjects. The second record shows his tracking during brief blanking off of the cursor (black bars). Tracking without visual feedback of the cursor was smooth but inaccurate. **B** and **C:** ramp tracking by a monkey during reversible inactivation of the dentate nucleus. **B:** average tracking errors ($\pm$ 1 S.E. of the mean) as the target is blanked off; cerebellar inactivation has little effect on performance. **C:** average tracking errors as the cursor is blanked off; cerebellar inactivation leads to great increase in errors.

The remainder of this chapter will present a neural network model of the Smith Predictor (Wolpert, 1992), and show that both its components -- a predictive model and a time delay model -- could be generated by the cerebellum.

We will consider the control of a planar two jointed arm, comprising two links hinged at a shoulder and elbow joint, and moving in the horizontal plane (Figure 3). Negative feedback control was used as the basic control system driving the arm, and the angular position error and angular velocity error for each joint were scaled by position and velocity gains to produce joint torques (box C in Figure 1). The way in which the desired trajectory of the arm is chosen for a particular movement will not be considered here; it will be assumed that the desired state for each point in time is available to the controller in appropriate co-ordinates. Likewise the problems of visual analysis of the target and of co-ordinate transformations are not examined in this model. Thus the model operates in a kinematic coordinate framework, and may be considered to be downstream of the motor cortex, within the intermediate cerebellum. We make the assumptions that the controller has access to the target's velocity and position as represented in joint angle co-ordinates, and that the arm starts off stationary but on-target. As there is no redundancy in the specifications of the tracking task, and with only two joints in the arm, the arm configuration required to match hand position and velocity to the target is unique. Figure 3a indicates the need for predictive control -- in this figure the arm was controlled with proportional-derivative (PD) feedback but without a Smith Predictor. Even though the gains were set to get good control with immediate feedback, the arm was unstable when a feedback delay of only 30msec was introduced. In the following simulations, we sought control of the arm with high gains and with 100msec delays.

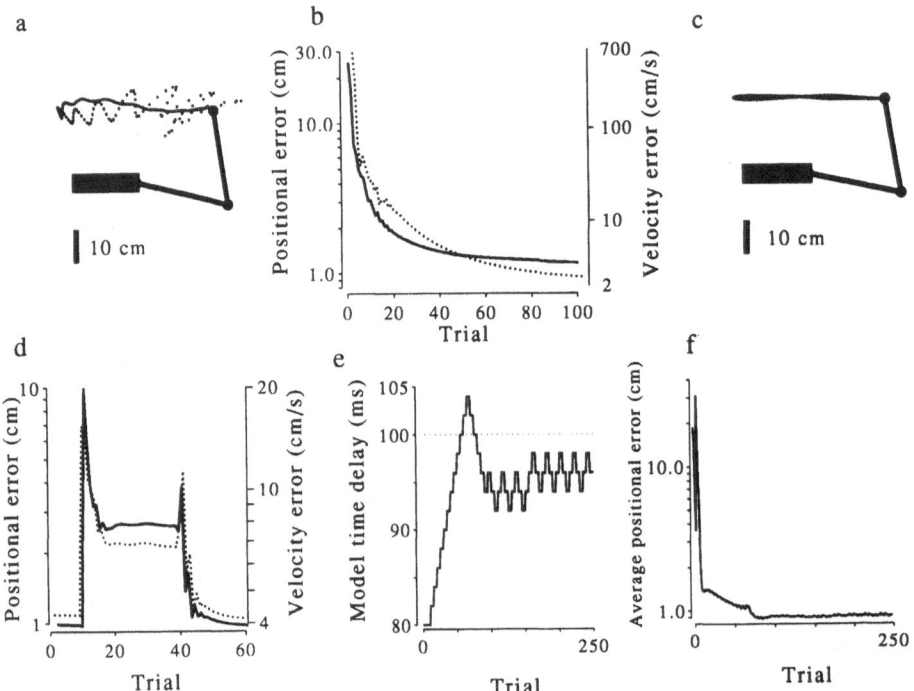

Figure 3. *A neural network model of the Smith Predictor:* tracking of a sinusoidal target using PD feedback. A target was moved with a 0.5 Hz cosine motion across the arm's workspace. The arm was modelled using rigid body dynamics of two freely pivoting links. The mass was assumed to be uniformly distributed along the links, which were modelled as thin cylinders. The dynamic equations of the arm (Jordan and Rumelhart, 1992) were simulated using a fourth order Runge-Kutta algorithm with a sampling frequency of 500 Hz. **a:** With the feedback delay set to only 30msec, the arm was unstable with open loop gains of 2.5 at the shoulder and elbow. **b:** Learning curves for the simulation when using a predictive neural network learning the arm's dynamics -- these were with a 100msec feedback delay and with open loop gains of 10 (shoulder) and 5 (elbow). The curves represent the position and velocity errors averaged over 10 networks. **c:** Smooth and stable tracking on the 100th trial despite the long delay and high gains. **d:** Learning curves as the model adapts to the addition and then the removal of a 7.5kg weight. **e:** Changes in the model delay time and in the tracking performance (**f**) during simultaneous learning of both the dynamic and time delay models.

Learning the Arm's Dynamics

The Smith Predictor must predict the current joint angles and velocities from the delayed feedback and from a copy of the motor command. Our neural network model therefore has six inputs: feedback of joint angles and velocities for each joint delayed by 100msec, and the outgoing joint torques for each joint. It had 4 outputs which were required to predict the two joint positions and velocities. A standard three layered feedforward neural network was used with the back-propagation algorithm to modify the synaptic weights after each iteration (Rumelhart and McClelland, 1986). Although this is not a physiological training rule, it serves to demonstrate that a solution can be found, and we can later concentrate on how more physiologically accurate networks and learning rules could achieve the same result.

The network's 6 input and 4 output units were linear; it had 50 logistic hidden units with asymptotes of ±1.0. Initially the weights were chosen randomly distributed between ±0.5.

The outputs of the network were then delayed by the model's delay, and compared to the actual state signals. The mismatch between the four delayed outputs (the estimated positions and velocities of each joint) and the corresponding delayed feedback signals were used as the error terms for the backpropagation algorithm (Comparator 3 in Figure 1). Hence, the output of the dynamic model will only be correct if it anticipates actual feedback by an amount equal to the time delay (Barto, 1990). For this simulation we simply delayed the signals by the known feedback delay without worrying about the delay mechanism (see section on "Fitting the model"). However, we will show how the value of this delay can be learned at the same time as the dynamic model.

To implement backpropagation the network's weights and activations are needed at the same time as the error term; since the error values were known only after a delay, the activities were stored and made available to the backpropagation algorithm. While storing the activation states of the neurones is not physiological impossible (Houk *et al.*, 1990), the exact mechanism that might be used is unknown, and we treat these simulations as a first attempt. The weights did not need to be stored, because the learning rate was very low and the weights changed little over the delay period. Finally, for perfect prediction of the current state of the arm, using delayed feedback, it would be necessary to retain in temporal order the previous 100msec of torques sent to the arm. However, the torques generated during human movement are unlikely to change dramatically over this time-scale because of the filtering properties of motor neurones and muscles, so we simply used a running average of the torques, using a 100msec moving average (MA in Figure 1).

The position and velocity gains of the PD controller were set to 10 for the shoulder and 5 for the elbow. The outer feedback loop had unity gain -- this was just within the stability margin with a 100msec feedback delay, so using gains of 10 and 5 put the controller well outside the stability region. To avoid gross instability while the network was naive (and therefore inaccurate) all the gains were initially set to a value of 0.5, and their values doubled at each trial up to their final values. This 'gain schedule' therefore lasts a maximum of 5 trials. Each trial consisted of two tracking periods of 2 seconds duration. During the first period backpropagation was used at each iteration, whereas during the second period the learning rate was set to zero, effectively turning off backpropagation and allowing us to assess the networks performance in the absence on intra-trial learning.

Figure 3b shows the learning curve for the network: this is the average from 10 runs with different random initialization of the network's weights. As usual with backpropagation learning there was an initial rapid decline in errors over the first few trials followed by a more gradual decline. Figure 3c shows the performance of the network on the 100th trial. The controller was producing near optimal trajectories with average positional errors of under 1 cm. All ten networks produced similar performance after 100 trials. This is the result of accurate predictions -- the network was operating with gains 4 times greater than that in Figure 3a, and with feedback delays of 100msec rather than 30msec. Figure 3d shows the model's ability to adapt to changes in the arm dynamics. After initial training, a mass of 7.5kg was added to the forearm -- the weight was added on trial 10 and removed on trial 40. The learning curves show that the weight significantly increased tracking errors. However, the network soon adapted, and errors reduced. The performance could not return to the original value because the PD feedback gains were unchanged. Thus the controller was driving a heavier mass, and could not be expected to do as well. However, when the weight was removed, the performance again briefly deteriorated as the network relearned the original unweighted dynamics. This indicates that it was indeed adapting to the arm's dynamic behaviour, rather than undergoing some non-specific changes in the face of the added weight.

Learning the Time Delay

Like the arm's dynamics, physiological feedback time delays are not fixed throughout life; they change due to growth and axonal enlargement, and are modality and stimulus dependant. It is therefore reasonable to postulate that the model delay within the Smith Predictor must be adaptable and there are really only two pieces of information which could be used. First, it would be possible to estimate the time between sending out a motor command and perceiving the response via the delayed feedback pathways. The second possibility is to continually change the model delay and try to optimise some performance criteria. It turns out that for these sorts of tracking tasks, the performance is a smooth unimodal function of the model delay relative to the actual delay. So it is possible to perform gradient descent on this error curve to find its minimum, and hence the best value of the model delay.

The model's delay was therefore changed by 2msec every 5 trials. The direction of the change was dependant on the improvement in mean performance over the last 5 trials. If the last change led to an increase in performance, the next change in the model delay was in the same direction; if performance had deteriorated, the next change was in the opposite direction. In this way a crude gradient descent scheme was implemented.

Each change in the model delay had two effects, as can be appreciated from Figure 1. The first is that the dynamic model will learn to predict ahead by an amount equal to the model delay, whether or not this is the same as the actual delay. The second is that the comparator will no longer compare the predictions of sensory feedback in correct synchrony with the actual feedback, and inappropriate adjustments to the dynamic model will therefore be made on the strength of this false error signal.

So it is a potentially difficult problem to learn both the dynamic and time delay models at the same time -- the two models interact and mutually interfere. Our solution, based on evidence from motor psychophysics (Miall et al.,1990), was to have two very different learning rates for these two systems so that one system can fully adapt while the other changes little. To test this, synchronous learning of both the dynamics and the delay was simulated. The feedback delay was 100msec, but the model delay was set initially to 80msec, and the dynamic model neural network was started as usual with randomised weights. Figure 3e shows the time course of the changes in the model delay and Figure 3f shows the concomitant positional errors during tracking. The curves show how the tracking performance rapidly improved as the dynamic network and the model delay became more accurate. The model delay then overshot, leading to a slight decrement in performance, before settling around a mean value of 96msec. Because of our crude gradient descent algorithm, it could not stay at any one delay value indefinitely, but alternated up and down one delay increment.

Fitting the Model into Cerebellar Physiology

Our hypothesis requires that there should be separate dynamic and time delay neural models. The models should receive as inputs an efferent copy of the motor command being sent to the limb, and also proprioceptive information about the current state of the body. The latter is needed for an accurate internal representation of the limb, as the arm's mechanical properties depend on its position and motion. Hence the internal dynamic model must be updated by proprioceptors. The models should lie on feedback loops, so that their output feeds back onto the input as indicated in Figure 1. Finally, there must be mechanisms to allow the models to be adapted to predict accurately the behaviour of the controlled object, i.e. neural learning mechanisms.

We believe that the cerebellum is an obvious candidate site for these neural models (Miall et al., 1993a). Hence, sufficient information should be available to allow the cerebellum not only to generate but also modify each neural model. The dynamic responses of a limb change greatly during growth, while delays can change either because of increased axonal lengths, or following changes in the sensory responses (Deno et al., 1989; Wolpert et al., 1993). So to lay down useful internal representations of the motor system requires that the controller actively explores the environment to assess the outcome of its actions (Barto, 1989). The responses received back from the environment tell the controller not only about its effects on the external world but also about the controlled object, i.e. the arm. Thus, there are two processes to be learned: an early estimate of the outcome of actions and an estimate of the delay before actual feedback will be received; these must be learned on the basis of delayed signals from the periphery.

Climbing fibre input from the inferior olive may provide a training input to the cerebellar cortex (see Ito, 1984; or Strata, 1989). The climbing fibres should therefore signal the need for adaptation, in other words signal back the fact of a mismatch in predicted and actual feedback. A mismatch could mean that the prediction was inaccurate, or the effector's behaviour had changed, but in either case the models would require adaptation. Gilbert and Thach (1977) showed that the average rate of climbing fibre activity increased as monkeys adapted their movements to a novel load; Gellman, Gibson and Houk (1985) and Andersson and Armstrong (1987) have shown that the most potent stimulus for climbing fibres is an unexpected sensory event, and that similar "reafferent" sensory stimuli resulting from the animal's own actions do not excite the climbing fibres. There is now strong evidence that coincidence of climbing fibre and parallel fibre inputs to Purkinje cells results in long term depression of the parallel fibre:Purkinje cell synapse (Ito, 1989; Crepel and Jaillard, 1991). This may allow the cerebellar cortex to learn or modify a neural representation of the limb dynamics. These data therefore suggest that the inferior olive signals differences between the prediction of sensory re-afference and the actual reafference. In other words, we suggest that the delayed prediction from the cerebellum is fed to the inferior olive for comparison with sensory feedback signals. There are indeed inhibitory pathways direct from the cerebellar nuclei to the inferior olive and also indirectly via the red nucleus (Weiss *et al.*, 1990). Thus the inferior olivary signals should modify the dynamic model, to ensure that the inner loop of the Smith Predictor accurately mimics actual performance. However, the climbing fibre inputs probably cannot signal the quantitative size of the mismatch. Instead, they may signal the event of a mismatch, and perhaps also the direction of the error (Houk, 1990). Barto (1990) reviews techniques to train neural networks with this sort of reinforcement signal.

Thus the major role that we would attribute to the inferior olive is the comparison between expected and actual sensory signals (Comparator 3 in Figure 1). This is vital to provide a teaching signal for the cerebellum.

The second model of the Smith Predictor requires an output which is delayed to match the returning visual feedback, perhaps 150 - 250msec later. The size of the feedback time delay could be estimated by measuring the delay between issuing a motor command and assessing its result. This would be most easy to do if the motor command were discrete (Miall *et al.*, 1993b), as the reafferent signal would then change abruptly. The parallel fibres within the cerebellar cortex could act as a "tapped delay line" (Braitenberg, 1961), thus allowing the encoding of a time interval; Desmond and Moore (1988) propose that a chains of pontine nuclear cells do the same thing. However, we prefer the idea of using a predictive neural network trained to predict backwards in time: back-prediction of a signal is equivalent to delaying it. Thus, we suggest that both models within the Smith Predictor consist of predictive neural networks; the first model makes a forward prediction of the outcome of the movement. The second model makes a backward prediction, based on the output of the first model, and results in a delayed copy of the controller's actions.

The fact that the Smith Predictor contains two separate models requires training signals for both. We propose that the inferior olive provides one, allowing the dynamic model to be learned. The second training signal could be provided by the diffuse noradrenergic and serotonergic inputs from the locus coerulus and raphe nucleus. These could provide a "performance measure", reporting to the cerebellum on the overall success of the behaviour (Gilbert, 1975). We imagine that this measure could be something like a running average of positional errors in a tracking task, or the retinal slip accumulated over a few minutes in a VOR task. In support of this view, Van Neerven, Pompeiano, Collewijn and Van der Steen (1990) have shown that beta-noradrenaline can interfere with VOR adaptation in the rabbit, while D'Ascanio, Manzoni and Pompeiano (1991) has shown than noradrenaline-blockers reduce the gain of vestibulo-spinal reflexes. Thus the time delay model of the Smith Predictor might be trained with reinforcement learning on the basis of non-specific performance criteria (Barto, 1990), while the more specific signals provided by climbing fibres train the dynamic model.

These two learning schemes could have very different learning rates. We have shown that humans (Miall *et al.*, 1990) and monkeys (unpublished data) are very much faster to adapt to changes in the gain or load of a tracking manipulandum than they are to a change in its feedback delay. This would suggest that the dynamic model is rapidly modified within the cerebellum, whereas the temporal delay model is much slower to adapt. Deno *et al.*,(1989) have shown that oculomotor adaptation to feedback delays does occur over several days; and long experience with delayed feedback in tracking paradigms certainly improves performance (unpublished data). Thus adaptation to time delays does take place in primates, as would be expected from an adaptive Smith Predictor, but is slow. Hence, we propose that the dynamic model would adapt rapidly, driven by inferior olivary input, whereas the time delay model would adapt more slowly, driven either by the same inferior olive signal or by the noradrenergic or serotonergic inputs. The difference in learning rates may be functionally unimportant, as in every-day experience it is the dynamic behaviour of the motor system that changes rapidly, for example when carrying heavy objects, rather than the feedback delays.

Finally, the model we have presented here operates in kinematic coordinates, and we propose this as a model of the intermediate cerebellum. There may actually be two independent Smith Predictors within the cerebellum, and we suggest the second one would operate in egocentric coordinates and be situated in the lateral cerebellum (Miall *et al.*, 1993a). The lateral cerebellum forms a link between visual association areas, especially the posterior parietal cortex, and the motor and premotor cortices. This cerebro-cerebellar pathway may well be the major route by which visual information reaches the cortical motor areas for the guidance of the limbs (Stein and Glickstein, 1992). If this route was to contain a Smith Predictor, it would serve to transform a movement command (an instruction to reach a desired goal) specified by the posterior parietal cortex in visual, egocentric coordinates into a motor control signal and transfer it to the motor cortex. This then provides a common computational role for both lateral and intermediate cerebellum, and can explain many of the symptoms of cerebellar damage which are so obvious during visually guided movement.

ACKNOWLEDGEMENTS. We would like to thank the Wellcome Trust, the Medical Research Council, and the McDonnell Pew Centre for Cognitive Neuroscience for their support.

REFERENCES

Andersson, G. and Armstrong, D.M. (1987). Complex spikes in Purkinje cells in the lateral vermis (b zone) of the cat cerebellum during locomotion. *Journal of Physiology (London)*, **385**, 107-134.

Barto, A.G. (1989). From chemotaxis to cooperativity: Abstract exercises in neuronal learning strategies. In R. Durbin, R.C. Miall, and G. Mitchison (Eds.), *The computing neuron* (pp. 73-98). Wokingham: Addison-Wesley.

Barto, A.G. (1990). Connectionist Learning for Control: An Overview. In T. Miller, R.S. Sutton, and P.J. Werbos (Eds.), *Neural networks for control* (pp. 5-58). Cambridge, Mass: MIT Press.

Braitenberg, V. (1961). Functional interpretation of cerebellar histology. *Nature*, **190**, 539-540.

Crepel, F. and Jaillard, D. (1991). Pairing of pre- and postsynaptic activities in cerebellar Purkinje cells induces long-term changes in synaptic efficacy in vitro. *Journal of Physiology (London)*, **432**, 123-141.

D'Ascanio, P., Manzoni, D., and Pompeiano, O. (1991). Changes in gain of vestibulospinal reflexes after local injection of Beta-adrenergic substances in the cerebellar vermis of decerebrate cats. *Acta Oto-Laryngologica*, **111**, 247-250.

Deno, D.C., Keller, E.L., and Crandall, W.F. (1989). Dynamical neural network organization of the visual pursuit system. *IEEE Transactions on Biomedical Engineering*, **36**, 85-92.

Desmond, J.E. and Moore, J.W. (1988). Adaptive timing in neural networks: The conditioned response. *Biological Cybernetics*, **58**, 405-415.

Gellman, R.S., Gibson, A., and Houk, J.C. (1985). Inferior olivary neurons in the awake cat: detection of contact and passive body displacement. *Journal of Neurophysiology*, **54**, 40-60.

Gilbert, P.F.C. (1975). How the cerebellum could memorise movements. *Nature*, **254**, 688-689.

Gilbert, P.F.C. and Thach, W.T. (1977). Purkinje cell activity during motor learning. *Brain Research*, **128**, 309-328.

Haggard, P.N., Miall R.C., Wade, D.T., Anslow, P., Renowden, S., Fowler, S. and Stein, J.F. (1994) Damage to cerebellocortical pathways following closed head injury: an MRI and behavioural study. *Neurosurgery and Psychiatry* (in Press).

Houk, J.C. (1990). Role of cerebellum in classical conditioning. *Society of Neuroscience Abstracts*, **16**, 205.8 (Abstract)

Houk, J.C., Singh, S.P., Fisher, C. and Barto A.G. (1990) An adaptive sensorimotor network inspired by the anatomy and physiology of the cerebellum. In T. Miller, R.S. Sutton, and P.J. Werbos (Eds.), *Neural networks for control* (pp. 301-348). Cambridge, Mass: MIT Press.

Ito, M. (1984). *The cerebellum and neural control*. New York: Raven Press.

Ito, M. (1989). Long-term depression. *Annual Review of Neuroscience*, **12**, 85-102.

Jordan, M.I. and Rumelhart, D.E. (1992). Forward models: Supervised learning with a distal teacher. *Cognitive Science*, **16**, 307-354.

Miall, R.C., Kerr, G.K., Wolpert, D.M., and Forsyth, D. (1990). Adaptation to task dynamics and visual feedback in human visually guided movements. *Neuroscience Letters Supplement*, **38**, S51.

Miall, R.C., Weir, D.J., Wolpert, D.M. and Stein, J.F. (1993a). Is the cerebellum a Smith Predictor? *Journal of Motor Behaviour*, **25**, 203-216.

Miall, R.C., Weir, D.J., and Stein, J.F. (1993b). Intermittency in human manual tracking tasks. *Journal of Motor Behavior*, **25**, 53-63.

Miall, R.C., Haggard, P.N., and Stein, J.F. (1993c) Visuo-motor pursuit movements in ataxic patients with cerebellar damage. *Journal of Physiology (London)*, **473**, 24P.

Rumelhart, D. and McClelland, J. (1986) *Parallel distributed processing*. MIT Press, Cambridge Mass.

Smith, O.J.M. (1959). A controller to overcome dead time. *ISA Journal*, **6**, 28-33.

Stein, J.F. and Glickstein, M. (1992) The role of the cerebellum in the visual guidance of movement. *Physiological Reviews*, **72**, 967-1017.

Strata, P. (1989). *The olivocerebellar system in motor control*. Berlin: Springer-Verlag.

Van Neerven, J., Pompeiano, O., Collewijn, H., and Van der Steen, J. (1990). Injections of beta-noradrenergic substances in the flocculus of rabbits affect adaptation of the VOR gain. *Experimental Brain Research*, **79**, 249-260.

Weiss, C., Houk, J.C., and Gibson, A.R. (1990). Inhibition of sensory responses of cat inferior olive neurons produced by stimulation of the red nucleus. *Journal of Neurophysiology*, **64**, 1170-1185.

Wolpert, D.M., Miall, R.C., Cumming, B.C. and Boniface, S.J. (1993). Retinal adaptation of visual processing time delays. *Vision Research*, **33**, 1421-1430.

Wolpert, D.M. (1992) *Overcoming time delays in visuo-motor control.* D.Phil. Thesis, Lincoln College, Oxford.

SIGNALLING PROPERTIES OF DEEP CEREBELLAR NUCLEI NEURONES

J.M. DELGADO-GARCIA and A. GRUART

Laboratorio de Neurociencia
Facultad de Biología
Universidad de Sevilla
41012-Sevilla, Spain

SUMMARY

The firing activity of identified deep cerebellar nuclei neurones was recorded in alert cats during experimentally-induced eyelid movements. *Type A* neurones increased their discharge rate coinciding with the beginning of reflex blinks, regardless of the stimulus modality applied (air puffs, flashes or tones). The increased activity was modulated by lid position during the blink. *Type B* neurones fired a brief burst of spikes before the blink, followed by a decrease in their firing rate.

An experimental simulation of afferent neural signals to nuclear areas was carried out by electrical stimulation of the appropriate areas of the pontine nuclei and the inferior olive. The amplitude of the synaptic field potentials induced in deep cerebellar nuclei following inferior olive electrical stimulation was modulated by conditioning stimuli in the pontine nuclei or by different sensory stimulations. A similar modulation of the synaptic field potential amplitude was observed during the acquisition of an eyelid response during a classical conditioning paradigm. The present results suggest the involvement of afferent inputs on cerebellar nuclear neurones during eyelid responses to novel stimuli.

INTRODUCTION

Deep cerebellar nuclei neurones are the main target of Purkinje cells and may contribute in a still unknown manner to modulate the activity of the overlying cortex (Ito, 1984; Thach *et al.*, 1992). Nuclear neurones are monosynaptically inhibited by Purkinje cells (Ito *et al.*, 1970) and excited from axon collaterals of mossy and climbing fibers projecting to the corresponding areas of the cerebellar cortex (Courville *et al.*, 1977; Llinás and Mühlethaler, 1988; Shinoda *et al.*, 1992). In turn, their axons represent the almost exclusive output of the cerebellum. Nuclear cells are thus a modulating link between cerebellar cortex and the rest of the central nervous system.

The eyelid/nictitating membrane response is a widely used paradigm in the study of the genesis and control of unconditioned and conditioned motor responses (see references in Welsh and Harvey, 1992). It has been proposed that putative interpositus neurones in the rabbit are activated by sensory cues (Berthier and Moore, 1990) or that their firing is correlated with the time-course of conditioned nictitating membrane responses (Berthier *et al.*, 1991). In contrast, other authors propose that interpositus neurones show spike activities time-locked to ongoing movements (Armstrong and Edgley, 1984; Thach *et al.*, 1992).

Although the rostral interpositus nucleus has been proposed as the site where the memory trace for classical conditioning is formed and stored (Thompson, 1988), basic information is still needed about the normal function of cerebellar nuclear neurones during reflex blinks as a

Neural Control of Movement, Edited by W.R. Ferrell
and U. Proske, Plenum Press, New York, 1995

basis to compare with the (expected) changes during the acquisition of conditioned eyelid/nictitating membrane responses. The first aim of the present work has been to study the discharge of identified nuclear cells during experimentally-induced reflex blinks in the alert cat. However, unitary recordings have some technical limitations, particularly for the long-term recordings required during the acquisition of new motor skills. Consequently, an attempt has been made to study the characteristics of the synaptic field potentials induced in nuclear areas by inferior olive stimulation during the induction of unconditioned and conditioned eyelid movements.

MATERIAL AND METHODS

Experiments were carried out in six adult female cats obtained from an authorized supplier. All manipulations were carried out according to the guidelines of the European Communities Council (86/609/EEC) and of the current Spanish legislation on the use of laboratory animals in chronic experiments.

Under general anaesthesia (35mg/kg of sodium pentobarbital and 0.4mg/kg of atropine sulfate), animals were implanted with search coils on the scleral margin of the left eye and into the lower margin of both upper lids. Following stereotaxic coordinates from Berman's atlas (1968), animals were also implanted with stimulating electrodes on the left VIth nerve and in the right divisions of the magnocellular red nucleus, dorsolateral pontine nucleus, medial longitudinal fascicle close to the oculomotor complex, restiform body and inferior olive. Animals were also provided with a head-holding system for stabilization of the unitary recordings. A hole was drilled in the left occipital bone to allow access to the recording sites through a transcerebellar approach. Recording sessions were carried out for 3h on alternate days, beginning two weeks after surgery. The animal was lightly restrained with an elastic bandage, mounted on the recording table and its head immobilized by attaching the head-holding system to a bar fixed to the table. Further details of this chronic preparation have been published elsewhere (Delgado-García et al., 1990; Gruart et al., 1993).

Extracellular unitary activity was recorded with glass microelectrodes filled with 2M NaCl of 3-6Mohms, while field potentials were recorded with micropipettes of a lower resistance (1-3Mohms). Neuronal electrical activity was filtered in a bandwidth from 10Hz to 10kHz. Electrical stimuli were cathodal 50μs square pulses of <0.15mA. Recorded units were identified by their antidromic activation from their projection sites. The collision test was used systematically to determine whether the recorded and the activated unit was the same. The electrical activity of the orbicularis oculi muscle was recorded with bipolar hook electrodes implanted close to the external canthus of the left eye. Eye and eyelid movements were recorded with the magnetic search-coil technique. Unitary activity and/or field potentials were recorded during presentation of the following stimuli: i) 20-100ms air puffs (0.5-3kg/cm^2) directed to the left cornea; ii) bright full-field xenon flashes; iii) 90 db tones of different pitch (600 or 6,000Hz) and duration (10-350ms); iv) horizontal optokinetic ramp stimuli at 10-30 deg/s; and v) vestibular sinusoidal stimulation in the horizontal plane at 0.5Hz. Classical conditioning of the eyelid response was achieved by either: i) the presentation of a 350ms tone of 600Hz and 90db followed 250ms later by a 100 ms air puff of 2kg/cm^2 directed to the left eye, or ii) the presentation of a 25ms air puff of 0.8Kg/cm^2 followed 250ms later by a 100ms air puff of 3kg/cm^2, both of them directed to the left eye.

Neuronal activity, eye and eyelid movements, square pulses corresponding to blink-evoking stimuli and planetarium and servo-controlled turntable position outputs were stored digitally on a video recording system for off-line analysis. Records corresponding to antidromic activations, collision tests and field potentials were printed on a X-Y plotter for latency and amplitude measurements of their voltage profiles. Recorded data were transferred to a computer for analysis. Computer programs were developed to display peri-stimulus time histograms of neural activity and the average of eye and eyelid position and velocity. Antidromic and synaptic field potentials were also averaged when needed.

At the end of the recording sessions, animals were deeply reanesthetized (50mg/kg of sodium pentobarbital) and electrolytic marks were placed at selected recording sites and at all the stimulation points. Routine histological procedures were followed to identify stimulation and recording sites.

RESULTS

Characteristics of Eyelid Movements

Eyelid movements in response to long (>25 ms) air puffs consisted of a succession of 1-4 downward movements. The early downward displacement presented a latency of 16.2 ±3 ms and lasted 20.1 ± 4ms. The late sags (see unconditioned responses in Fig. 1B) lasted for about 40 ms with a constant latency in their beginning as they could still be noticed in averaged records (Fig. 1A). Blink responses to flashes and tones showed longer latencies (52.5 ± 5.3 and 50.3 ± 6.1ms, respectively) and a random presence of the late sags.

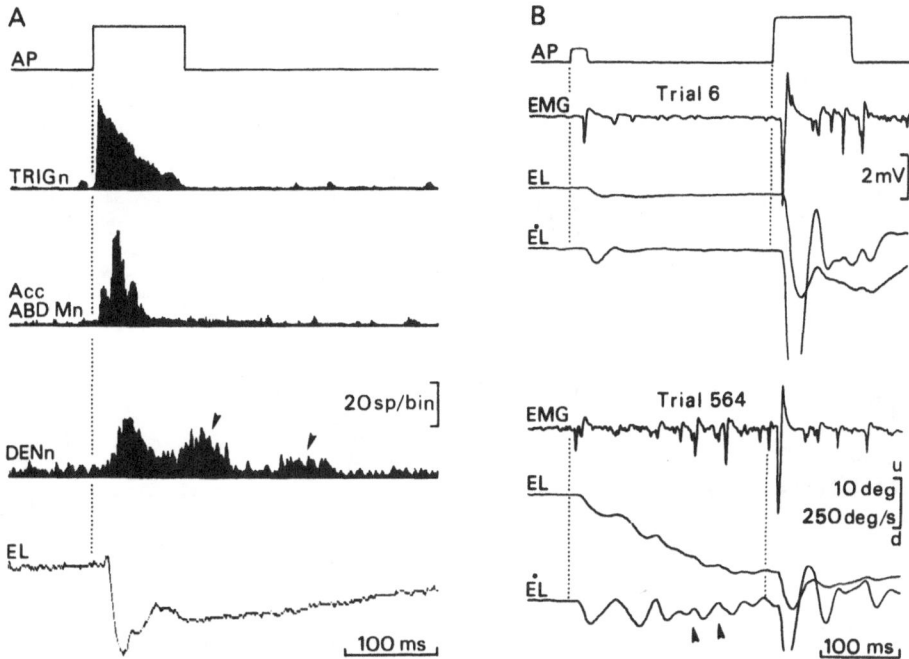

Figure 1. A. Peri-stimulus time histograms of the spike activity of a spinal trigeminal neurone (TRIGn), an accessory abducens motoneurone (Acc ABD Mn) and a type A dentate neurone (DENn) in response to repeated (n=50) corneal air puff (AP, 100ms, 2kg/cm^2) stimuli. The averaged eyelid movement (EL) is also shown. Arrowheads point to the late discharge components of the DENn. Records were obtained in different recording sessions. B. Evolution of an eyelid conditioned response following repeated presentations of a short (25ms), weak (0.8kg/cm^2) AP followed by a stronger corneal stimulus (100ms; 3Kg/cm^2). From top to bottom are shown single EMG records of the orbicularis oculi muscle and eyelid position (EL) and velocity (E'L) corresponding to the indicated trials. Note the oscillatory nature (arrowheads) of the acquired motor response.

Unitary Activity

Nuclear *type A* neurones (n=55) were activated antidromically from the red nucleus (50%), the restiform body (30%) or the medial longitudinal fascicle (20%) and increased their spike activity in coincidence with the beginning of eye blinks regardless of the stimulus modality, i.e., air puffs, flashes or tones. Their electrical response mirrored the variations of eyelid movements with the presentation of different blink-inducing sensory stimuli. The variations in latency and/or lid movement produced by repeated presentation of the same stimulus were followed by similar changes in the latency and/or intensity of the neuronal discharge. As shown in Fig. 1A, the discharge profile of type A neurones was different from that presented by putative second order trigeminal neurones, as the latter fired in a continuous but decreasing manner in response to a sustained puff of air, with no activity after stimulus removal. In contrast, identified accessory abducens motoneurones fired during the early stages of eyelid movements, indicating that the late downward movements of the upper lid was not the (indirect) result of the action of the retractor bulbi muscle, but mainly that of the orbicularis oculi muscle (see EMG traces in Fig. 1B).

Nuclear *Type B* neurones (n=18) were activated antidromically from the red nucleus (60%), restiform body (25%) or the medial longitudinal fascicle (15%) and decreased their discharge rate during reflexly-induced blinks. Some of these neurones presented a sharp increase in firing slightly preceding the blink, followed by the indicated decrease in their discharge rate. No further modulation in the electrical activity of type B neurones during the late components of the eyelid response was noticed. This neuronal type was not found in the fastigial nucleus.

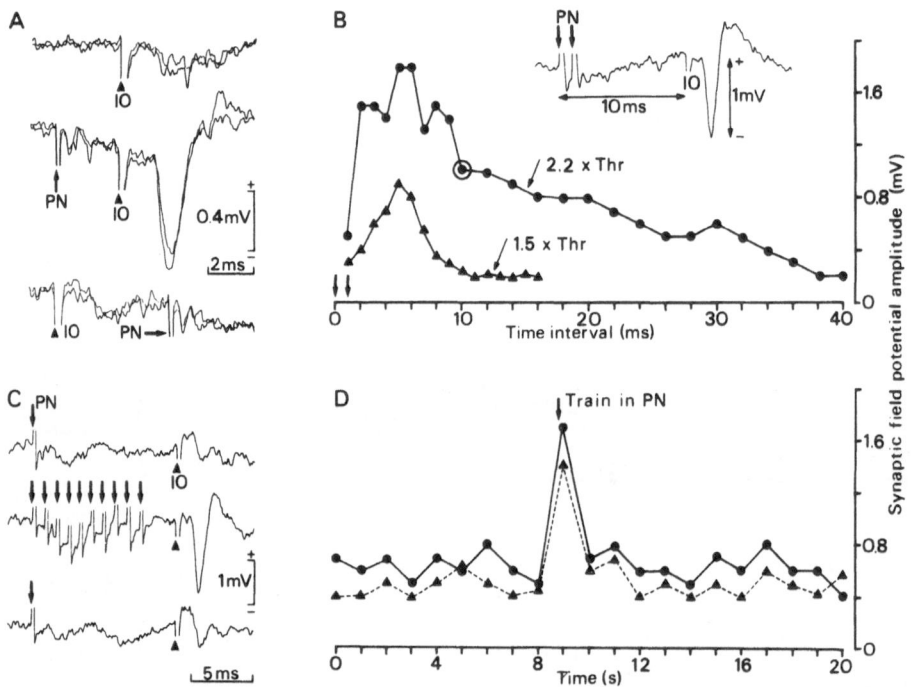

Figure 2. A. Top records: Synaptic field potentials induced in the fastigial nucleus following inferior olive (IO) stimulation. Middle records: Effects of a single conditioning stimulus applied in the pontine nuclei (PN). Bottom records: Result of reversing the order of the applied stimuli. B. Effects of double pulse conditioning stimuli in PN on the amplitude of the synaptic field potentials induced by IO stimulation in the fastigial nucleus. The record in the inset corresponds to the encircled dot. C. Effects of a train of stimuli applied in PN at 1.5 x Thr. on the amplitude of the synaptic field potential induced by IO stimulation at the fastigial nucleus. The top record was obtained 1s before and the bottom record 1s after the train was applied (middle record). D. Session evolution for the experimental paradigm shown in C.

Field Potential Analysis

As long-term recordings of identified nuclear cells presented several technical difficulties, we decided to study the variability in the effects of inferior olive stimulation on the activation of deep cerebellar nuclear neurones during different experimental paradigms involving the presence of unconditioned and conditioned eyelid responses.

Control field potentials induced in cerebellar nuclei by inferior olive stimulation consisted of two negative waves at 0.5-1 ms and 2-3 ms followed by a late (4-6) positive wave. Field potential deep profile analysis and unitary recordings showed that the two negative waves corresponded, respectively, to the antidromic and synaptic activation of cerebellar nuclei neurones, while the delayed positivity was the result of IPSPs generated by Purkinje cells on subjacent nuclear neurones (see also Llinás and Mühlethaler, 1988). The first negative component was not present in all the recording sites in which the second negativity was still recorded (Fig. 2A, top). The presentation of a novel stimulus (silhouettes, flashlights, sounds) to the alert animal increased by 5-10 times the amplitude of the second negativity and, although less noticeably, that of the delayed positivity. This increase in synaptic field potential amplitude disappeared following repeated presentation of the (same) stimulus. It was noticed that during 1/s presentation of flashes of light the increase in amplitude of evoked synaptic field potentials was restricted to the 50-80 ms following the stimulus, i.e., during the time period in which the eyelid response to the flash was being produced. In the fastigial nucleus, the amplitude of the second negativity was modulated 120-130 deg in advance to eye position during sinusoidal horizontal stimulation of the animal at 0.5 Hz, increasing for contralateral and decreasing for ipsilateral head rotation. The opposite results were obtained during optokinetic stimulation, i.e., an increase in the amplitude of the second negativity during planetarium rotation toward the ipsilateral side and a decrease (or no change) during stimulation toward the contralateral side. The discharge of single identified nuclear units in the cat follows the populational firing profile suggested by field potential analysis (Gruart and Delgado-García, in preparation). Simultaneous recording at different cerebellar nuclear sites during sensory stimulation in presence of inferior olive stimuli at 1 Hz demonstrated that the reported variability in evoked field potential amplitude was specific for both stimulus modality and recording site.

Interestingly, a similar increase in the amplitude of the evoked synaptic field potential was induced in the alert animal by the presentation of a 50 μs conditioning stimulus in the pontine nuclei prior to inferior olive stimulation (Fig. 2A, B). The effect of the conditioning stimulus was removed by a low dose (5 mg/kg) of Ketamine anaesthesia. The conditioning effects of pontine nucleus stimulation were in the range of ms (Fig. 2B). We did not succeed in maintaining the increased synaptic field potential amplitude following trains of up to 1,000 Hz applied in pontine nuclei at different time intervals prior to inferior olive stimulation (Fig. 2C, D).

As shown in Fig. 3 (A, C and E), both the second negativity and the delayed positivity induced in the fastigial nucleus following single pulses applied to the inferior olive increased up to 6 times in amplitude during the acquisition of a conditioned eyelid response in a classical conditioning paradigm. This increase in synaptic field potential amplitude was restricted to the 200 ms time-window preceding presentation of the unconditioned stimulus. As shown in Fig. 3 (B, D and F), when the inferior olive stimulus was applied outside this time-window no significant changes were observed in the amplitude of the second negativity of the evoked field potential. Finally, the conditioned responses disappeared and the amplitude of the evoked synaptic field potential returned to control values during the extinction of the learned motor response (Fig. 3G).

DISCUSSION

The present results demonstrate that the discharge of identified cerebellar nuclear neurones is related to the characteristics of ongoing movements of the eyelid. Changes in the modality, intensity, frequency and duration of applied stimuli were accompanied by changes in neuronal electrical responses coinciding with changes in the evoked eyelid movements. The reciprocal firing properties of type A and B neurones support an agonistic-antagonistic interplay of neural messages shaping the activation profiles of motoneuronal units involved in the actual eyelid displacement. These findings also support previous reports suggesting that although some dentate neurones in the monkey are stimulus-related and may participate in the initial triggering mechanisms involved in motor responses to sensory cues (Chapman et al., 1986), most types of nuclear neurones present discharge rates time-locked to ongoing movements (Armstrong and Edgley, 1984; see further ref. in Thach et al., 1992).

The downward excursion of the eyelid induced by long air puffs was shown here to be composed of successive steps of a constant latency (Gruart et al., 1993). Whether type A neurone responses during the late sags of reflex blinks are the result of a damping mechanism to decrease terminal tremor (Thach et al., 1992) or a contributing element of the eyelid position signal generation system remains to be elucidated. In any case, both the discharge of type A neurones and the actual trajectory of reflexly-induced blinks can be considered as examples of the non-continuous nature of movement execution (Llinás, 1991). Repeated presentation of the same stimulus at 1 Hz to an alert animal may synchronize the olivo-cerebellar system (Llinás and Sasaki, 1989) timing the activity of nuclear neurones, i.e., the main output of the cerebellum. Late oscillations evoked in reflex blinks were not observed during spontaneous blinks, because these movements are programmed centrally and because

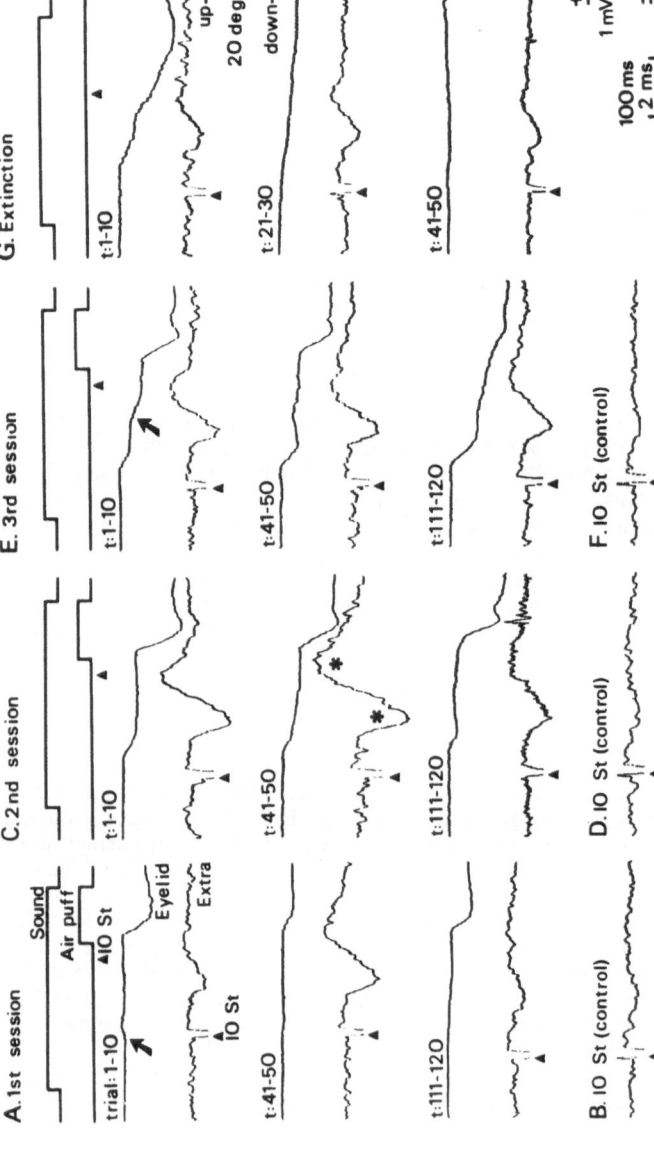

Figure 3. Evolution of the motor eyelid response during a classical conditioning paradigm. Tone was 600Hz at 90db. Air puff was 2 kg/cm^2 directed to the left cornea. Records in **A,C** and **E** were obtained from the first three conditioning sessions. Eyelid movements and field potentials recorded in the fastigial nucleus were averaged (n=10) during the trials indicated. Time calibrations for blinks (100ms) and field potentials (2ms) are shown. Inferior olive (IO) stimulation was applied at the moment indicated by arrowheads. **B, D** and **F** correspond to averaged (n=10) records obtained in response to IO stimuli at random times between conditioning trials. **G:** Data obtained during an extinction session. Arrows point to the beginning (A) and to the appearance of a second time-fixed component (E) of the eyelid conditioned response. Asterisks indicate the moment of maximum amplitude in the evoked synaptic field potential.

they are not synchronized by repetitive stimuli. The strength of the late components of reflex blinks at the neuromuscular level (Fig. 2B) was similar to that of conditioned eyelid responses and may explain why they are more easily removed following cerebellar lesions (Welsh and Harvey, 1992). The lower variability and the higher strength of the early downward eyelid response suggest that it is generated by brain stem circuits, independent of direct cerebellar control.

According to the present results, and besides the well-known inhibitory inputs from Purkinje cells, excitatory inputs from collaterals of mossy and climbing fibre afferents have powerful regulatory effects on the firing properties of deep cerebellar nuclear neurones. The conditioning effects of mossy fibre afferents on subsequent stimuli applied to the inferior olive suggest the presence at the nuclear cell membrane of specific postsynaptic mechanisms able to recognize the temporal order of arrival of electrical signals through both afferent systems. The ability of nuclear neurones to generate rebound spike bursts mostly when the cells are hyperpolarized, i.e., following IPSPs from Purkinje axon terminals has been described (Llinás and Mühlethaler, 1988). However, it is still possible that such low-threshold spike bursts can also be triggered by other parameters not easily controlled in an *in vitro* preparation.

As sensory cues of different modalities can successfully substitute for the direct electrical stimulation of pontine nuclei it can be suggested that the powerful excitatory mechanism present in nuclear cells has to be of a definite functional relevance in the genesis of output signals from cerebellar nuclei to the rest of the central nervous system. The increase of the excitatory influence of inferior olive inputs on nuclear cells during attentive and orienting movements as well as during the acquisition of new motor skills further reinforces the hypothesis of the participation of olivary-cerebellar circuits in the generation and control of movements.

REFERENCES

Armstrong, D.M. and Edgley, S.A. (1984). Discharges of nucleus interpositus neurones during locomotion in the cat. *Journal of Physiology* **351**, 411-432

Berman, A.L. (1968). *The brain stem of the cat. A cytoarchitectonic atlas with stereotaxic coordinates.* The University of Wisconsin Press, Madison (Wisconsin)

Berthier, N.E., Barto, A.G. and Moore, J.W. (1991). Linear systems analysis of the relationship between firing of deep cerebellar neurones and the classically conditioned nictitating membrane response in rabbits. *Biological Cybernetics* **65**, 99-105

Berthier, N.E. and Moore, J.W. (1990). Activity of deep cerebellar nuclear cells during classical conditioning of nictitating membrane extension in rabbits. *Experimental Brain Research* **83**, 44-54

Chapman, C.E., Spidalieri, G. and Lamarre, Y. (1986). Activity of dentate neurones during arm movements triggered by visual, auditory and somesthetic stimuli in the monkey. *Journal of Neurophysiology* **55**, 203-226

Courville, J., Augustine, J.R. and Martel, P. (1977). Projections from the inferior olive to the cerebellar nuclei in the cat demonstrated by retrograde transport of horseradish peroxidase. *Brain Research* **130**, 405-419

Delgado-García, J.M., Evinger, C., Escudero, M. and Baker, R. (1990) Behavior of accessory abducens and abducens motoneurones during eye retraction and rotation in the alert cat. *Journal of Neurophysiology* **64**, 413-422

Gruart, A., Zamora, C. and Delgado-García, J.M. (1993). Response diversity of pontine and deep cerebellar nuclei neurones to air puff stimulation on the eye in the alert cat. *Neuroscience Letters* **152**, 87-90

Ito, M. (1984). *The cerebellum and neural control.* Raven Press, New York

Ito, M., Yoshida, M., Obata, K., Kawai, W and Udo, M. (1970). Inhibitory control of the intracerebellar nuclei by the Purkinje cell axons. *Experimental Brain Research* **10**, 64-80

Llinás, R. (1991). The noncontinuous nature of movement execution. In: "Motor Control: Concepts and Issues" eds. D.R. Humphrey and H.-J. Freund, 223-242. John Wiley and Sons, New York

Llinás, R. and Mühlethaler, M. (1988). Electrophysiology of guinea-pig cerebellar nuclear cells in the *in vitro* brain stem-cerebellar preparation. *Journal of Physiology* **404**, 241-258

Llinás, R. and Sasaki, K. (1989). The functional organization of the olivo-cerebellar system as examined by multiple Purkinje cell recordings. *European Journal of Neuroscience* **1**, 587-602

Shinoda, Y., Sugiuchi, Y., Futami, T. and Izawa, R. (1992). Axon collaterals of mossy fibers from the pontine nucleus in the cerebellar dentate nucleus. *Journal of Neurophysiology* **67**, 547-560

Thach, W.T., Kane, S.A., Mink, J.W. and Goodkin, H.P. (1992). Cerebellar output: multiple maps and modes of control in movement coordination. In: *The cerebellum revisited* eds. R. Llinás and C. Sotelo, 283-300. Springer-Verlag, New York

Thompson, R.F. (1988). The neural basis of basic associative learning of discrete behavioral responses. *Trends in Neuroscience* **11**, 152-155

Welsh, J.P. and Harvey, J.A. (1992). The role of cerebellum in voluntary and reflexive movements: History and current status. In: *The cerebellum revisited* eds. R. Llinás and C. Sotelo, 301-334. Springer-Verlag, New York

VISUAL INPUT TO THE LATERAL CEREBELLUM

D.E. MARPLE-HORVAT

Department of Physiology
University of Bristol
Bristol, BS8 1TD, England

SUMMARY

We have recorded the activity of lateral cerebellar neurones (both cortical and nuclear) in cats resting and walking on a horizontal ladder. A substantial proportion (40%) of the cells tested are visually responsive which suggests that the lateral cerebellum is important in visuomotor control. Furthermore, the visual responses of cerebellar neurones can sometimes be context dependent, and the same cell can show later discharge changes which are motor in character. This means that the cerebellar contribution can be more complex than just simply signalling visual events, perhaps indicating something about the significance of the event and whether a motor response is required, and if so how long there is to make it.

INTRODUCTION

When we talk about visuomotor control - how visual inputs give rise to motor outputs - clearly there are at least two dimensions. There is an anatomical dimension and a physiological dimension. I am not going to attempt to add anything to Professor Glickstein's review of the anatomy. The question I shall address is this: when we come to the cerebellum and make recordings of the neurones in awake, behaving animals, what do their discharges represent, what are they saying, situated as they are maybe halfway along the computational chain from visual inputs to motor outputs. In this report I shall describe single unit recordings recently obtained in awake, behaving cats, and compare them with earlier work with monkeys trained to perform a visually guided tracking task.

DISTRIBUTION OF VISUALLY RESPONSIVE NEURONES

Figure 1A, which is taken from Snider and Stowell (1944) shows where they managed to record visually evoked potentials in response to a brief flash of light in anaesthetised cats. The circle in this Figure shows where we made our first recordings, and we have since ranged more widely through most of lobulus simplex, crus I and the underlying lateral cerebellar nucleus. In our awake cats we tested 112 cells with one of two visual stimuli, either a brief full field flash of light or a rung which moves as the walking cat approaches. The cat subsequently steps up onto and over this displaced rung.

Of the 112 cells that we tested 45 gave a short latency response to one of these two stimuli. The stereotaxic locations of these cells are shown in Figure 1B. These have been calculated from the known coordinates of the microelectrode penetration at the surface of the cerebellum, the angle of the track and the depth at which the cell was recorded. The main finding is that about four-fifths of these visually responsive neurones were more than 7mm

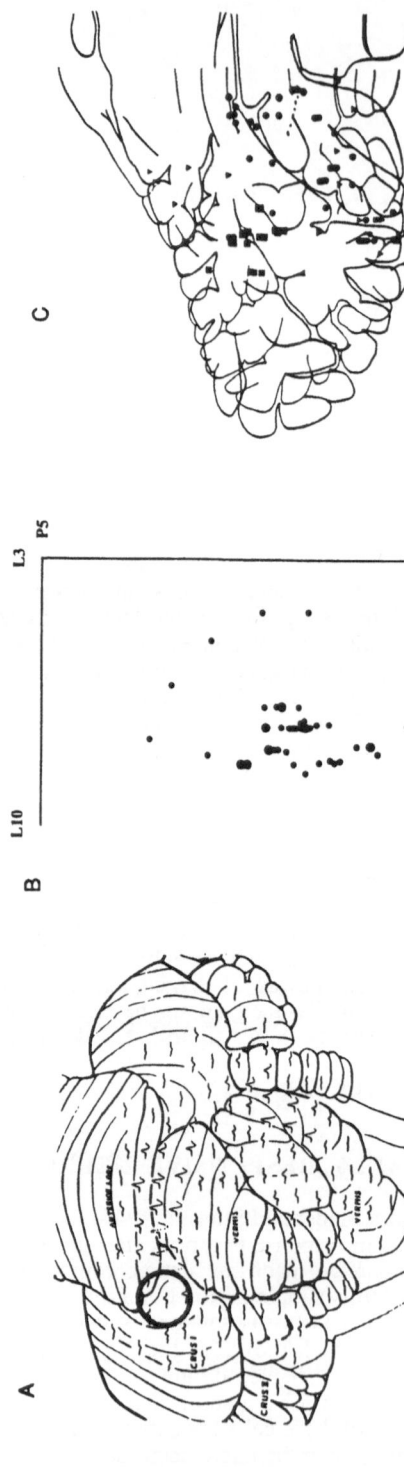

Figure 1. A: Visual evoked potentials in anaesthetised cats, with (superimposed) location of our initial recordings in the awake animal. **B:** Stereotaxic locations of visually responsive neurones (triangles) in the monkey plotted against two coronal cerebellar sections.

234

away from the midline, which means that the cortical cells were hemispheral (rather than paravermal) and the nuclear cells within the lateral cerebellar nucleus (rather than interpositus). We have attempted to find visual responses more medially in this and a parallel study but have found very few. These cells are both cortical and nuclear, these have not been distinguished in this Figure. We can compare this distribution in the cat with the monkey cerebellum which is shown in Figure 1C. Different symbols represent cells with different functional characteristics; the triangles are visually responsive cells which are distributed fairly widely but in highest concentration in dorsal paraflocculus. We have not looked at this cortical area in the cat but from this and the anatomical evidence it seems that it would be a very fruitful place to look for visually responsive cells. In addition, Mushiake and Strick (1992) recently reported preliminary findings recording in dentate nucleus where as many as three quarters of dentate neurones whose discharges were identified as 'task dependent' showed altered activity during the reaction time in monkeys performing a visually guided tracking task (see also Chapman, Spidalieri and Lamarre, 1986).

RESPONSES TO FULL FIELD FLASH ILLUMINATION

Figure 2A shows 12 cells that were recorded in a fairly circumscribed area of cortex in lobulus simplex. Although there are some variations in these responses - two at least of them show decreased discharge as the main component - there are nevertheless some features in common. We have drawn a dotted line at a latency of 50ms after the stimulus (delivered at the time of the downwards arrow) and the maximum modulation tended to be close to this line. We can see the common features more clearly in Figure 2B. This is the pooled discharge histogram of those 12 cells and the onset of modulation is at 25ms with the peak being at 45-50ms. Taking a control period 200ms prior to delivery of the flash and calculating the mean and standard deviation values for the discharge level, the peak is about 16 standard deviations away from that mean level, so this is a strong response and represents a very clear signal. The response is also very brief, perhaps reflecting the fact that the stimulus was essentially instantaneous, the flash lasting about 10 microseconds. The response is on average either 1 or 2 spikes per stimulus, and the mean for all the cells that responded to the flash was 1.3 spikes per trial. The cell of Figure 2C could be another example from the cat, but in fact there is more cerebellum here and this is one of the monkey recordings whose cortical location is shown (right) by the spot (Marple-Horvat and Stein, 1990). The discharge of this cell was again modulated fairly early, starting around 25ms, and the peak is around 35ms, so it is similar to our recordings in the cat.

RESPONSES TO RUNG MOVEMENT

The discovery of such visual responses lends support to the candidacy of the lateral cerebellum as a visuomotor control structure; the flash proved very useful in probing for visual input and it was in fact an effective stimulus for cat visual pontine cells which Baker, Gibson, Glickstein and Stein (1976) recorded, which was part of the reason why we used it. But we also wanted to look at the responsiveness of neurones to a visual event to which a motor response was required. We have a horizontal circular ladder (Amos, Armstrong and Marple-Horvat 1987) and train the cats to walk around the rungs of this ladder which are set a comfortable distance apart (circa 20cm). Most of the rungs are fixed, but there are two special ones which we call pre-displaceable rungs. These can be made to move as the cat approaches, triggered by interruption (by either forelimb) of an infra-red beam in one of two locations (first sensor or second sensor). The cat is either three steps or two steps away from the pre-displaceable rung when it moves. What do we see in response to this visual stimulus to which the cat has to respond by stepping up or down (3 or 6cm) onto the displaced rung when it has moved?

We found 16 cells of 25 tested (a smaller number than the 99 tested against flash because it takes much longer to accumulate enough trials) - sixteen cells that responded to rung movement. An example is shown in Figure 3A. The response is slightly longer latency than to flash, at 65ms, and has a longer duration, 80ms. Although the initial increased firing rate takes the eye this represents only half a spike extra whereas the silencing of the neural discharge following it represents a more important loss of about 4 spikes per encounter. The more usual response seen was an increased discharge. Figure 3B shows the response of a

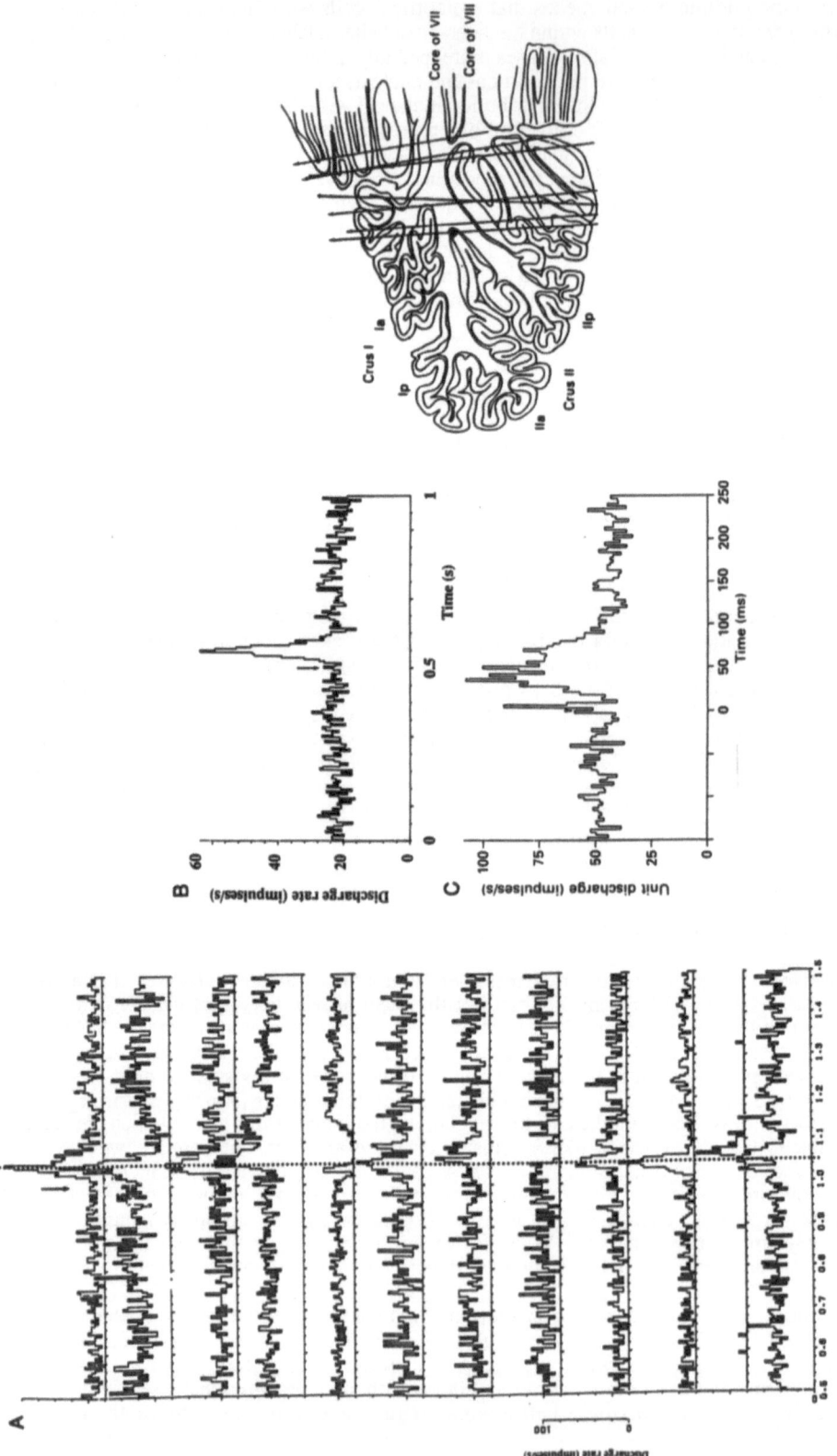

Figure 2. A: Response of twelve cortical neurones to a brief flash of light. **B:** Pooled discharge histogram of the cells shown in A. **C:** Similar response to flash obtained in the monkey (stimulus delivered at time zero, arrowed; stimulus artifact in next three bins).

cortical cell to flash (again all stimuli are delivered at the time of the downwards arrow). The response to flash represents two spikes per delivery. The next discharge histogram (3C) drawn to the same vertical scale (in impulses per second) to assist in making comparisons shows the response of this cell when the pre-displaceable rung moves as the cat approaches it, when he is closer to it at the second sensor, only two steps (40cm) away. The latency is broadly similar but the evoked activity lasts longer and in fact there is much more of it. There are now roughly 9 spikes per encounter with this moving rung. The third histogram for this cell (3D) is obtained by subtracting (B) from (C) to show the difference between the two. What the difference histogram shows is that just the earlier part of the response to rung movement has been reduced in size, but the peak is the same height and all of the later discharge is completely unaffected. So the discharge to rung movement is both bigger and lasts longer (perhaps because the stimulus - rung movement - lasts longer, about 150ms). We also tested this cell to rung movement when the cat was further away at the first sensor location. Figure 3E shows the response when he was further away, 3 steps (60cm) and this represents 6 spikes per encounter. Finally, the bottom histogram is a difference obtained by subtracting in this case 3E from 3D and it shows how much larger is the response when the cat is closer; again the difference is basically 3 spikes, and statistics applied to the difference histogram show that the difference is significant.

It is clear, therefore, that this cell and others like it can give a different response to the same event; the rung moves in exactly the same way but the context is different; the cat is either closer to or further away from the rung. The sensory context is different in that the image of the rung and the visual angle subtended by its movement will both be larger when the cat is closer; the motor context is different in that less time is available to modify the normal locomotor pattern. The former (size of visual angles) may well be the best information available to indicate the latter (distance away and hence time available to respond) when approaching a familiar object of known size.

In addition, some of these 16 cells also showed a much later modulation accompanying and related to the change in trajectory of the limb to step up or down onto the displaced rung. Arrival at the rung was approximately half a second at minimum after the rung had finished moving so there is a sizeable gap between these early latency visual responses and the subsequent modification of limb trajectory. So we have some recordings which show both visual character and motor character at different times.

This type of response where the same cell can react differently to the same visual event has been seen on a few occasions in the monkey as well. We used two ways of looking at the susceptibility of a cerebellar cortical neurone to a flash of light delivered at different times when a monkey was performing a visually guided tracking task (see Marple-Horvat and Stein 1990, Figure 7). At time 0 a target circle stepped from one location to another, and then over a one second period the monkey would make his response and move a manipulandum to bring a spot back inside the circle at its new location. We could deliver a flash at different times with different delays from the target movement and whether we looked at the number of spikes or how high the peak frequency went and how significant this was in terms of standard deviations away from the mean discharge level prior to the flash, we found a pattern of greater susceptibility at some times than others. At about 250ms after target movement the flash was most effective in provoking a response and away from this interval the response became progressively smaller.

There is a great deal of interest in complex spikes, but we have encountered problems in trying to say something about what the complex spike discharges of Purkinje cells are doing in walking cats. Even in a neurone which we held for forty five minutes (which is the longest that we have managed) we have experienced difficulty in discriminating the complex spike discharges (which are not easily separable from the accompanying simple spike train). In a cell which clearly showed a decrease in simple spike output following the flash with a short latency, we wondered whether there was any significant increase in the accompanying complex spike activity. Using our normal criterion for significance of two adjacent bins more than two standard deviations away from the mean, this was not the case. Modulation of complex spikes was only significant at the lower level of one standard deviation. With more trials this modulation might have become more clear. The situation when walking was even worse; several hundred deliveries were made of the flash but there were something like 20 or 30 encounters with the displaceable rung. Although it looked like there was a suggestion of some increased complex spike discharge in response to that rung movement again it failed on our usual criterion of two adjacent bins being significant; and it was not at all clear whether the simple spike train showed any decreased frequency around the same time. We remain

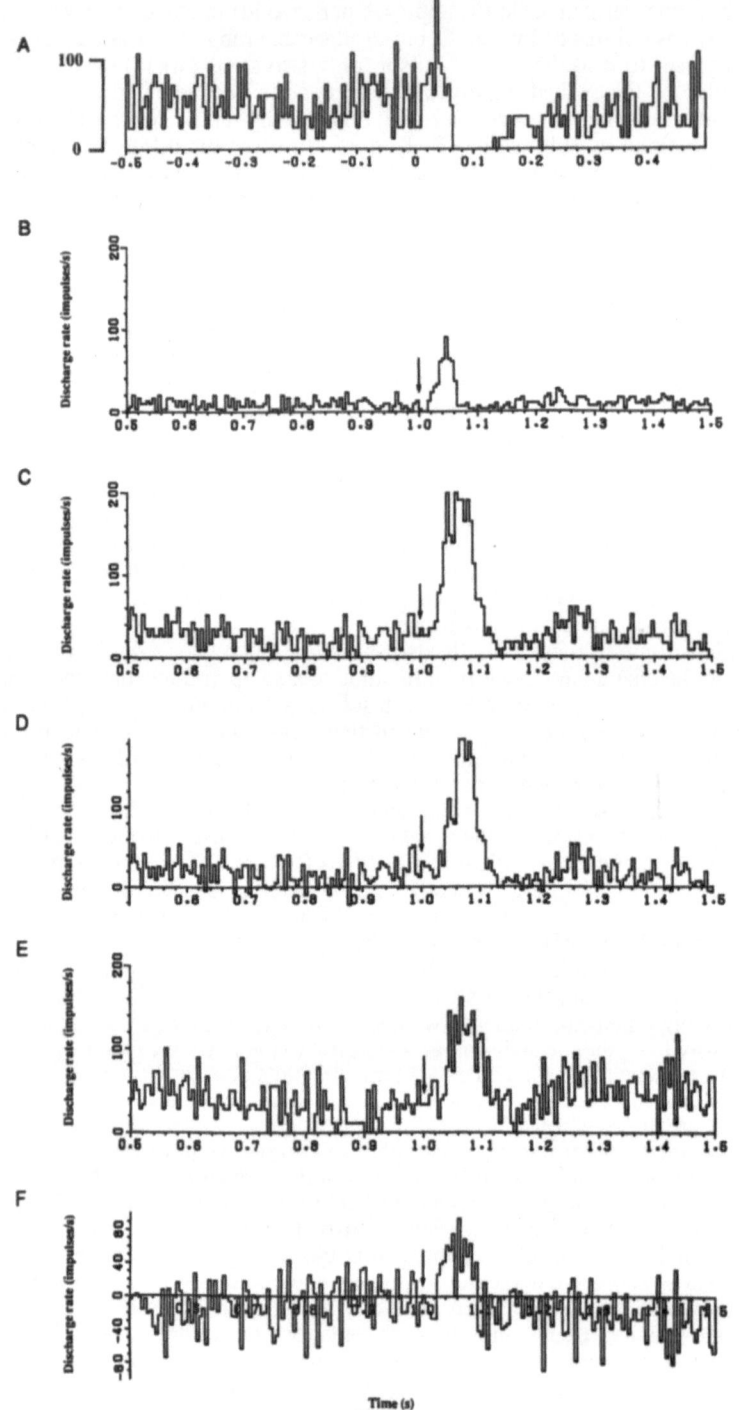

Figure 3. A: Response of a cerebellar neurone to rung movement (which began at time zero and finished approx. 100ms later). **B:** Response of a cortical neurone to flash. Response of the same cell to rung movement when close is shown in **C**, and the difference between the two (C-B) in **D**. Response to rung movement when further away is shown in **E**, and **F** is the difference (D-E).

cautious therefore in trying to say anything about what the complex spikes are doing in these circumstances.

CONCLUSION

In conclusion, we have found that a substantial proportion of lateral cerebellar neurones, 40%, are visually responsive which does suggest to us that the lateral cerebellum is important in visuomotor control (this finding is to some extent predicted by anatomical experiments involving injection of tritiated amino acids or WGA-HRP into the ventromedial pontine visual area. Terminal labelling, though most concentrated in the dorsal paraflocculus, was also distributed over most of the cerebellar hemisphere: see Robinson, Cohen, May, Sestokas and Glickstein, 1984). Nobody doubts the importance of the cerebellum for motor control and I think it is becoming ever clearer that it is involved in visuomotor control specifically. Secondly, the visual responses of cerebellar neurones can sometimes be context dependent, and the same cell can show later discharge changes which are motor in character. This means that the cerebellar contribution can be more complex than just simply signalling visual events, perhaps indicating something about the significance of the event and whether a motor response is required, and if so how long there is to make it. Finally I think these findings correspond nicely with the recordings that we and other people have made in the lateral cerebellum of rhesus monkeys performing visually guided tracking tasks.

ACKNOWLEDGEMENTS. The single unit recordings from the cerebellum described here were obtained in cats working together with David Armstrong and Jose Criado who was a visiting fellow from the University of Salamanca. The monkey recordings are from experiments performed jointly with John Stein at the Physiology Laboratory, Oxford. In both studies Steve Gilbey provided expert technical assistance. The work was funded by the Medical Research Council of Great Britain.

REFERENCES

Amos, A.J., Armstrong, D.M. & Marple-Horvat, D.E. (1987). A ladder paradigm for studying skilled and adaptive locomotion in the cat. *Journal of Neuroscience Methods* **20**, 323-340.

Baker, J., Gibson, A., Glickstein, M. & Stein, J.F. (1976). Visual cells in the pontine nuclei of the cat. *Journal of Physiology* **255**, 415-433.

Chapman, C.E., Spidalieri, G. & Lamarre, Y. (1986). Activity of dentate neurones during arm movements triggered by visual, auditory and somesthetic stimuli in the monkey. *Journal of Neurophysiology* **55**, 203-226.

Marple-Horvat, D.E. & Stein, J.F. (1990). Neuronal activity in the lateral cerebellum of trained monkeys, related to visual stimuli or to eye movements. *Journal of Physiology* **428**, 595-614.

Mushiake, H. & Strick, P.L. (1992). Activity of dentate neurones during sequential movements. *Society for Neuroscience Abstracts* **18**, 1046.

Robinson, F.R., Cohen, J.L., May, J., Sestokas, A.K. & Glickstein, M. (1984). Cerebellar targets of visual pontine cells in the cat. *Journal of Comparative Neurology* **223**, 471-482.

Snider, R.S. & Stowell, A. (1944). Receiving areas of the tactile, auditory, and visual systems in the cerebellum. *Journal of Neurophysiology* **7**, 331-357.

INFERIOR OLIVE AND THE SACCADIC NEURAL INTEGRATOR

P. STRATA, F. ROSSI and F. TEMPIA

Department of Human Anatomy and Physiology
University of Turin, C.so Raffaello 30
10125 Turin, Italy

SUMMARY

Following a lesion of the inferior olive, spontaneous saccades are characterised by a backward postsaccadic drift. When performed in light, this drift has a time constant of 100-150ms and after a few tens of ms, it gradually slows down to reach a steady level under a drive of the optokinetic reflex. In dark, this drift is followed by a slower drift due to the leakage of the neural integrator. When calculated on saccades of 10° of amplitude and ending near the midline the time constant is 0.9s. The amplitude of the postsaccadic drift (y) in light depends on the saccadic amplitude (x) and on the eccentricity (z), according to the equation y = 0.23 x + 0.24 z. This means that the gain of the pulse to step transformation is 0.77 at all saccadic amplitudes. Following flocculus-paraflocculus lesion the time constant of the neural integrator is 0.9s and the equation relating the amplitude of the postsaccadic drift to the saccadic amplitude and to the eccentricity is y = 0.21 x + 0.23 z. Thus, there is a striking similarity between the inferior olive and flocculus-paraflocculus lesion.

There is anatomical evidence that the pathway from the inferior olive to the flocculus projects to the prepositus hypoglossi nucleus which in turn sends fibres to the inferior olive. Such a loop is superimposed on that of the brain stem neural integrator formed by the medial vestibular and prepositus hypoglossi nuclei. The interruption of the loop at either the inferior olive or flocculus levels would lead to similar impairment of the neural integrator. We propose the hypothesis that the olivocerebellar loop is part of the neural integrator with the function of improving the dynamic performance of the integrator and of being also responsible for its adaptive capabilities.

INTRODUCTION

Saccades are fast ballistic eye movements which rotate the eyeball in the orbit in order to shift the sight line. This phasic activity is followed by a tonic one with the aim in maintaining the eye in the acquired position and to stabilize the image on the retina. The phasic component of the saccadic performance is due to a burst of action potentials of the oculomotor neurones innervating the eye muscles, called *pulse* of innervation. The pulse amplitude determines the saccade peak velocity, whereas the pulse area (product of duration and amplitude) determines the saccadic amplitude. The tonic component is due, instead, to a tonic discharge of the same motoneurones named *step* of innervation. The amplitude of the step component defines the steady-state position of the eye after the saccade (Robinson, 1975, 1981; see Fuchs *et al.*, 1985). A third component of the motoneurone discharge is called *slide* and it has the function to help the transition between the pulse and the step (Optican and Miles, 1985).

The phasic discharge is generated by a burst of activity of neurones located in the paramedian pontine reticular formation (pulse generator) (Keller, 1974; see Fuchs *et al.*, 1985). This activity, in addition to being forwarded to the oculomotor neurones, is also transmitted to a neural integrator that transforms the phasic activity into a tonic one (Robinson, 1974, 1975). Due to an intrinsic leakiness of the neural integrator, in darkness the tonic activity decreases with time so that the eye slowly drifts towards the centre of the orbit. In the light, an optokinetic reflex compensates for the integrator leakiness and prevents the eye from drifting towards the centre of the orbit. If the pulse area and the step amplitude are appropriately tuned with the time constant of the eye dynamics, the steady-state position will exactly match the displacement due to the pulse, so that the eye will be perfectly stable after the saccade and this will provide an optimal stabilization of the image on the retina. If not, the eye will drift until it reaches the position defined by the step component of the motor command or until an optokinetic reflex, even impaired, is able to maintain a steady position. Such a matching requires a continuous parametric regulation of both these saccadic components.

The cerebellum plays a major role in motor function and it is involved in both the dynamic and the adaptive aspects of the saccadic system. Two areas are mainly involved in such control. The dorsal vermis, with the fastigial nucleus, is implicated in the accuracy of the saccadic amplitude. Lesioning of these structures leads to saccadic dysmetria. When only the cerebellar cortex is lesioned, the dysmetria depends on saccadic direction; centripetally directed saccades tend to overshoot the target, whereas centrifugally directed ones tend to undershoot (Richtie, 1976). In contrast, when the lesion includes the fastigial nucleus, saccades are usually hypermetric (Optican and Robinson, 1980). The vestibulocerebellum, and more precisely the flocculus-paraflocculus area, is involved in the gaze-holding process. In the monkey, following a flocculus-paraflocculus lesion, the position reached by the eye at the end of the saccadic shift is not maintained and the eye presents a drift which has been described as having more often an onward direction. Such a drift has been attributed to a pulse-step mismatch (Zee *et al.*, 1981). After this lesion the monkey is also unable to adaptively modify a pulse-step mismatch induced by a partial eye muscle tenotomy (Optican *et al.*, 1986).

An adaptive modification of the step requires a change in the performance of the neural integrator. Therefore, in order to understand the mechanism of the adaptive modifications of the step component a necessary prerequisite is to identify the location of the neural integrator and to study the reciprocal influences with the cerebellum. It is known that cerebellar performance depends on the functional interactions between the activity of the main two inputs to the cerebellar cortex, the mossy fibres and the climbing fibres (Eccles *et al.*, 1967; Ito, 1984). The latter ones originate from the inferior olive nucleus and they have often been suggested to be of basic importance for the motor adaptive mechanisms (see Ito, 1984; Strata, 1989). Therefore, it is also of interest to direct more specifically our attention to the importance of the inferior olive nucleus in controlling the activity of the neural integrator.

The neural integrator is likely located in the brain stem and more precisely in the region of the nucleus prepositus hypoglossi and the medial vestibular nucleus. In fact, following a lesion of these two nuclei in the monkey, the gaze holding abilities are greatly impaired or lost (Cheron *et al.*, 1986a,b; Cheron and Godaux, 1987; Cannon and Robinson, 1987; Kaneko and Fuchs, 1991; Kaneko, 1992) and a lesion of the rostral area of the prepositus hypoglossi nucleus is sufficient to disrupt the integrator function (Cheron *et al.*, 1986b). In addition, cells with a tonic discharge related to eye position have been found in both these structures (Fuchs and Kimm, 1975; Keller and Daniels, 1975; Lopez-Barneo *et al.*, 1982; McFarland, 1988) including neurones of the prepositus hypoglossi nucleus projecting to the abducens nucleus (Escudero *et al.*, 1992). These two structures are reciprocally connected (Baker and Berthoz, 1975b; Balaban, 1983) and they may be the circuit acting as a positive feedback for the integration process. Finally, the paramedian pontine reticular formation (the pulse generator) projects to the ipsilateral prepositus hypoglossi nucleus and mainly to its rostral part (Graybiel, 1977). Direct projections exist also from the prepositus hypoglossi to the oculomotor nuclei (Graybiel and Hartweig, 1974; Baker and Berthoz, 1975a) (see Fig. 4).

The flocculus is closely related to the neural integrator. Some of its Purkinje cells impinge upon the rostral part of the prepositus hypoglossi nucleus (Kotchabhakdi, 1977; Alley, 1977; Graybiel, 1977; Yingharoen and Rinvik, 1983) and they show an eye position related activity (Noda and Suzuki, 1979a, b). Its climbing fibre input comes from the dorsal cap and the lateral outgrowth of the inferior olive (Ruigrok *et al.*, 1992).

The inferior olive is the source of climbing fibres to the cerebellar cortex and sends collateral fibres to the intracerebellar and vestibular nuclei (see Eccles *et al.*, 1967; Ito, 1984; Strata, 1989), the other main afferent pathway being the mossy fibre input ending on granule cells. The region of the inferior olive which is involved in the control of eye movements is the dorsal cap. This may be divided into three regions: a caudal area related to the horizontal plane, and the two more rostrally located regions which respond to images moving around horizontal axes (Simpson *et al.*, 1981; Leonard *et al.*, 1988). Although the anatomical organization of the olivocerebellar system is well known, the functional significance in motor control and in its plasticity is still a controversial issue (Ito, 1984; Strata, 1989).

Recently the main dynamic characteristics of the spontaneous saccades of the pigmented rat have been described. They are performed essentially on the horizontal plane and have a main sequence which reveals a rather high peak velocity. Instead, the gaze-holding capacity is rather poor (Chelazzi *et al.*, 1989). We have analysed the effect of inferior olive lesion on the spontaneous saccades performed by the pigmented rat both in dark and in light (Strata *et al.*, 1990, 1992). The inferior olive was almost completely lesioned by means of 3-acetylpyridine. Recording took place at least one month after the lesion. Eye movements were recorded by means of a phase detection coil system.

The mean amplitude of the saccadic shift both in the dark and in the light did not show any significant change, and neither did the peak velocity or duration. In contrast, there was a clear deficit in holding the eye in the position achieved by the saccadic shift. Such a deficit has been quantitatively analysed and may be described as follows.

First, we measured the time constant of the neural integrator by analysing the time course of the drift towards the midline in the dark. When calculated on saccades starting from the midline and ending at 10° of eccentricity, the time constant in intact rats was 4.2s, whereas after inferior olive lesion the value was 0.9s.

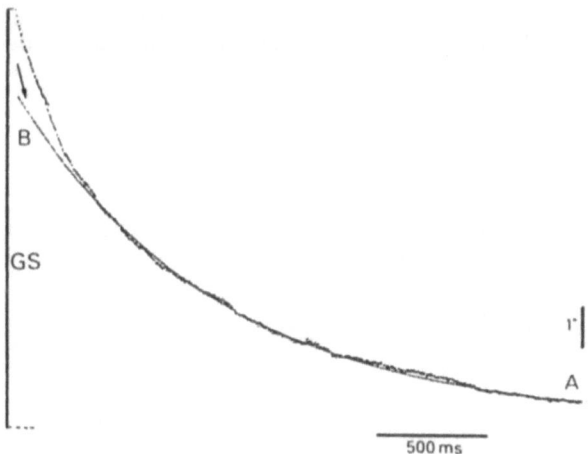

Figure 1. Time course of postsaccadic eye position obtained by averaging 20 saccades of 10° of amplitude and ending at 10° of eccentricity performed in the dark (A). B is the curve which best fits the segment of the drift which starts 400 ms after the gaze shift (GS). The arrow indicates the eye position that would be determined by the step which starts at 7.6°. (Modified from Strata *et al.*, 1990).

Another typical feature of the saccades performed by the rat after the inferior olive lesion was the consistent presence of a backward postsaccadic drift. When performed in the light, this drift had a time constant around 100-150ms and after a few tens of ms it gradually slowed down to reach a steady level under the drive of the optokinetic reflex. In the dark this drift was followed by a much slower drift which may be attributed to the integrator leakage. It is maintained that the fast drift is the consequence of a mismatch between the size of the pulse and that of the step. In this case, since we have shown that the amplitude of the saccade was unchanged after an inferior olive lesion, we have to draw the conclusion that the step was smaller relative to the pulse. However, in light, because of the very short time constant of the rat neural integrator, particularly after inferior olive lesion, at the time the eye reached the step value, this had further decreased from its initial value. In addition, we have shown that following the inferior olive lesion the initial rise of the velocity of the reflex is slowed down (Hess *et al.*, 1988). Consequently, it takes several hundreds of ms before the gaze becomes stabilized by the optokinetic reflex. Therefore, the amplitude of the postsaccadic drift in the light for eccentric saccades is due in part to a pulse-step mismatch and in part to the leakage of the neural integrator. We have thus investigated the contribution of these two components of the postsaccadic drift for saccades of different amplitudes and eccentricities.

Fig. 1 shows the average record of 20 saccades of 10° of amplitude and ending at 10° of eccentricity performed in dark (A). Since the time constant of the fast drift was around 100-150ms, the time course of the slow drift after 400ms represented the time course of the step due to the leakage of the neural integrator. Line B shows the best fitting exponential of this part of the drift and it was thus possible to calculate its starting point which was 76% of the saccadic amplitude. Therefore, for a saccade of 10° of amplitude, the initial size of the step was 7.6°. This means that the gain of the pulse to step transformation was 0.76.

A second method to calculate this gain value was to determine the relationship between the amplitude of the postsaccadic drift in the light and the saccadic amplitude for classes of saccades of similar eccentricities. In each class, the leakage of the neural integrator is constant and therefore the increase of drift with the increase of the amplitude provided a measure of the gain of the pulse to step transformation. In this experimental condition there was a linear relationship between the two parameters and the average gain was 0.79.

The same issue was addressed with a third method. When classes of saccades of similar amplitudes and different eccentricities were studied, it was possible to obtain a plot where one could extrapolate the relationship between the value of the postsaccadic drift and the saccadic amplitude for saccades ending near the midline. In this condition the postsaccadic drift was due only to the pulse-step mismatch. Fig. 2A shows schematically these different classes of saccades ending near the midline and Fig. 2B a plot of the experimental values. The amplitude of the postsaccadic drift was linearly related to the saccadic amplitude. The regression line had a slope of 0.24 and a Y intercept at 0.49. Therefore, the gain of the pulse to step transformation was 0.76. If we disregard the Y intercept, which is a fraction of a degree, and we take into consideration the average of the three gain values of the pulse to step transformation, which is 0.77, we may write the following equation:

$$y' = 0.23 \ x \qquad\qquad (1)$$

where y' is the amplitude of the postsaccadic drift (due only to a pulse-step mismatch) and x the saccadic amplitude. This means that at all saccadic amplitudes the amount of drift was 0.23° for each degree of amplitude.

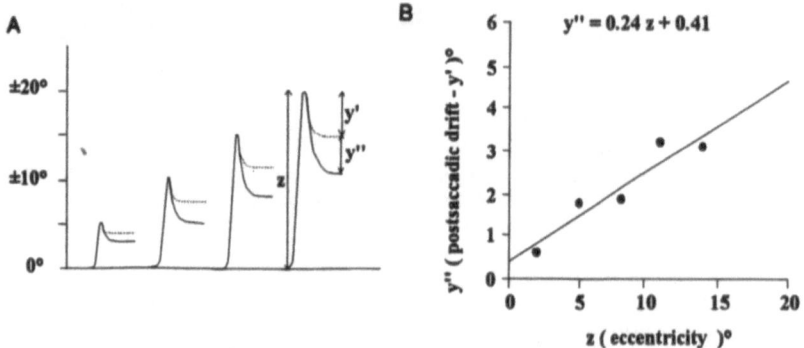

Figure 2. Relationship between postsaccadic drift and saccadic amplitudes for saccades of 0° of eccentricity performed in the light. **A** shows schematically 4 different classes of saccades of different amplitudes. **B** shows the experimental values. Correlation coefficient r=0.96. See text. (B is modified from Strata *et al.*, 1990).

In order to study the postsaccadic drift due to the leakage of the neural integrator, we have calculated the drift value for classes of saccades of different amplitudes and different eccentricities. Fig. 3A shows schematically 4 different classes of saccades. The amplitude of the postsaccadic drift (y) is the sum of the drift due to the mismatch (y'), and calculated in Fig. 2B, and of the drift due to integrator leakage (y") which depends on the eccentricity (z).

Fig. 3B shows the plot of the experimental values. The amount of drift due to integrator leakage was linearly related to the eccentricity. The regression line had a slope of 0.24. By disregarding the Y intercept of 0.41, we may write:

$$y'' = 0.24 \, z \qquad (2)$$

By combining (1) and (2) we may write:

$$y = 0.23 \, x + 0.24 \, z \qquad (3)$$

where y is the sum of the postsaccadic drift due to the pulse-step mismatch (y') and of that due to the leakage of the neural integrator (y").

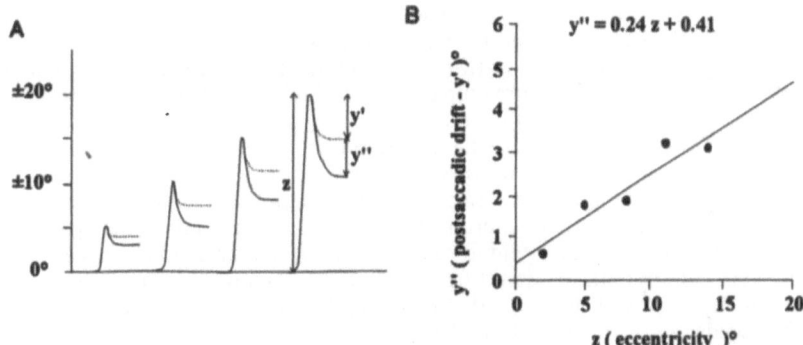

Figure 3. Relationship between postsaccadic drift and saccadic amplitudes for saccades of different amplitudes and eccentricities performed in the light. A shows schematically 4 different classes of saccades. The dotted lines indicate the postsaccadic drift for the classes of saccades at 0° of eccentricities displayed in Fig. 2A, so that y' is the drift due to the pulse-step mismatch; y" is the additional drift due to the eccentricy. B shows y" as a function of eccentricity. Correlation coefficient r=0.94. (B is modified from Strata *et al.*, 1990).

EFFECT OF FLOCCULUS-PARAFLOCCULUS LESION ON THE GAZE HOLDING CHARACTERISTICS OF THE SACCADES

A quantitative analysis similar to that just described for rats with inferior olive lesions has also been performed following flocculus-paraflocculus lesions (Chelazzi *et al.*, 1990). Surprisingly, all the values obtained in the two experimental conditions were not significantly different. In fact, the time constant of the neural integrator was 0.9s. In addition, the equation relating the amplitude of the postsaccadic drift (y), due to both the pulse-step mismatch and to the leakage of the neural integrator, to saccadic amplitude x and eccentricity z was:

$$y = 0.21 \, x + 0.23 \, z \qquad (4)$$

This means that the amount of the postsaccadic drift due to the pulse to step transformation was 0.21° for each degree of saccadic amplitude and the amount due to the eccentricity was 0.23° for each degree of eccentricity.

RELATIONSHIP BETWEEN THE INFERIOR OLIVE AND THE NEURAL INTEGRATOR

Recent investigations provide evidence that the saccadic neural integrator is localized in the rostral region of the nucleus prepositus hypoglossi and in the medial vestibular nucleus (see above). Both areas are reciprocally connected with the flocculus (Blanks, 1988; McCrea, 1988). Therefore, it is not surprising that a lesion to this cerebellar region affects the activity of the neural integrator. The inferior olive projects to the cerebellar cortex and to the intracerebellar nuclei, with some collaterals to the vestibular nuclei (Ito, 1984) and the dorsal cap is the region which projects to the flocculus. Therefore, it is very likely that it is the dorsal cap which controls the activity of the integrator through the flocculus. In addition, the inferior olive exerts a strong control on the electrical activity of the Purkinje cells by means of a phasic excitatory (Eccles *et al.*, 1966) and a tonic inhibitory action (Montarolo *et al.*, 1982). Although these data explain why the lesion of either the inferior olive or of the flocculus-paraflocculus affects the function of the neural integrator, it was unexpected to find that both produce astonishingly similar effects on the integrator time constant and on the gain of the pulse to step transformation.

A simple explanation for such a similarity would be that following the lesion of the inferior olive the inhibition exerted by the Purkinje cells of the cerebellar cortex on their target neurones in the intracerebellar nuclei is lost or greatly reduced. Therefore, a lesion of the inferior olive would be equivalent to a functional inactivation of the Purkinje cells. A remarkable loss of Purkinje cell inhibition following the inferior olive lesion has been documented and has been attributed to the lack of a trophic factor released by the climbing fibres (Ito *et al.*, 1979; Ito, 1984). However, such a loss is not very large (Montarolo *et al.*, 1981; Karachot *et al.*, 1987), and could be due, at least in part, to an irreversible damage of the Purkinje cells (Rossi *et al.*, 1987) because of their intense hyperactivity which follows their climbing fibres deafferentation (Montarolo *et al.*, 1981). If such a loss is partial, it is unlikely that the removal of the entire output of the flocculus-paraflocculus corresponds to the partial suppression of the activity of the Purkinje cells.

An alternative explanation is that the ablation of the flocculus-paraflocculus induces a retrograde degeneration of the inferior olive cells (Ito *et al.*, 1980) with the loss of the collaterals to the vestibular nuclei, an effect which would mimic the inferior olive lesions. This explanation would attribute to the inferior olive and to the collaterals to the vestibular nuclei the entire job of controlling the integrator function, leaving no role to the circuit through the Purkinje cells. This hypothesis could be tested by destroying the flocculus-paraflocculus Purkinje cells by means of kainic acid which would spare the inferior olive cells from the retrograde degeneration (Ito *et al.*, 1980; Rossi *et al.*, 1993).

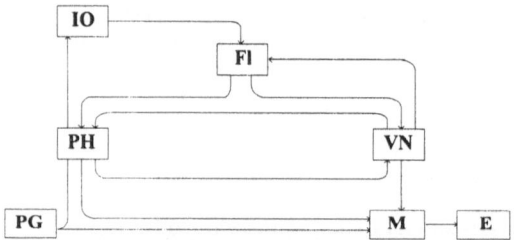

Figure 4. Hypothetical block diagram illustrating the structures involved in the saccadic neural integrator. The nucleus prepositus hypoglossi (PH) and the medial vestibular nucleus (VN) are the structures for the basic integration. The dorsal cap of the inferior olive (IO) and the flocculus (Fl) are additional structures which form a loop with the PH to improve the integration process and to subserve plasticity. PG: pulse generator; M: ocular motoneurones; E: eye.

Here we propose the hypothesis that the dorsal cap of the inferior olive and the flocculus are part of the saccadic neural integrator. This part enhances its capabilities and it is responsible for its plasticity. Fig. 4 illustrates the possible circuit relating the dorsal cap of the inferior olive (IO) and the flocculus (Fl) to the brain stem integrator located in the rostral region of the nucleus prepositus hypoglossi (PH) and in the medial vestibular nucleus (VN). The nucleus prepositus hypoglossi has a consistent projection to the dorsal cap of the inferior olive with both excitatory and inhibitory GABAergic fibres (De Zeeuw *et al.*, 1993). This olivary region is known to project to the flocculus which in turn projects to the medial vestibular neurones, and also to the rostral region of the nucleus prepositus hypoglossi (Kotchabhakdi, 1977; Alley, 1977; Graybiel, 1977; Yingharoen and Rinvik, 1983) which should thus be regarded as a subcerebellar nucleus. The medial vestibular nucleus is directly connected with the flocculus and with the inferior olive, but not with the dorsal cap (Gerrits, 1985). The latter does not respond to vestibular stimulation (Barmack and Hess, 1980).

According to our hypothesis, that part of the olivocerebellar system which includes the dorsal cap and the flocculus (cerebellar loop), would form a loop through the nucleus prepositus hypoglossi which is superimposed on the loop of the classical brain stem neural integrator (brain stem loop). When either the inferior olive or the flocculus are lesioned, the cerebellar loop is interrupted and the integrator activity is partially impaired with the consequent reduction of the time constant and of the gain of the pulse to step transformation. In addition, any plasticity of the integrator is lost as shown in the monkey following a flocculus-paraflocculus lesion (Optican et al., 1986). A long-term parametric regulation of the integrator is thus possible only when the olivocerebellar loop is intact. The sensitivity of the dorsal cap of the inferior olive to retinal slip velocity (Barmack and Hess, 1980) is necessary for the parametric regulation of the integrator in order to match its performance to that of the pulse.

A number of observations are consistent with the idea that the plastic changes which occur during motor adaptive behaviour are located at the synapses between the parallel fibers and the Purkinje cell in the cerebellar cortex (see Ito, 1984, 1991 for review). Other experimental evidence is, instead, in favour of the idea that the cerebellum is important in inducing modifications of synapses at the brain stem level (see Lisberger, 1988 for review). Our experiments show that the integrity of the inferior olive is necessary for proper gaze-holding in the rat and they confirm the important role of the flocculus. In addition, they are congruent with the hypothesis that the olivocerebellar system has a basic role in the plasticity of the saccadic neural integrator, but the site of the plastic changes remains to be determined.

REFERENCES

Alley, K. (1977). Anatomical basis for interaction between cerebellar flocculus and brain stem. In Developments in Neuroscience, Vol 1, Control of gaze by brain stem neurons, ed. Baker, R. and Berthoz, A., pp 109-117. Elsevier, Amsterdam.

Baker, R. and Berthoz, A. (1975a). Is the prepositus hypoglossi nucleus the source of another vestibulo-ocular pathway?. Brain Research 86, 121-126.

Baker, R. and Berthoz, A. (1975b). Vestibular input to the prepositus hypoglossi nucleus. Federation Proceedings 34, 439.

Balaban, C.D. (1983). A projection from nucleus reticularis tegmenti pontis of Bechterew to the medial vestibular nucleus in rabbits. Experimental Brain Research 51, 304-309.

Barmack, N.H. and Hess, D.T. (1980). Multiple-unit activity evoked in dorsal cap of inferior olive of the rabbit by visual stimulation. Journal of Neurophysiology 43, 151-164.

Blanks, R.H.I. (1988). Cerebellum. In Neuroanatomy of the oculomotor system, ed. Büttner-Ennever, A., pp 225-272. Elsevier, Amsterdam.

Cannon, S.C. and Robinson, D.A. (1987). Loss of the neural integrator of the oculomotor system from brain stem lesion in monkey. Journal Neurophysiology 57, 1383-1409.

Chelazzi, L., Ghirardi, M., Rossi, F., Strata. P. and Tempia, F. (1990). Spontaneous saccades and gaze holding abilities in the pigmented rat: II. Effect of localized cerebellar lesions. European Journal of Neuroscience 2, 1085-1094.

Chelazzi, L., Rossi, F., Tempia, F., Ghirardi, M. and Strata, P. (1989). Saccadic eye movements and gaze holding in the head-restrained pigmented rat. European Journal of Neuroscience 1, 639-646.

Cheron, G., Gillis, P. and Godaux, E. (1986a). Lesions in the cat prepositus complex: effects on the optokinetic system. Journal of Physiology 372, 95-111.

Cheron, G., Godaux, E., Laune, J.M. and Vanderkelen. B. (1986b). Lesions in the cat prepositus complex: effects on the vestibulo-ocular reflex and saccades. Journal of Physiology 372, 75-94.

Cheron, G., and Godaux, E. (1987) Disabling of the oculomotor neural integrator by kainic acid injections in the prepositus-vestibular complex of the cat. Journal of Physiology 394, 267-290.

De Zeeuw, C.I., Wentzel, P. and Mugnaini, E. (1993). Fine structure of the dorsal cap of the inferior olive and its GABAergic and non-GABAergic input from the nucleus prepositus hypoglossi in rat and rabbit. Journal of Comparative Neurology 327, 63-82.

Eccles, J.C., Ito, M. and Szentágothai, J. (1967). The cerebellum as a neuronal machine. Springer, New York.

Eccles, J.C., Llinás, R. and Sasaki, K. (1966). The excitatory synaptic action of climbing fibres on the Purkinje cells of the cerebellum. Journal of Physiology 182, 268-296.

Escudero, M., De la Cruz, R.R. and Delgado Garcia, J.M. (1992). A physiological study of vestibular and prepositus hypoglossi neurones projecting to the abducens nucleus in the alert cat. Journal of Physiology 458, 539-560.

Fuchs, A.F., Kaneko, C.R.S. and Scudder, C.A. (1985). Brainstem control of saccadic eye movements. Annual Review of Neuroscience 8, 307-337.

Fuchs, A.F. and Kimm, J. (1975). Unit activity in vestibular nucleus of the alert monkey during horizontal angular acceleration and eye movement. *Journal of Neurophysiology* **38**, 1140-1161.

Gerrits, N.M. (1985). Brainstem control of the cerebellar flocculus. Doctoral thesis. Leiden: Krips Repreo-Meppel.

Graybiel, A.M. (1977). Direct and indirect preoculomotor pathways of the brain stem: an autoradiographic study of the pontine reticular formation in the cat. *Journal of Comparative Neurology* **175**, 37-78.

Graybiel, A.M. and Hartweig, E.A. (1974). Some afferent connections of the oculomotor complex in the cat: an experimental study with tracer techniques. *Brain Research* **81**, 543-551.

Hess, B. J. M., Savio, T. and Strata, P. (1988). Dynamic characteristics of optokinetically controlled eye movements following inferior olive lesions in the brown rat. *Journal of Physiology* **397**, 349-370.

Ito, M. (1984). *The cerebellum and neural control*. Raven Press, New York.

Ito, M. (1991). The cellular basis of cerebellar plasticity. *Current Opinion in Neurobiology* **1**, 616-620.

Ito, M., Jastreboff, P.J. and Miyashita, Y. (1980). Retrograde influence of surgical and chemical flocculectomy upon dorsal cap neurons of the inferior olive. *Neuroscience Letters* **20**, 45-48.

Ito, M., Nisimaru, N. and Shibuki, K. (1979). Destruction of the inferior olive induces rapid depression in synaptic action of cerebellar Purkinje cells. *Nature* **277**, 568-569.

Kaneko, C.R.S. (1992). Effects of ibotenic acid lesions of nucleus prepositus hypoglossi on optokinetic and vestibular eye movements in the alert, trained monkey. In *"Sensing and controlling motion"*. Annals of the New York Academy of Sciences **656**, 408-427.

Kaneko, C.R.S. and Fuchs, A.F. (1991). Saccadic eye movement deficits following ibotenic acid lesions of the nuclei raphe interpositus and prepositus hypoglossi in monkey. *Acta Otolaryngology Supplementum* **481**, 213-216.

Karachot, L., Ito, M. and Kanai, Y. (1987). Long-term effects of 3-acetylpyridine-induced destruction of cerebellar climbing fibers on Purkinje cells inhibition of vestibulospinal tract cells of the rat. *Experimental Brain Research* **66**, 229-246.

Keller, E.L. (1974). Participation of medial pontine reticular formation in eye movement generation in monkey. *Journal of Neurophysiology* **37**, 316-332.

Keller, E.L. and Daniels, P,.D. (1975). Oculomotor related interaction of vestibular and visual stimulation in vestibular nucleus cells in alert monkey. *Experimental Neurology* **46**, 187-198.

Kotchabhakdi, N. (1977). Cerebellar projections from the perihypoglossal nuclei. In *Developments in Neuroscience, Vol 1, Control of gaze by brain stem neurons*, ed. Baker, R. and Berthoz, A., pp 119-130. Elsevier, Amsterdam.

Leonard, C.S., Simpson, J.I. and Graf, W. (1988). Spatial organization of visual messages of the rabbit's cerebellar flocculus. I. Typology of inferior olive neurons of the dorsal cap of Kooy. *Journal of Neurophysiology* **60**, 2073-2090.

Lisberger, S.G. (1988). The neural basis for learning of simple motor skills. *Science* **242**, 728-735.

Lopez-Barneo, J., Darlot, C., Berthoz, A. and Baker, R. (1982). Neuronal activity in prepositus nucleus correlated with eye movements in the alert cat. *Journal of Neurophysiology* **47**, 329-352.

McCrea, R.A. (1988). The nucleus prepositus. In *Neuroanatomy of the oculomotor system*, ed. Büttner-Ennever, A. pp 203-225. Elsevier, Amsterdam.

McFarland, J.L. (1988). The role of the nucleus prepositus hypoglossi and the adjacent medial vestibular nucleus in the control of horizontal eye movement in the behaving monkey. PhD Thesis. University of Washington, Seattle, Washington.

Montarolo, P.G., Palestini, M. and Strata, P. (1982). The inhibitory effect of the olivocerebellar input on the cerebellar Purkinje cells in the rat. *Journal of Physiology* **332**, 187-202.

Montarolo, P.G., Raschi, F. and Strata, P. (1981). Are the climbing fibres essential for the Purkinje cells inhibitory action?. *Experimental Brain Research* **42**, 215-218.

Noda, H. and Suzuki, D.A. (1979a). The role of the flocculus of the monkey in saccadic eye movements. *Journal of Physiology* **294**, 317-334.

Noda, H. and Suzuki, D.A. (1979b). The role of the flocculus of the monkey in fixation and smooth pursuit eye movements. *Journal of Physiology* **294**, 335-348.

Optican, L.M. and Miles, F.A. (1985). Visually induced adaptive changes in primate saccadic oculomotor control signals. *Journal of Neurophysiology* **54**, 940-958.

Optican, L.M. and Robinson, D.A. (1980). Cerebellar dependent adaptive control of primate saccadic system. *Journal of Neurophysiology* **44**, 1058-1079.

Optican, L.M., Zee, D.S. and Miles, F.A. (1986). Floccular lesions abolish adaptive control of post-saccadic ocular drift in primates. *Experimental Brain Research* **64**, 596-598.

Richtie, L. (1976). Effects of cerebellar lesions on saccadic eye movements. *Journal of Neurophysiology* **39**, 1246-1256.

Robinson, D.A. (1974) The effect of the cerebellectomy on the cat's vestibuloocular integrator. *Brain Research* **71**, 195-207.

Robinson, D.A. (1975). Oculomotor control signals. In *Basic Mechanisms of Ocular Motility and Their Clinical Implications*, ed. Lennerstand, G. and Bach-y-Rita, P., pp. 337-374. Pergamon, New York.

Robinson, D.A. (1981). Control of eye movements. In *Handbook of Physiology. The nervous system II*, ed. Brooks, V.B., pp. 1275-1320. American Physiological Society, Bethesda.

Rossi, F., Cantino, D. and Strata, P. (1987). Morphology of the Purkinje cell axon terminals in intracerebellar nuclei following inferior olive lesion. *Neuroscience* **22**, 99-112.

Rossi, F., Borsello, T., Vaudano, E. and Strata, P. (1993). Regressive modifications of climbing fibres following Purkinje cell degeneration in the cerebellar cortex of the adult rat. *Neuroscience* **53**, 759-778.

Ruigrok, T.J.H., Osse, R.J. and Voogd, J. (1992). Organization of inferior olivary projections to the flocculus and ventral paraflocculus of the rat cerebellum. *Journal of Comparative Neurology* **316**, 129-150.

Simpson, J.I., Graf, W. and Leonard, C.S. (1981). The coordinate system of visual climbing fibers to the flocculus. In *Progress in oculomotor research*, ed. Fuchs, A. and Becker, W., pp. 475-484. Elsevier, Amsterdam.

Strata, P. (1989) (ed.) *The olivocerebellar system in motor control*. Springer, New York.

Strata, P., Chelazzi, L., Ghirardi, M., Rossi, F. and Tempia, F. (1990). Spontaneous saccades and gaze holding abilities in the pigmented rat: I. Effects of inferior olive lesion. *European Journal of Neuroscience* **2**, 1074-1084.

Strata, P., Chelazzi, L., Tempia, F., Rossi, F. and Ghirardi, M. (1992). Cerebellar control of saccadic eye movements in the pigmented rat. In *The cerebellum revisited*, ed. Llinás, R. and Sotelo, C., pp. 215-225. Springer, New York.

Yingharoen, K. and Rinvik, E. (1983). Ultrastructural demonstration of a projection from the flocculus to the nucleus prepositus hypoglossi in the cat. *Experimental Brain Research* **51**, 192-198.

Zee, D.S., Yamazaki, A., Butler, P.H. and Gücer, G. (1981) Effect of ablation of flocculus and paraflocculus on eye movements in primate. *Journal of Neurophysiology* **46**, 878-899.

COMPARATIVE STUDIES

PRESYNAPTIC GAIN CONTROL IN A LOCUST PROPRIOCEPTOR

M. BURROWS, T. MATHESON AND *G. LAURENT

Department of Zoology
University of Cambridge
Downing Street
Cambridge CB2 3EJ, England

*Division of Biology
California Institute of Technology
Pasadena
California 91125, USA

SUMMARY

A common finding in both vertebrates and invertebrates is that the central terminals of their mechanosensory neurones receive synaptic inputs. These inputs are often caused by neurones that release GABA, and change the chloride conductance which, at the normal resting potential, causes a depolarisation. These inputs in turn reduce the efficacy with which the sensory spikes release transmitter onto postsynaptic neurones. The consequence of this presynaptic inhibition is that the information coded in the spikes of the sensory neurones may not be reliably transmitted to postsynaptic neurones. This paper suggests a new function for some of these presynaptic inputs, from an analysis of proprioceptive afferents in a locust. It proposes that they can form part of an automatic gain control mechanism that limits the efficacy of the proprioceptive signals, when activated by the spikes in other afferents from the same sense organ responding to the same movement. The result is that the actions of one sensory neurone are interpreted by the central nervous system only in the context of the network actions of all the other sensory neurones responding to the same stimulus.

INTRODUCTION

Presynaptic inhibition of afferent neurones is generally considered to have several possible functions. In cutaneous afferents of vertebrates, for example, lateral inhibitory interactions may improve spatial discrimination (Janig et al., 1968; Schmidt, 1971). Interactions between signals from muscle afferents could enhance the input to one set of motor neurones while decreasing that to another, perhaps antagonistic, set (Rudomin, 1990). Similarly the inputs from cercal hairs of a locust are modified by signals from a receptor monitoring cercal movements, perhaps so that their output is related to voluntary movements (Bernard, 1987; Boyan, 1988). Proprioceptive afferents in crayfish also receive GABAergic synaptic inputs that can reduce the strength of their inputs to postsynaptic motor neurones (Cattaert et al., 1992; El Manira and Clarac, 1991). Interactions between the cercal hair afferents of cockroaches could sharpen the directional sensitivity of particular interneurones (Blagburn and Sattelle, 1987). Many sensory neurones also receive a synaptic

Neural Control of Movement, Edited by W.R. Ferrell
and U. Proske, Plenum Press, New York, 1995

input that is linked to a fictive locomotory rhythm (Gossard *et al.*, 1990, 1991; El Manira *et al.*, 1991) and could be part of the mechanism that modifies the effectiveness of sensory transmission according to the phase of the movement (Sillar and Skorupski, 1986; Gossard *et al.*, 1990).

Few studies of presynaptic modification of sensory signals have, however, considered the possible interactions of receptors from the same sense organ. In the crab, interactions between sensory neurones from a proprioceptor could extend the dynamic range of the postsynaptic sensory neurone (Wildman and Cannone, 1991). In the crayfish, electrical coupling between the sensory neurones from one proprioceptor may facilitate the transmission of their sensory signals to postsynaptic motor neurones (El Manira *et al.*, 1993). In spiders (Foelix, 1975) and in some proprioceptive afferents in the legs of *Drosophila* (Shanbhag *et al.*, 1992) synapses occur between the axons of sensory neurones in the nerves before they enter the central nervous system, but the function of these synapses and therefore their likely effects have not been established.

A Locust Proprioceptor

The movements and position of the femoro-tibial joint of a locust leg are monitored by several different types of receptor, prominent among which is a chordotonal organ (the femoral chordotonal organ, FCO). This consists of about 100 sensory neurones suspended in an elastic ligament and linked by an apodeme to the insertion of the extensor tibiae muscle (Fig. 1A). The arrangement is such that the organ is stretched when the tibia is flexed and relaxed when it is extended. This convenient anatomy means that joint movements can be simulated experimentally by grasping the apodeme of the organ with forceps driven by a function generator. This stimulates only the sensory neurones of this organ, and not any of the other joint receptors. About 50 of the FCO sensory neurones respond to these movements, some being excited by flexion movements, others by extension (Matheson, 1990, 1992; Zill, 1985a). The various features of the velocity, acceleration, direction of movement, and position of the joint are thereby coded in a number of parallel sensory channels. The axons of these sensory neurones then project to the segmental metathoracic ganglion (Fig. 1B) where they make a series of distributed and parallel connections with particular spiking and nonspiking local interneurones, with intersegmental interneurones and with motor neurones (Burrows, 1987, 1988; Burrows *et al.*, 1988; Laurent and Burrows, 1988). All the connections are excitatory. As a result of this pattern of connections, imposed movements of the proprioceptor produce resistance reflexes in the muscles that move the femoro-tibial and adjacent joints (Field and Burrows, 1982; Field and Rind, 1981; Zill, 1985b). In specific conditions these reflexes can be both powerful and reliable, but at other times can be modified, most notably during active movements when they reverse in sign to assist the motor commands.

Each movement of the joint excites many sensory neurones and their convergence in the central nervous system could saturate the responses of particular interneurones and motor neurones. There is no efferent control of the organ itself to regulate the initiation of the sensory spikes in accord with the desired motor response. This raises the question of whether there is any mechanism within the central nervous system that could regulate the input to the central neurones from the sensory neurones of this organ.

Feedback Control of the Sensory Neurones

Recordings made from the axons of the sensory neurones, close to their output terminals within the central nervous system, reveal a wealth of depolarising potentials, especially when the femoro-tibial joint moves, in addition to the spikes that are initiated at the sense organ itself (Fig. 1C,D). For example, imposed flexion and extension movements of the FCO apodeme evoke a burst of spikes in many of the FCO afferents (Fig. 1C). An intracellular recording from one of the sensory neurones shows that it spikes during each movement, but that these spikes are superimposed on a depolarisation which is correlated with the spikes in the other afferents excited by these movements. The depolarisation also occurs at the same time, and has the same general waveform, as the depolarisation of a flexor tibiae motor neurone that responds in the resistance reflex. The depolarising inputs in the sensory neurones are also generated during voluntary movements of the joint (Fig. 1D). For example, each spike in the slow extensor tibiae motor neurone causes a small extension

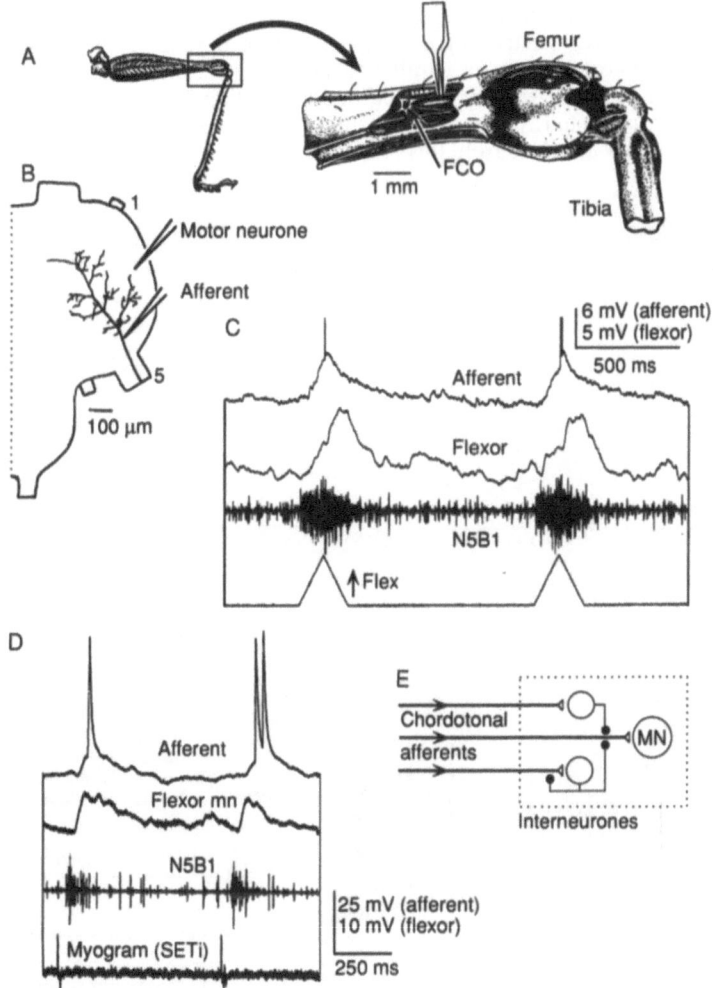

Figure 1. **A :** The location of the femoral chordotonal organ (FCO) in the distal femur of a hind leg. Its apodeme is grasped by forceps so that tibial movements can be mimicked. **B:** Its afferent neurones project to the metathoracic ganglion. The drawing shows the branches of one afferent and the sites where recordings were made from sensory and motor neurones. **C:** Movements of the FCO apodeme (lower trace) excite many afferents recorded in nerve 5B1, and evoke a depolarisation of one of the afferents recorded intracellularly, and a depolarisation of a tibial flexor motor neurone. Both neurones are hyperpolarised to emphasise the amplitude of these inputs. The sensory spikes (truncated for display purposes), are superimposed on a barrage of depolarising inputs. **D:** An afferent also receives synaptic inputs during voluntary movements of the tibia caused by spikes in the slow extensor motor neurone (SETi). The spikes of the sensory neurone signalling the movement are preceded by depolarising potentials. A flexor motor neurone is also depolarised. **E:** A simplified diagram of the possible pathways for presynaptic inhibition of the afferent terminals. Chordotonal (FCO) afferents make direct connections with motor neurones (MN) and an unidentified population of interneurones. The interneurones make central synapses back onto the afferent neurones, presynaptic to the motor neurones.

movement which is signalled by spikes in many FCO afferents. One of these afferents spikes once or twice on each movement, but also receives a depolarising input that precedes the sensory spikes.

These potentials are generated in the central nervous system and could not simply be reflections of generator potentials conducted electrotonically from the periphery. There do not appear to be direct synaptic or electrotonic connections between the afferents. The simplest explanation for the depolarising inputs is that they are synaptic potentials caused by other chordotonal afferents activating interneurones through reliable and high gain pathways (Fig. 1E). These interneurones, or parallel sets of interneurones, can also be activated by inputs from other sense organs, by interneurones signalling inputs from other legs, and by the neurones that generate rhythmic motor patterns. The most potent source of activation is nevertheless the other afferents from this proprioceptor.

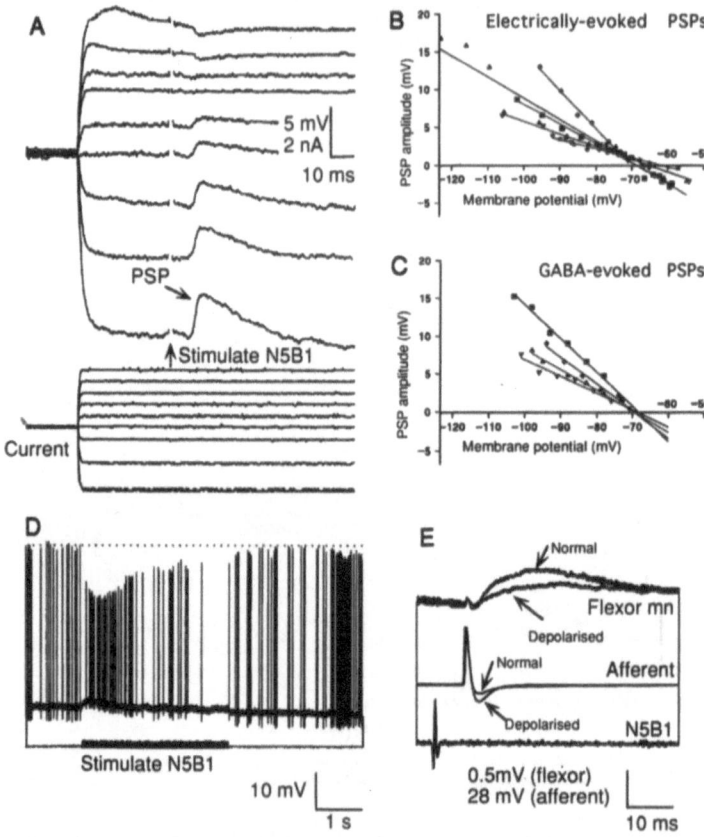

Figure 2. Properties and effects of the depolarising potentials in the afferent terminals. **A:** An afferent is held at different membrane potentials under current clamp while the nerve (N5B1) that contains other FCO afferents is stimulated electrically to evoke a PSP. The PSPs reverse a few millivolts more positive than resting potential, indicating that they are depolarising inhibitory potentials. **B:** Plotting the amplitude of electrically-evoked PSPs against membrane potential for 5 sensory neurones indicates that the reversal potential is near -68mV. **C:** Pressure injection of GABA into the neuropil where the afferents terminate evokes a depolarisation that reverses at a similar membrane potential. The responses of 4 afferents are plotted. B, C are based on Burrows and Laurent (1993). The lines represent least-squares linear regressions. All the r^2 values exceed 0.93, and all the slopes are significantly ($P < 0.01$) different from zero. **D:** Electrical stimulation of N5B1 causes a barrage of central synaptic inputs in an impaled afferent. These reduce the amplitude of the sensory spikes. **E:** Injecting current (+1.5nA) into an FCO afferent reduces the amplitude of the EPSP that it causes in a postsynaptic flexor motor neurone, as compared with normal membrane potential. The afferent spike is also monitored in nerve 5B1.

The Inputs are Caused by GABAergic Interneurones

The nature of these depolarising inputs has been determined by current clamping the afferents in an isolated ganglion, and by applying agonists and antagonists to mimic and block their effects (Burrows and Laurent, 1993). Electrical stimulation of the nerve that contains the axons of the FCO afferents evokes a depolarising potential associated with a conductance change in the terminals of an impaled afferent. These potentials are caused only by the chordotonal afferents and not by other afferents in this nerve (Burrows and Matheson, 1993). This evoked potential reverses at a membrane potential a few millivolts more positive than the normal resting potential, and thus usually appears to be depolarising (Fig. 2A). GABA, or its agonist muscimol, mimic the conductance changes associated with the natural or evoked potentials, and reduce the amplitude of the potentials evoked by movements of this receptor. GABA injected into the neuropil containing the terminals of the afferents also generates potentials that reverse at the same potential as the naturally occurring ones (Fig. 2B,C). Both the natural and electrically evoked potentials are blocked reversibly by picrotoxin. All this indicates that the potentials are depolarising inhibitory potentials of synaptic origin, probably caused by interneurones that release GABA, and not by synapses from afferents, which probably release acetylcholine. This conclusion is confirmed by electron microscopy which shows GABA-like immunoreactivity in the majority of input synapses to chordotonal afferents (Watson *et al.*, 1993). The fact that there are some input synapses which do not show GABA-like immunoreactivity suggests that the presynaptic control of the afferents may be more complex than the pharmacology has so far indicated.

The Inputs Reduce the Efficacy of Transmission from the Afferents to Motoneurones

The depolarising inhibitory synaptic input to the afferent terminals has at least two effects. First, it changes the excitability of the terminals by reducing their ability to support spikes (Burrows and Laurent, 1993). Second, it reduces the amplitude of the spikes that invade the terminals (Fig. 2D). A consequence of these changes is that the amplitude of excitatory synaptic potentials evoked by the afferents in postsynaptic motor neurones is reduced (Burrows and Matheson, 1993). This reduction in the efficacy of transmission will then reduce the gain of the reflex pathway. These effects are mimicked when the membrane potential of an afferent is altered by injecting current into its terminals; when the afferent is depolarised, the amplitude of the EPSP in a motor neurone is reduced (Fig. 2E).

These observations clearly show that the effectiveness of one afferent in signalling to a postsynaptic neurone can be regulated by spikes in other afferents from the same sense organ. Why should the afferent signals be regulated by other afferent signals in this way, and under what circumstances is the regulation expressed? The answer to these questions emerges from experiments that relate the response properties of a particular afferent to the timing and pattern of its synaptic inputs. The possible number of interactions is large as there may be as many as 30 different response types within the population of FCO afferents (Matheson, 1992).

The Inputs to an Afferent are Greatest when the Afferent itself Spikes

The picture that is emerging so far can be summarised in the following way. First, phasic afferents may receive tonic synaptic inputs. For example, moving the joint at a velocity below that needed to evoke a spike in a velocity-sensitive afferent reveals a tonic synaptic input at the new joint position (Fig. 3A). Second, phasic afferents may receive both phasic and tonic synaptic inputs. Moreover, the tonic input may be dependent on the range of angles over which the joint is moved. For example, a second velocity-sensitive afferent receives a phasic synaptic input each time the joint is moved to a more flexed position (Fig. 3B). As the joint is progressively flexed the depolarisation increases. At the same time, the afferent produces more spikes at the more flexed positions of the joint. In addition to the phasic input, the afferent also receives a tonic input that is dependent on the joint position, increasing in amplitude toward full flexion. Both the phasic and the tonic inputs are therefore maximal when the tibia is moved in the preferred direction and within the preferred range of joint angles of this afferent. Third, afferents that respond preferentially to a particular velocity of joint movement may receive phasic inputs during all velocities of movement (Fig. 3C). This means that the spike response to a preferred movement will always be superimposed on a synaptic input, whereas other velocities of movement will evoke a synaptic input alone. Fourth, tonic afferents may receive a phasic input during movements and a tonic input at particular positions of the joint.

In all these examples therefore, the synaptic inputs are greatest for the stimulus that also evokes the greatest spike response of the particular afferent. This means that the effectiveness of the strongest spike signals of these afferents will be reduced by the convergent, presynaptic action of the spikes in the other afferents that are excited by the same stimulus. The spike-coded information of one afferent is thus only read by its postsynaptic neurones in the context of the summed activity of many other afferents. Furthermore, the synaptic inputs to a particular afferent are caused by specific sets of afferents, and not by all neurones of the population, and occur only during certain movements and at certain positions of the joint. This points to specific mechanisms for regulating the effectiveness of an afferent in particular circumstances.

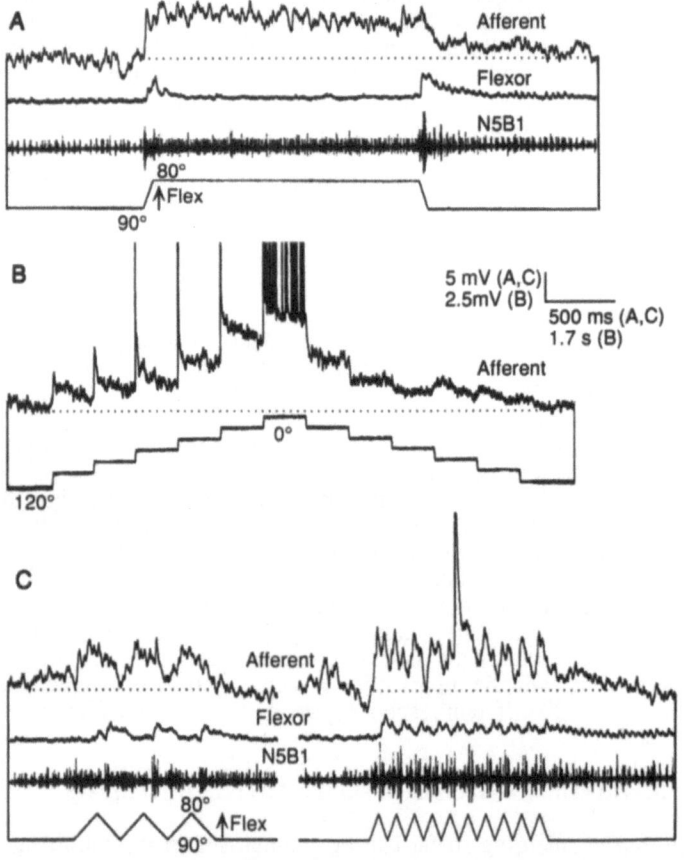

Figure 3. Timing of the synaptic inputs to an FCO sensory neurone. **A:** Tonic synaptic inputs are evoked in a phasic, flexion-sensitive afferent when the FCO apodeme is moved to simulate a tibial flexion. Many other afferents also respond to the movement, and a flexor motor neurone is transiently depolarised during the movements. The movement here was below the velocity threshold of this afferent. The afferent is hyperpolarised to emphasise the synaptic inputs. **B:** A phaso-tonic (flexion-sensitive) FCO afferent receives both phasic and tonic synaptic inputs as the apodeme is moved in 20° steps over the full range of possible tibial movements. The phasic inputs occur only during flexion movements, and the tonic inputs are greatest at more flexed angles, at which the afferent produces its greatest spike response. **C:** A velocity-sensitive afferent receives synaptic inputs at different velocities of imposed movements. Low velocity movements evoke a clear synaptic input (but no spikes) in an impaled afferent, that is associated with the spikes in other FCO afferents. At velocities close to its velocity threshold, the afferent produces a spike on one cycle of the movement that is superimposed on a large synaptic input produced by each cycle of the movement. A flexor motor neurone is also depolarised by each imposed movement.

The Central Inputs Form an Automatic Gain Control Mechanism

The presynaptic inputs to the afferents suggest a mechanism whereby the gain of the sensory effects on central neurones, and hence on the movements produced, is automatically regulated by other afferents from the same proprioceptor. The net effect would be to limit the gain of the synaptic transmission to postsynaptic neurones. Only the fastest conducting afferents will be able to exert their effects at full gain on postsynaptic neurones. For all the afferents, the gain of their effects will be reduced in accord with the other afferents that are also activated by a particular movement, and which excite the feedback loop. This control mechanism will prevent saturation of the input to postsynaptic neurones when many sensory neurones are active at the same time. This will happen during most movements, but in particular during high velocity movements, and during those when the tibia is close to full flexion. The gain control mechanism may also be capable of quite subtle effects because each afferent receives a specific pattern of synaptic inputs, generated by a specific array of afferents, that could allow its reflex effects to be changed independently. Furthermore, not all the terminals of the same sensory neurone may be regulated to the same extent, because it may be necessary to preserve the pathway to the interneurones that mediate the feedback at full gain while reducing the effectiveness of those terminals that contact motor neurones.

The existence of such a gain control mechanism in one proprioceptor inevitably raises the question of whether this mechanism is limited to the specialisations of this particular receptor, or whether it is a more widespread phenomenon characteristic of many mechanoreceptors. Movements of the joints of most animals are signalled by many receptors acting in parallel and in ways similar to those that have been described here. It is reasonable to expect therefore that similar automatic gain control mechanisms will be found. This of course must now be answered by searching for such mechanisms in proprioceptors of other animals.

ACKNOWLEDGEMENT. This work was supported by grants from NIH (NS 16058), SERC (UK), HFSP and the Wellcome Trust.

REFERENCES

Bernard, J. (1987) Effectiveness of the cercal chordotonal inhibitory organ in the cockroach. Synaptic activity during imposed cercal movements. *Comparative Biochemistry and Physiology* **87A**, 53-56.

Blagburn, J.M. and Sattelle, D.B. (1987) Presynaptic depolarization mediates presynaptic inhibition at a synapse between an identified mechanosensory neurone and giant interneurone 3 in the first instar cockroach, *Periplaneta americana*. *Journal of Experimental Biology* **127**, 135-157.

Boyan, G.S. (1988) Presynaptic inhibition of identified wind-sensitive afferents in the cercal system of the locust. *Journal of Neuroscience* **8**, 2748-2757.

Burrows, M. (1987) Parallel processing of proprioceptive signals by spiking local interneurones and motor neurones in the locust. *Journal of Neuroscience* **7**, 1064-1080.

Burrows, M. (1988) Responses of spiking local interneurones in the locust to proprioceptive signals from the femoral chordotonal organ. *Journal of Comparative Physiology* **164**, 207-217.

Burrows, M and Laurent, G. (1993) Synaptic potentials in the central terminals of locust proprioceptive afferents generated by other afferents from the same sense organ. *Journal of Neuroscience* **13**, 808-819.

Burrows, M., Laurent, G.J. and Field, L.H. (1988) Proprioceptive inputs to nonspiking local interneurones contribute to local reflexes of a locust hindleg. *Journal of Neuroscience* **8**, 3085-3093.

Burrows, M. and Matheson, T. (1994) A presynaptic gain control mechanism among sensory neurons of a locust leg proprioceptor. *Journal of Neuroscience* **14**, 272-282.

Cattaert, D., El Manira, A. and Clarac, F. (1992) Direct evidence for presynaptic inhibitory mechanisms in crayfish sensory afferents. *Journal of Neurophysiology* **67**, 610-624.

El Manira, A. and Clarac, F. (1991) GABA-mediated presynaptic inhibition in crayfish primary afferents by non-A, non-B-GABA receptors. *European Journal of Neuroscience* **3**, 1208-1218.

El Manira, A., Cattaert, D., Wallén, P., DiCaprio, R.A. and Clarac, F. (1993) Electrical coupling of mechanoreceptor afferents in the crayfish: a possible mechanism for enhancement of sensory signal transmission. *Journal of Neurophysiology* **69**, 2248-2251.

El Manira, A., DiCaprio, R.A., Cattaert, D. and Clarac, F. (1991) Monosynaptic interjoint reflexes and their central modulation during fictive locomotion in crayfish. *European Journal of Neuroscience* **3**, 1219-1231.

Field, L.H. and Burrows, M. (1982) Reflex effects of the femoral chordotonal organ upon leg motor neurones of the locust. *Journal of Experimental Biology* **101**, 265-285.

Field, L.H. and Rind, F.C. (1981) A single insect chordotonal organ mediates inter- and intra-segmental leg reflexes. *Comparative Biochemistry and Physiology* **68A**, 99-102.

Foelix, R.F. (1975) Occurrence of synapses in peripheral sensory nerves of Arachnids. *Nature*, **254**, 146-148.

Gossard, J-P., Cabelguen, J-M and Rossignol, S. (1990) Phase-dependent modulation of primary afferent depolarization in single cutaneous primary afferents evoked by peripheral stimulation during fictive locomotion in the cat. *Brain Research* **537**, 14-23.

Gossard, J-P., Cabelguen, J-M and Rossignol, S. (1991) An intracellular study of muscle primary afferents during fictive locomotion in the cat. *Journal of Neurophysiology* **65**, 914-926.

Janig, W., Schmidt, R.F. and Zimmermann, M. (1968) Two specific feedback pathways to the central afferent terminals of phasic and tonic mechanoreceptors. *Experimental Brain Research* **6**, 116-129.

Laurent, G. and Burrows, M. (1988) A population of ascending intersegmental interneurones in the locust with mechanosensory inputs from a hind leg. *Journal of Comparative Neurology* **275**, 1-12.

Matheson, T. (1990) Responses and locations of neurones in the locust metathoracic femoral chordotonal organ. *Journal of Comparative Physiology* [A] **166**, 915-927.

Matheson, T. (1992) Range fractionation in the locust metathoracic femoral chordotonal organ. *Journal of Comparative Physiology* [A] **170**, 509-520.

Rudomin, P. (1990) Presynaptic inhibition of muscle spindle and tendon organ afferents in the mammalian spinal cord. *Trends in Neuroscience* **13**, 499-505.

Schmidt, R.F. (1971) Presynaptic inhibition in the vertebrate central nervous system. *Ergebnisse der Physiologie* **63**, 20-101.

Shanbhag, S.R., Singh, K. and Naresh Singh, R. (1992) Ultrastructure of the femoral chordotonal organs and their novel synaptic organization in the legs of *Drosophila melanogaster* Meigen (Diptera: Drosophilidae). *International Journal of Insect Morphology and Embryology* **21**, 311-322.

Sillar, K.T. and Skorupski, P. (1986) Central input to primary afferent neurones in crayfish, *Pacifastacus leniusculus* is correlated with rhythmic output of thoracic ganglia. *Journal of Neurophysiology* **55**, 678-688.

Watson, A.H.D., Burrows, M. and Leitch, B. (1993) GABA-immunoreactivity in processes presynaptic to the terminals of afferents from a locust leg proprioceptor. *Journal of Neurocytology* **22**, 547-557.

Wildman, M.H. and Cannone, A.J. (1991) Interaction between afferent neurones in a crab muscle receptor organ. *Brain Research* **565**, 175-178.

Zill, S.N. (1985a) Plasticity and proprioception in insects. I. Responses and cellular properties of individual receptors of the locust metathoracic femoral chordotonal organ. *Journal of Experimental Biology* **116**, 435-461.

Zill, S.N. (1985b) Plasticity and proprioception in insects. II. Modes of reflex action of the locust metathoracic femoral chordotonal organ. *Journal of Experimental Biology* **116**, 463-480.

CENTRAL AND REFLEX RECRUITMENT OF CRAYFISH LEG MOTONEURONES

P. SKORUPSKI

Department of Physiology
University of Bristol
Bristol BS2 8EJ, UK

SUMMARY

The reflex pathways of the crayfish thoracocoxal (leg to body) joint are reviewed as a model system for studying central and feedback control of movement. Movement of this joint can evoke both positive and negative feedback reflexes via parallel reflex pathways. Transmission in these pathways depends on central factors, such as the phase of centrally generated motor output. In addition, reflex responses are not uniform within a pool of motoneurones: positive feedback reflexes are restricted to subgroups of a motor pool. Thus, selective recruitment of motoneurones by the CNS could potentially bring about different reflex effects. Finally, selective neuromodulation of reflex pathways is described. The amine, octopamine, abolishes positive feedback reflexes, but leaves negative feedback reflexes relatively unaffected.

INTRODUCTION

In a jointed limbed animal the typical response to perturbation of a joint is a muscular action to resist the imposed movement. This type of negative feedback reflex may be regarded as essentially homeostatic in function, reflecting the stability of the resting CNS. It has been suggested, however, that animals achieve movement and manouverability by exploiting *instability* that the CNS can control (Hasan and Stuart, 1988). In keeping with this notion there is now evidence that positive feedback reflexes are present during locomotion (Andersson and Grillner, 1981; Bässler, 1986; Elson, Sillar and Bush, 1992; Skorupski and Sillar, 1986). In arthropods the term 'assistance reflex' is often used, because such reflexes assist rather than resist the imposed movement (DiCaprio and Clarac, 1981).

Since assistance reflexes are recorded about the same joints in response to the same movements where, in other circumstances, (negative feedback) resistance reflexes are evoked, a reflex reversal must be taking place in the CNS. What are the neural control mechanisms that regulate transmission in alternative reflex pathways in different behavioural contexts? This question is most commonly considered in terms of alternative inputs to the motoneurones, together with mechanisms for presynaptic control. The reflex pathways of the crayfish basal limb region have proved a useful model system for studying such mechanisms, at the level of the primary afferent-motoneurone synapse (Cattaert, El Manira and Clarac, 1992; Skorupski, 1992) and in terms of alternative reflex pathways (Skorupski, 1992).

An additional mechanism in crayfish is reflex specialization of the motor pool: different motoneurones within a pool are involved in different reflex responses (Skorupski, Rawat and Bush, 1992). The motoneurones underlying an assistance reflex in a given muscle may not

Neural Control of Movement, Edited by W.R. Ferrell
and U. Proske, Plenum Press, New York, 1995

be exactly the same as those underlying a resistance reflex in the same muscle. Thus the CNS could potentially modulate the final reflex effect by selectively recruiting different subsets of motoneurones to the same muscle.

Background

The crayfish thoracocoxal (TC) joint of each walking leg is operated by two antagonistic muscles. The coxal promotor rotates the leg forwards (as would occur during the swing phase of forward walking) and the coxal remotor muscle rotates the leg backwards (as would occur during the stance phase). Movement and angular position of the TC joint are monitored by two proprioceptors, the thoracocoxal muscle receptor organ (TCMRO), which signals leg remotion, and the thoracoxal chordotonal organ (TCCO), which signals leg promotion (Alexandrowicz and Whitear, 1957; Bush, 1981). These two receptor strands span the joint in parallel, but the TCMRO is innervated by two nonspiking afferent neurones sensitive to receptor *lengthening*, whereas the TCCO is innervated by up to 60 spiking afferents that respond to receptor *shortening* (see Fig. 3). Although the TCMRO is a muscle receptor, with its own efferent innervation, the experiments to be reviewed here were all done under open loop conditions (i.e. the receptor-motor nerve was cut). Stretching the two receptors together (during which TCMRO-mediated reflex effects predominate) mimics a remotion signal, and shortening them (which elicits mainly TCCO-mediated reflex effects) mimics a promotion signal (Skorupski *et al.* 1992).

Thoracic ganglia that are isolated from the periphery, apart from the TCMRO and TCCO, produce spontaneous motor output (Sillar and Skorupski, 1986). In many preparations this consists of alternating bursts of promotor and remotor activity with a rather long cycle period of 20-25 seconds. In others promotor activity is more or less continuous, though varying in intensity and occasionally interrupted by a few remotor nerve spikes.

Assistance and Resistance Reflexes

Motor responses to TC joint movement (i.e. to lengthening and shortening of the TCMRO and TCCO) may be classified as either resistance or assistance reflexes. *Assistance reflexes* involve: 1) excitation of promotor motoneurones by a promotion signal (i.e. by receptor shortening), and 2) excitation of remotor motoneurones by a remotion signal (i.e. by receptor stretch). *Resistance reflexes* involve: 3) excitation of promotor motoneurones by a remotion signal (i.e. by receptor stretch), and 4) excitation of remotor motoneurones by a promotion signal (i.e. receptor shortening).

To a considerable extent assistance and resistance reflexes are restricted to different subgroups of the promotor and remotor motor pools. However, in a given motoneurone the reflex response may vary with the state of a preparation, or the phase of activity in a bursting preparation (Skorupski and Sillar, 1986; Skorupski *et al.*, 1992; Skorupski, 1992). This means there are potentially two mechanisms contributing to the net reflex effect recorded from a muscle nerve: 1) the prevailing pattern of synaptic inputs (from among alternative reflex pathways) to an individual motoneurone, and 2) the particular subset of motoneurones activated. When the multi-unit reflex responses vary, how may one distinguish between modulation of inputs to a given motoneurone and selective recruitment of motoneurones with different inputs?

One method is tracking multiple units, using a combination of intra- and extracellular recording, and comparing their simultaneous reflex responses. For example, the remotor nerve does not normally display any tonic activity, but in bursting preparations three units may typically be discerned, firing in alternation with promotor bursts. Intracellular recording can reveal subthreshold activity in additional remotor motoneurones. Fig. 1A compares the responses of one promotor, and four remotor motoneurones, to stretch and release of the TCMRO and TCCO in parallel. Responses elicited by ramp stretch-hold-release stimulation during promotor bursts (left) and during remotor bursts (right) were averaged separately. During remotor bursts three units in the extracellular remotor nerve recording responded to the stimulation, while subthreshold responses were recorded intracellularly from a fourth remotor motoneurone. Two of the units respond to stretch, a TCMRO-mediated assistance reflex, while the third responds to release, a TCCO-mediated resistance reflex. A subthreshold depolarizing response to release, representing the compound input from a number of TCCO

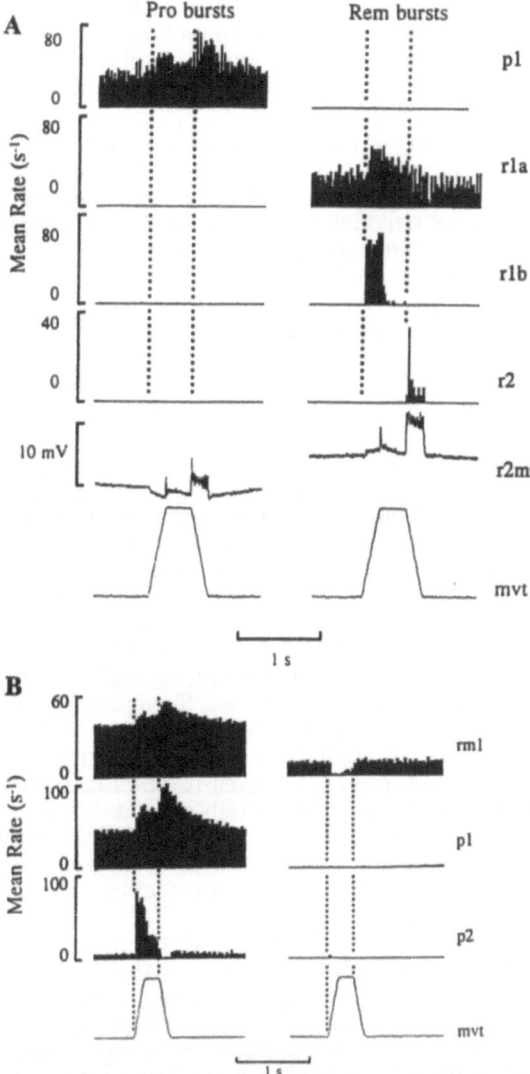

Figure 1. Reflex specialisation in promotor and remotor motoneurones, demonstrated by ramp stretch-hold-release length changes applied to the TCMRO and TCCO in parallel. A. Peristimulus-time histograms comparing responses of one promotor (p1) and three remotor units (r1a, r1b, r2); a fourth remotor motoneurone (r2mn) was recorded intracellularly and its responses averaged simultaneously. Responses evoked during promotor bursts (left, n=26) and remotor bursts (right, n=15) are displayed separately. Stimulus monitor trace (mvt): stretch (=remotion) is upwards. B. Reflex responses of the TCMRO receptor-motoneurone (rm1) and two promotor motoneurones (p1, p2), plotted separately for promotor bursts (left, n=101) and remotor bursts (right, n=129). Modified, with permission, from Skorupski et al.,1992 (A), and Skorupski and Bush, 1992 (B).

afferents, can also be seen in the fourth remotor motoneurone. Thus, within the same pool of motoneurones, different units receive different inputs, which correspond either to assistance or resistance reflexes.

A similar subdivision occurs in the promotor pool, although there is more overlap. In Fig. 1B reflex responses of two promotor units and the receptor motoneurone that exclusively innervates the TCMRO are compared (the TCMRO, unlike the TCCO, is a muscle receptor organ with an efferent innervation; unit *rm1* is analogous to a mammalian *gamma-motoneurone*). The assistance reflex, which now corresponds to excitation generated by receptor release (i.e. by a promotion signal), is selectively distributed within the promotor pool. Some promotor motoneurones, as well as the TCMRO receptor-motoneurone, rm1, are excited by receptor shortening, while others, recorded at the same time, clearly lack this input. In both the promotor and the remotor pools, then, motoneurones may be divided into two feedback groups. In each case *group 1* is an assistance reflex group, consisting of promotor motoneurones excited by a TCCO-mediated promotion signal and remotor motoneurones excited by a TCMRO-mediated remotion signal. *Group 2* motoneurones are members of resistance reflex groups: promotor motoneurones excited by a TCMRO-mediated remotion signal and remotor motoneurones excited by a TCCO-mediated promotion signal.

Thus, different motoneurones within the pool innervating the same muscle can have different proprioceptive inputs, which can generate reflex responses of opposite sign. These different units have characteristic spontaneous activity patterns that can be recognized from preparation to preparation. Typically, several group 1 motoneurones are active, with 1-3 promotor and 1-2 remotor units firing in a reciprocal pattern. Group 2 motoneurones are generally only recruited during more intense bursts of motor activity (compare background firing levels in the histograms of Figs. 1 and 2).

Proprioceptive input to any given motoneurone is phase-dependent. Usually intracellular recordings are required to reveal this, since reflex responses are subthreshold during the reciprocal phase of bursting. This, inhibitory phase of centrally generated bursting activity is probably the simplest mechanism whereby the CNS modulates reflex effects: a neurone may simply be prevented from responding at an inappropriate time in the cycle by powerful postsynaptic inhibition. Moreover, excitatory proprioceptive inputs are generally smaller during this phase (see Fig. 1A, r2mn). The shunting effect of postsynaptic inhibition may contribute to this (Skorupski *et al.,*1992), although presynaptic modulation of the input pathway may also occur (Cattaert *et al.,*1992; Sillar and Skorupski, 1986).

A reflex pathway may be modulated during centrally generated activity so that the synaptic input to a motoneurone is reduced in amplitude, or abolished, during the reciprocal phase of activity. In some cases proprioceptive input even reverses in sign, from excitation to inhibition, according to the prevailing level of centrally generated activity. This may be seen in the stretch response of unit rm1 in Fig. 1B. Since this unit is not normally silenced during remotor bursts, but continues firing at a lower rate (Skorupski and Bush, 1992), phase-dependent inhibition may be detected in extracellular recordings. Note also the absence of release-evoked excitation during this phase.

In both groups of promotor motoneurones a variable degree of excitation is seen on stretch. This is a direct input from the dynamically-sensitive TCMRO afferent, the T fibre. This input, however, is overridden by di- or polysynaptic inhibition during centrally generated remotor bursts. Among group 1 (assistance group) remotor motoneurones, one or more are also excited monosynaptically by the T fibre. Such motoneurones make inhibitory central outputs, which contribute to the indirect inhibition of promotor motoneurones by TCMRO stretch during remotor bursts (Skorupski, 1992; see Fig. 3).

In the crayfish walking system proprioceptive inputs to leg motoneurones can be centrally modulated. In addition, different individual motoneurones within a pool receive different patterns of centrally modulated input. Conceivably, then, there are two levels at which the CNS can select its feedback responses: 1) central regulation of the reflex pathways and 2) selective recruitment of subgroups of motoneurone with different inputs. To what extent does selective recruitment contribute to modulation of the net reflex response during normal behaviour? This is not known at present, although experiments with the amine neuromodulator, octopamine, provide circumstantial evidence that selective recruitment is possible.

Neuromodulation of Reflex Pathways

Octopamine abolishes any ongoing bursting activity and at higher concentrations inhibits any tonic promotor nerve activity (and sometimes induces tonic remotor nerve activity, a pattern never seen in normal saline). It also changes the prevailing reflex pattern in response to TCMRO/TCCO stimulation, so that resistance, rather than assistance, reflexes are favoured.

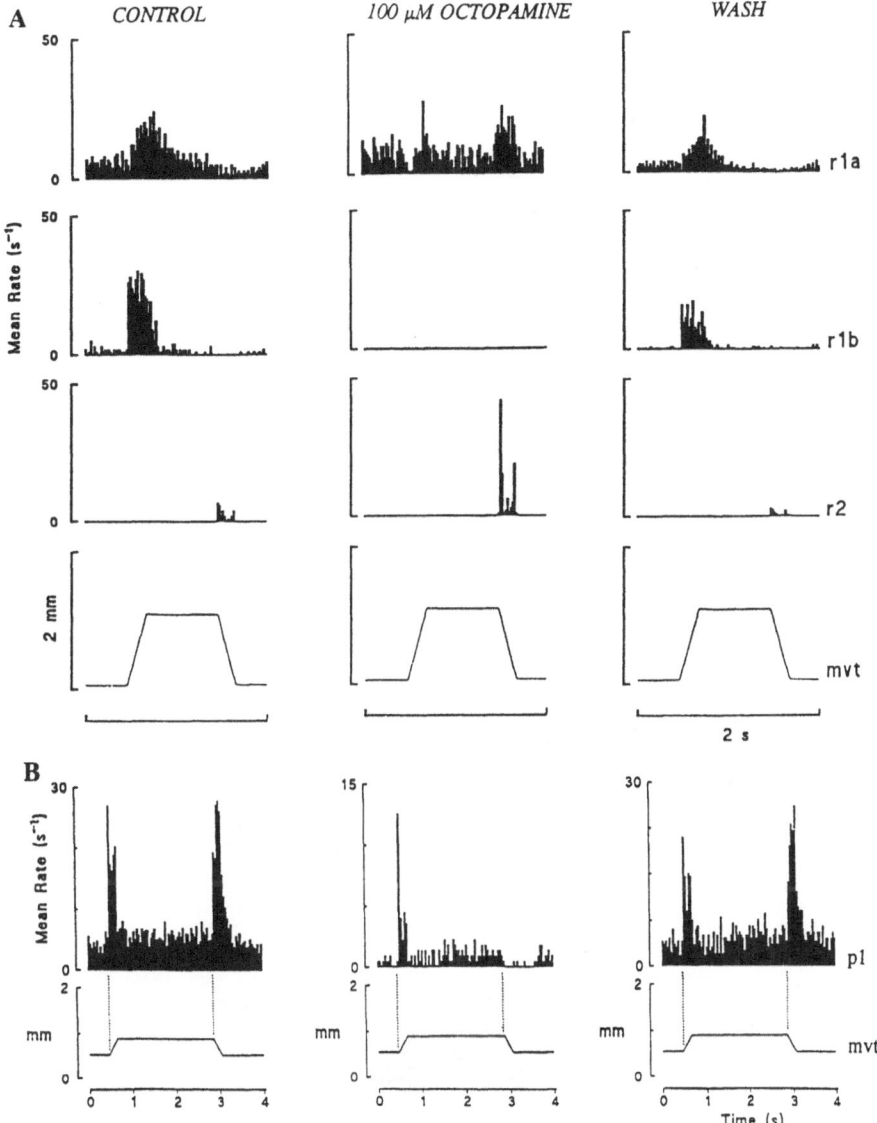

Figure 2. Octopamine selectively abolishes assistance reflexes. Net reflex responses (i.e. averaged without respect to phase of the motor output) of three remotor units (r1a, r1b, r2) compared before (left, n=50), during (middle, n=50) and after (right, n=48) bath application of 100μM octopamine. B. Net reflex responses of a promotor unit compared before, during and after application of 10μM octopamine (n=50 in each case). Skorupski, unpublished material.

The effect of octopamine in the remotor pool is illustrated in Fig. 2A. Reflex responses of three units, two group 1 and one group 2, are compared before, during and after octopamine application. While the two units classified as group 1 respond vigorously to stretch in normal saline, in the presence of octopamine this response is dramatically altered. In one case the motoneurone is inhibited and ceases to respond, while in the other the response changes from that typical of group 1 remotors (excitation on stretch) to that typical of group 2 (excitation on release). The response of the motoneurone classified as group 2 in normal saline is enhanced in octopamine. Octopamine also abolishes the assistance reflex in group 1 promotor motoneurones, while leaving any resistance reflex components qualitatively unaffected (Fig. 2B). Thus, in both promotor and remotor motor neurons, the overall effect of octopamine is to favour resistance over assistance reflexes.

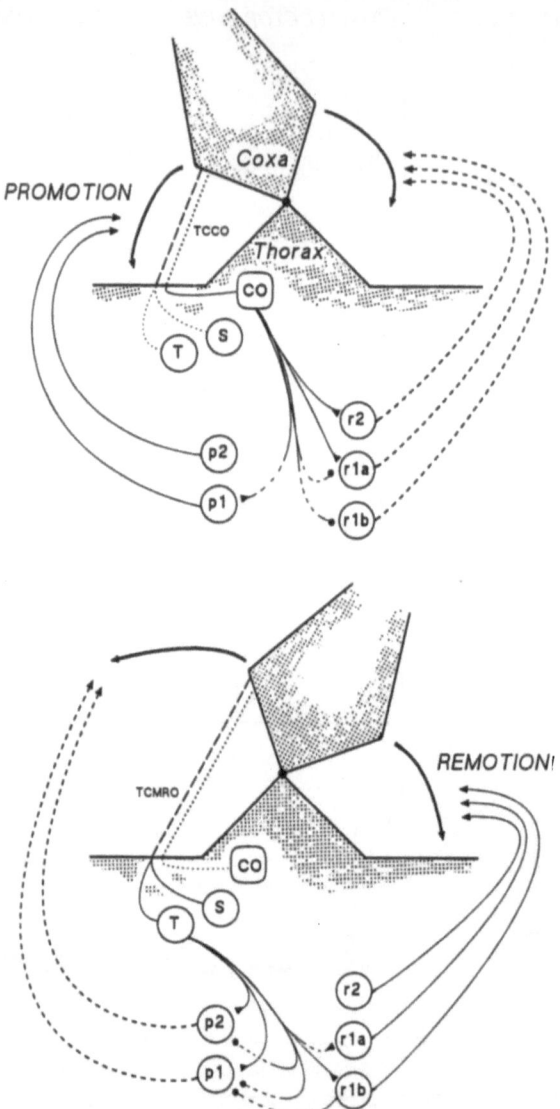

Figure 3. Diagrammatic representation of the crayfish thoracocoxal joint and reflex pathways considered in this review. *Upper diagram* shows the pathways activated by leg promotion and consequent receptor shortening, i.e. from the multiple, spiking, dynamically-sensitive TCCO afferents (CO), to the different subgroups of promotor (p) and remotor (r) motoneurones. *Lower diagram* shows pathways activated by leg remotion and receptor lengthening, from the single nonspiking, dynamically-sensitive T fibre of the TCMRO (reflex connections from the statically-sensitive S fibre of the TCMRO are omitted). Triangles represent excitatory synaptic actions, filled circles inhibitory. Sensori-motor connections thought *not* to be direct are indicated by broken lines.

Functional Implications and Comparison with other Systems

Fig. 3 summarizes the reflex pathways from TCMRO and TCCO afferents to promotor and remotor motoneurones. Positive feedback, assistance reflexes can be generated by either direction of movement: remotion, via the dynamically-sensitive T fibre of the TCMRO, can excite remotor motoneurones. Similarly, promotion, via dynamically-sensitive TCCO afferents, can excite promotor motoneurones. In each case the synaptic input underlying the positive feedback reflex is restricted to a subgroup of the motor pool.

Normally the assistance group (group 1) motoneurones are spontaneously active, with promotor and remotor units firing in alternating bursts. Oscillatory central drive preferentially recruits motoneurones involved in positive feedback loops, and assistance reflexes can readily be demonstrated. Octopamine, however, reverses this pattern, so that resistance reflexes are facilitated, or relatively unaffected, while assistance reflexes are abolished. This is accomplished in two ways. First, individual group 1 motoneurones are selectively inhibited so that assistance reflexes remain subthreshold. Second, a motoneurone's configuration of inputs can change so that it 'switches allegiance' between feedback groups of the motor pool, responding in the group 2 resistance, rather than group 1 assistance reflex modes. This is reminiscent of the stomatogastric system, where the activity of a neurone may be switched from one motor circuit to another by a neuromodulator (see Katz, this volume).

To what extent is the reflex specialization demonstrated in crayfish promotor and remotor motoneurones a matter of general interest? In crayfish it seems that movement-sensitive positive and negative feedback systems are operating, to a considerable extent, in parallel. Group 1 motoneurones, which, during rhythmic activity, are the earliest to be recruited by central drive, would be reinforced by feedback from the movement they generate. Group 2 motoneurones, which receive roughly in-phase central drive, would be able to respond vigorously to any perturbation of the ongoing movement.

The output of these two systems is summed in the periphery. The two systems are not segregated into separate muscular 'compartments'; instead, individual muscle fibres can be polyneuronally innervated by a mixture of group 1 and group 2 units (Skorupski, Vescovi and Bush, 1994). This may be related to the fact that crayfish leg motoneurones have extensive *central* outputs, both excitatory and inhibitory (Skorupski and Sillar, 1988; Chrachri and Clarac, 1989; Skorupski, 1992). As a consequence, these motoneurones themselves may be significantly involved in sensorimotor integration. For example, reflex reversal in crayfish may be mediated at least partly by these leg motoneurones since, by virtue of their central synaptic outputs in conjunction with their selective patterns of proprioceptive input, they can bias the net reflex effect, and thus determine whether a given sensory input results in reflex excitation or inhibition (Skorupski and Sillar, 1988; Skorupski, 1992). In other systems, on the other hand, reflex reversal may involve alternative, interneuronal, modes of processing (see chapters by Hultborn; and McCrea, Shefchyk and Pearson, this volume). This suggests that the subdivision of crayfish motor pools into feedback groups may not be directly comparable to other types of specialisation within a motor pool (e.g. subdivision of the motor pool of a bifunctional muscle into separate 'task groups': Loeb, 1985; see also Windhorst, Hamm and Stuart, 1989). The conceptual equivalent of a feedback group in other systems may be represented by different combinations of local interneurones and motoneurones.

REFERENCES

Alexandrowicz, J.S. and Whitear, M. (1957). Receptor elements in the coxal region of Decapoda Crustacea. *Journal of the Marine Biological Association of the United Kingdom* **36**, 603-628.

Andersson, O. and Grillner, S. (1981). Peripheral control of the cat's step cycle. I. Phase dependent effects of ramp-movements of the hips during fictive locomotion. *Acta Physiologica Scandinavica* **113**, 89-101.

Andersson, O. and Grillner, S. (1983). Peripheral control of the cat's step cycle. II. Entrainment of the central pattern generators for locomotion by sinusoidal hip movements during fictive locomotion. *Acta Physiologica Scandinavica* **118**, 229-239.

Bässler, U. (1986). Afferent control of walking movements in the stick insect *Cuniculina impigra*. II. Reflex reversal and the release of the swing phase in the restrained foreleg. *Journal of Comparative Physiology* **158**, 351-362.

Bush, B.M.H. (1981). Non-impulsive stretch receptors in crustaceans. In *Neurones without Impulses: their Significance for Vertebrate and Invertebrate Nervous Systems*. Society for Experimental Biology Seminar Series **6**. ed. Roberts, A. and Bush, B.M.H., pp.147-176. Cambridge University Press, Cambridge.

Cattaert, D., El Manira, A., and Clarac, F. (1992). Direct evidence for presynaptic inhibitory mechanisms in crayfish sensory afferents. *Journal of Neurophysiology* **67**, 610-624.

Chrachri, A. and Clarac, F. (1989). Synaptic connections between motor neurons and interneurons in the fourth thoracic ganglion of the crayfish, *Procambarus clarkii*. *Journal of Neurophysiology* **62**, 1237-1250.

DiCaprio, R.A. and Clarac, F. (1981). Reversal of a walking leg reflex elicited by a muscle receptor. *Journal of Experimental Biology* **90**, 197-203.

Elson, R.C., Sillar, K.T. and Bush, B.M.H. (1992). Identified proprioceptive afferents and motor rhythm entrainment in the crayfish walking system. *Journal of Neurophysiology* **67**, 530-546.

Hasan, Z. and Stuart, D.G. (1988). Animal solutions to problems of movement control: the role of proprioceptors. *Annual Review of Neuroscience* **11**, 199-223.

Hultborn, H. (1994). "Autogenetic" excitation of extensors during locomotion. *This volume.*

Katz, P.S. (1994). Neuromodulation and motor pattern generation in the crustacean stomatogastric system. *This volume.*

Loeb, G.E. (1985). Motor neuron task groups: coping with kinematic heterogeneity. *Journal of Experimental Biology* **115**, 137-146.

McCrea, D., Shefchyk, S. and Pearson, K.G. (1994). Activation of golgi tendon organ afferents produces disynaptic excitation and not inhibition of synergists during fictive locomotion. *This volume.*

Sillar, K.T. and Skorupski, P. (1986). Central input to primary afferent neurons in crayfish, *Pacifastacus leniusculus*, is correlated with rhythmic motor output of thoracic ganglia. *Journal of Neurophysiology* **55**, 678-688.

Sillar, K.T., Skorupski, P., Elson, R.C. and Bush, B.M.H. (1986). Two identified afferent neurones entrain a central locomotor rhythm generator. *Nature* **323**, 440-443.

Skorupski, P. (1992). Synaptic connections between nonspiking afferent neurons and motor neurons underlying phase-dependent reflexes in crayfish. *Journal of Neurophysiology* **67**, 664-679.

Skorupski, P. and Bush, B.M.H. (1992). Parallel reflex and central control of promotor and receptor motorneurons in crayfish. *Proceedings of the Royal Society* series B **249**, 7-12.

Skorupski, P., Rawat, B.M. and Bush, B.M.H. (1992). Heterogeneity and central modulation of feedback reflexes in crayfish motor pool. *Journal of Neurophysiology* **67**, 648-663.

Skorupski, P. and Sillar, K.T. (1986). Phase-dependent reversal of reflexes mediated by the thoracocoxal muscle receptor organ in the crayfish, *Pacifastacus leniusculus*. *Journal of Neurophysiology* **55**, 689-695.

Skorupski, P. and Sillar, K.T. (1988). Central synaptic coupling of walking leg motor neurones in the crayfish: implications for sensorimotor integration. *Journal of Experimental Biology* **140**, 355-380.

Skorupski, P., Vescovi, P.J. and Bush, B.M.H. (1994). Integration of positive and negative feedback loops in a crayfish muscle. *Journal of Experimental Biology* **187**, 305-313.

Windhorst, U., Hamm, T.M. and Stuart, D.G. (1989). On the function of muscle and reflex partitioning. *Behavioural and Brain Science* **12**, 629-681.

PRIMITIVE ROLE FOR GABAERGIC RETICULOSPINAL NEURONES IN THE CONTROL OF LOCOMOTION

A. ROBERTS

Biological Sciences
University of Bristol
Bristol, BS8 1UG, U.K.

SUMMARY

All animals need to be able to stop locomotion as well as start it. In hatchling embryos of the clawed toad *Xenopus*, swimming locomotion can be stopped by pressure on the head or cement gland. The receptors for this response are trigeminal neurones with free nerve endings which project centrally into the brainstem. Our evidence suggests that these receptors excite GABAergic reticulospinal neurones which project to both sides of the spinal cord to inhibit spinal motoneurones and interneurones and in this way turn off the spinal locomotor pattern generator to terminate swimming locomotion. Similar reticulospinal GABAergic pathways have been shown anatomically in mammals.

INTRODUCTION

An extensive system of neurones projecting from the brainstem to the spinal cord is a feature of all vertebrates. These neurones, characterised anatomically by widespread dendritic fields and axonal projections, form the reticulospinal system (Kuypers, 1981). The evidence has suggested that they have rather general and non-specific roles in the activation or inhibition of spinal motor systems (Magoun, 1963). More recently it has been shown that some stain for inhibitory transmitters such as GABA and glycine (Holstege, 1991; Holstege and Bongers, 1990). While direct evidence is lacking on the behavioural role of these neurones in adult vertebrates, there is strong circumstantial evidence from a very simple vertebrate, the *Xenopus* embryo, that GABAergic reticulospinal neurones are responsible for terminating swimming locomotion in response to sensory stimulation. This paper will review this evidence and suggest that GABAergic reticulospinal neurones in adult vertebrates could serve a similar function.

The *Xenopus* Embryo and its Behaviour

After two days of development the *Xenopus* embryo is about 5mm long and ready to hatch (Fig. 1A). When released from the egg it spends most of its time hanging from a strand of mucus secreted by a cement gland on the front of the head. However, if it is touched anywhere on the body its usual response is to swim off and continue swimming until its head contacts support. It then stops and becomes attached by mucus secreted by the cement gland (Fig. 1B). A fictive correlate of all this behaviour can be recorded from the ventral roots of embryos immobilised in a neuromuscular blocking agent such as α-bungarotoxin and this has allowed detailed investigation of the neural mechanisms underlying the behaviour (Roberts, 1990). The basic motor output for swimming, like that for rhythmic activities in so many other animals (Roberts and Roberts, 1983), appears to be produced by a largely spinal central pattern generator and the simplicity of the embryo spinal cord has allowed us to identify and study the roles and functions of most of the neurones involved in the chain of events from a sensory stimulus to the generation of swimming.

If during fictive swimming pressure is slowly applied to the head skin or cement gland then swimming very often stops. On the other hand, if the trunk skin is stimulated, this usually leads to an increase rather than a decrease in the swimming frequency (Sillar and Roberts, 1992). It is the pathways mediating the first of these two responses, the stopping response, that implicates inhibitory reticulospinal neurones and the first question is: what are the sensory receptors for this response?

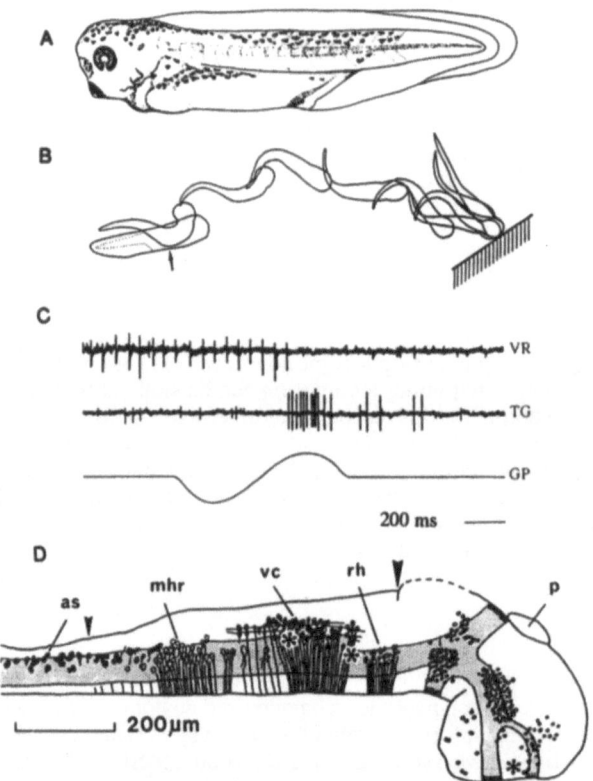

Figure 1. A: *Xenopus* embryo at stage of hatching is 5mm long. **B:** Frames from a video at 200fps show swimming when the embryo is touched on the trunk (at arrow) which stops when the embryo bumps into the side of the dish (hatched; K.M. Boothby, unpublished). **C:** After neuromuscular block, swimming activity can be recorded in ventral roots (VR). A gentle pressure stimulus to the cement gland (when GP trace moves up) leads to impulse discharge in trigeminal sensory neurones (TG) and termination of swimming (Boothby and Roberts, 1992a). **D:** Lateral view of right side of the brain showing neurones and axons tracts (stippled) with GABA-like immunoreactivity. In the hindbrain (between arrowheads) there are three groups all with decussating commissural axons: "midhindbrain reticulospinal neurones" (mhr) which are the only group to project axons into the spinal cord at this stage; vestibular commissural (vc); and rostral hindbrain (rh). The spinal cord has ascending neurones (as) with ascending axons. p is the pineal (Roberts *et al., 1987*).

The Sensory Receptors for the Stopping Response

The head skin and cement gland of the *Xenopus* embryo are innervated by neurones in the trigeminal ganglia (Hayes and Roberts, 1983). These can be divided into two clear populations on the basis of their extracellularly recorded impulse discharge responses to deformation of the skin or cement gland surface (Roberts and Blight, 1975; Roberts, 1980). The first class could loosely be called "pressure" receptors as they discharge repeatedly during slow, broad deformation such as might occur when an embryo swims into a solid object (Fig. 1C). The second class are rapidly adapting transient deformation or "touch" receptors which discharge very few impulses to the most rapid of local indentations of the skin surface. Clearly the first class of "pressure" receptors which are particularly dense in the cement gland itself and are absent from the trunk skin are the candidate receptors for the stopping response especially as the kinds stimuli which excite them, if given at rest, do not initiate swimming whereas "touch" stimuli do.

Inhibitory Neurones in the *Xenopus* Embryo Brainstem

How could trigeminal "pressure" receptors stop swimming? The most obvious proposal is that their central descending axons, which project through the brainstem but only rarely reach the spinal cord, could excite inhibitory reticulospinal neurones. Inhibitory neurone populations have been revealed in the *Xenopus* embryo brainstem by using antibodies to the glutaraldehyde fixation products of the inhibitory transmitters glycine and GABA. Only two anatomical classes of brainstem inhibitory neurone with spinal projections were uncovered at the hatching stage of development, both lying in the caudal half of the hindbrain. The first class which we have called "commissural interneurones" showed glycine-like immunoreactivity. These neurones have a ventral decussating axon and other anatomical features which suggested that they were the brainstem extension of the longitudinal column of spinal reciprocal inhibitory interneurones (Dale *et al,* 1986; Roberts *et al,* 1988). The second class, which we have called "midhindbrain reticulospinal neurones", showed GABA-like immunoreactivity (Fig. 1D) and were unusual in having descending projections to both sides of the spinal cord (Roberts *et al.,* 1987). Both these classes of neurone had dendrites in a position to contact the descending axons of trigeminal sensory receptors and axons reaching down into the spinal cord, so both could be part of the stopping pathway.

Pharmacology of the Stopping Response

To resolve whether the stopping response was dependent on glycinergic or GABAergic inhibition we examined the pharmacology of the response. Stopping behaviour and fictive stopping were both blocked by the $GABA_A$ antagonist bicuculline ($10\mu M$) (Boothby and Roberts, 1992b). To approach the mechanisms more directly we made intracellular recordings from spinal motoneurones and gave slow "pressure" stimuli to the cement gland to specifically excite the "pressure" receptors. These stimuli lead to large compound IPSPs in the motoneurones which were blocked by $10\mu M$ bicuculline (Fig. 2A) but were unaffected by the glycine antagonist strychnine ($2\mu M$). Similar compound IPSPs were seen in motoneurones when fictive swimming was stopped by cement gland stimulation (Fig. 2B). The IPSPs in motoneurones could also be blocked by the excitatory amino acid antagonist kynurenic acid (0.5mM).

These experiments suggested that the trigeminal "pressure" receptors excited GABAergic interneurones in the hindbrain and that these in turn produced GABA mediated inhibition of spinal neurones which could stop swimming. The "midhindbrain reticulospinal neurones" were therefore the obvious candidates. The effects of kynurenic acid implied that the trigeminal afferents or some other neurones in a more indirect pathway exciting the GABAergic interneurones could release a transmitter like glutamate.

Lesion Studies on the Stopping Response

The pharmacological experiments suggested a direct pathway from trigeminal afferent axons to the GABAergic "midhindbrain reticulospinal neurones" which would then project to both sides of the spinal cord to turn off swimming. We therefore performed lesion experiments to test whether input from one trigeminal ganglion was sufficient to stop swimming on each side of the spinal cord (Boothby and Roberts, 1992a; Fig. 2C to F). The stopping response was still reliable after cutting all commissural connections in the hindbrain and also when the only commissural connections to the opposite side were in the region of the "midhindbrain reticulospinal neurone" commissural axons (Fig. 2D, E). These experiments also showed that mid- and fore-brain were not needed for the stopping response. However, further lesions where the hindbrain was divided along the midline except in the most rostral region showed that there was an additional commissural pathway in the rostral hindbrain (Fig. 2F).

271

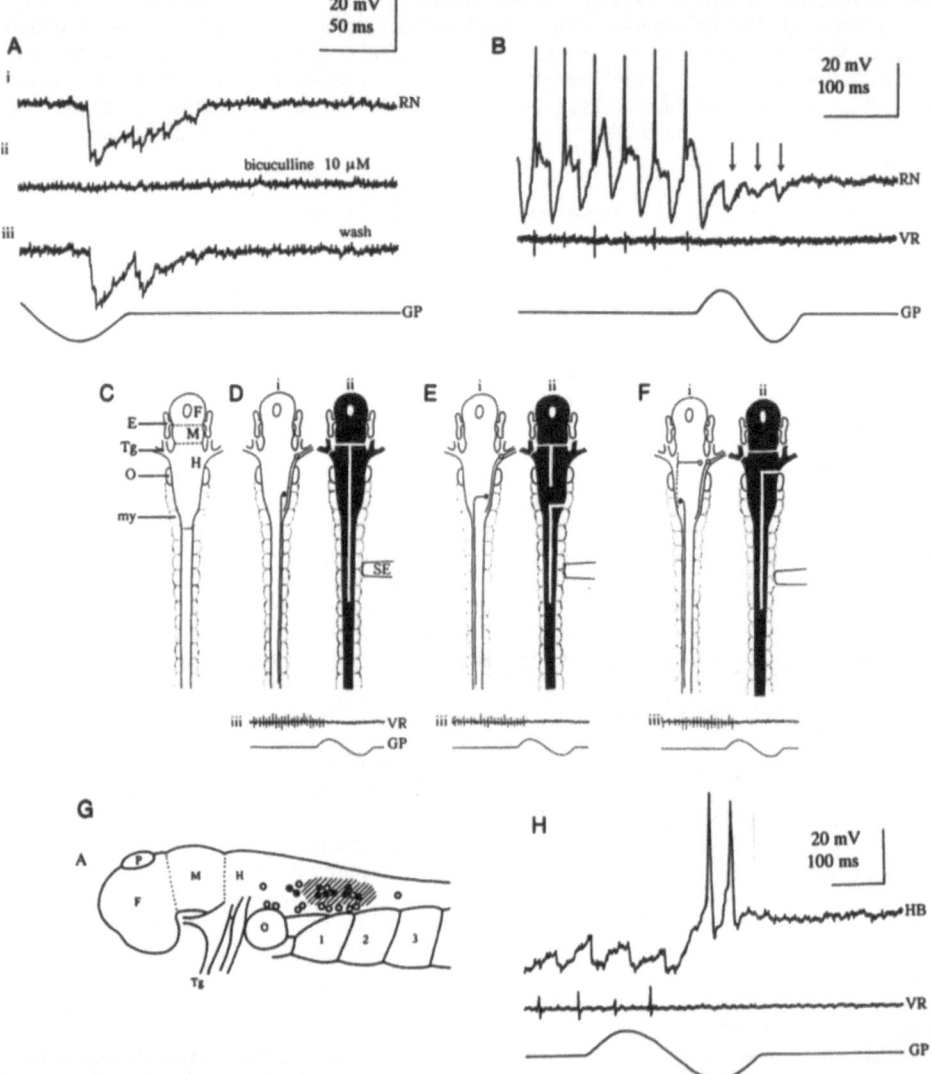

Figure 2. A: Pressure on the cement gland (GP trace moves down) leads to compound IPSP, seen in intracellular recording from a spinal rhythmic neurone (RN), which is reversibly blocked by bicuculline. **B:** The same stimulus given during fictive swimming recorded in ventral root (VR) and intracellularly in a rhythmic neurone (RN), evokes IPSPs (at arrows) and stops swimming. **C to F:** Lesions which still permit the stopping response shown in diagrams of the nervous system in dorsal view (F, forebrain; M, midbrain; H, hindbrain; E, eye; Tg, trigeminal ganglion; O, otic capsule; my, myotome; SE, suction electrode). For each lesion (i) shows intact pathway, (ii) shows lesion where left trigeminal is always cut and (iii) shows pressure to cement gland (GP) stopping swimming (seen in VR; Boothby and Roberts, 1992a). **G:** Side view of brain to show location (filled circles) of recorded neurones responding to cement gland stimulation which coincides with area of "midhindbrain reticulospinal neurones" (hatched). Open circles were rhythmically active neurones. **H:** Example of record from a hindbrain neurone inhibited during swimming (HB) and excited to fire by cement gland stimuli (GP trace moves down) which stopped swimming (seen in VR). A, B, G and H from Boothby and Roberts (1992b).

Recordings from Hindbrain Neurones

If GABAergic "midhindbrain reticulospinal neurones" are involved in the stopping response then one ought to be able to record from them and show that they are silent at rest and during swimming but are excited by the stimuli that usually stop swimming like slow pressure to the cement gland. Recordings from hindbrain neurones showed some neurones that were rhythmically active during swimming but in the region where "midhindbrain reticulospinal neurones" are located a number of neurones were recorded which had different properties (Boothby and Roberts, 1992b; Fig. 2G, H). These neurones were silent at rest and inhibited rhythmically during fictive swimming, like spinal sensory interneurones (Clarke and Roberts, 1984; Roberts and Sillar, 1990). They were not excited by stimuli which normally initiate swimming like local touch to the trunk skin or dimming the illumination (Roberts, 1990) but were excited, sometimes enough to fire impulses, by slow pressure to the cement gland. Furthermore these neurones could also be excited by pulling on the mucus strand secreted by the cement gland. This is a very specific stimulus for the "pressure" receptor sensory neurones innervating the cement gland (Roberts and Blight, 1975). When firing was induced in these neurones by intracellular current injection swimming was not initiated but neither was it stopped.

These recordings show that in the region of the brainstem where "midhindbrain reticulospinal neurones" lie there are neurones with many of the properties which would be expected from the inhibitory neurones of the stopping pathway. However, without dye injection to reveal anatomy or a demonstrated inhibitory effect on swimming, this evidence can only be circumstantial.

Conclusion on the *Xenopus* Embryo

I have sketched an outline of the evidence that swimming locomotion in *Xenopus* embryos can be terminated by a specific pathway which includes a population of GABAergic reticulospinal neurones and is shown in Figure 3. Stimuli to the head which stop swimming excite a specific population of trigeminal sensory neurones with unmyelinated axons and free nerve endings in the head skin and cement gland. These neurones project ipsilateral central axons caudally into the brainstem or hindbrain where the pharmacology suggests they release an excitatory amino acid transmitter. We have concluded that they then excite at least two classes of hindbrain interneurone. Firstly they excite a compact group of GABAergic interneurones with bilateral spinal projections which we have called "midhindbrain reticulospinal neurones". This conclusion depends on a number of lines of evidence: pharmacological experiments have shown that the stopping response and compound IPSPs evoked in spinal motoneurones by pressure on the cement gland are blocked by GABA antagonists and are therefore dependent on GABAergic inhibition. The "midhindbrain reticulospinal neurones" are the only hindbrain neurones with GABA-like immunoreactivity and descending spinal projections. Intracellular recordings in the region where these interneurones are located show some neurones excited by stimuli which stop swimming. Lesions which preserve the connections of the "midhindbrain reticulospinal neurones" do not affect the stopping response. Secondly, the lesion studies show that a stopping pathway crosses the hindbrain rostrally and well removed from the commissural axons of "midhindbrain reticulospinal neurones". This suggests that the pressure sensitive trigeminal afferents may also excite interneurones in the rostral hindbrain (E in Fig. 3) which can relay excitation across the nervous system to excite "midhindbrain reticulospinal neurones" on the opposite side.

We assume that the "midhindbrain reticulospinal neurones" make inhibitory synapses onto all classes of spinal interneurone that are involved in the generation of swimming activity. By producing similar IPSPs to those that we have recorded in motoneurones they would turn off rhythmic activity. The fact that each "midhindbrain reticulospinal neurone" has two descending axons going to each side of the spinal cord makes them unusual and particularly suited for a general inhibitory role.

GENERAL IMPLICATIONS

We are all aware of the calming effects of slow mechanical stimulation of the skin or massage. Our experiments on *Xenopus* embryos suggest that this pleasure may originate from a primitive system which is concerned with the termination of locomotion. Such systems are found in a wide range of animals from leeches and sea-slugs to fishes and mammals (see Boothby and Roberts, 1992a and b). Among the mammals it has been proposed that some reticulospinal pathways have a general inhibitory role in decreasing the excitability of spinal motoneurones (Holstege and Kuypers, 1982). There are a few cases

273

where evidence has been presented for direct inhibitory connections from the brainstem to spinal neurones (Magoun and Rhines, 1946; Llinas and Terzuolo, 1964; Holstege and Bongers, 1990). Recently, Holstege (1991) has shown one such connection in the rat to be GABAergic. This evidence suggests that there could be close parallels between the descending inhibitory pathways in mammals and the presumably primitive pathways described here in the *Xenopus* embryo. However, there are undoubtedly also other mechanisms like those described by Mori (1987) where brainstem excitatory neurones terminate locomotion by exciting spinal inhibitory neurones.

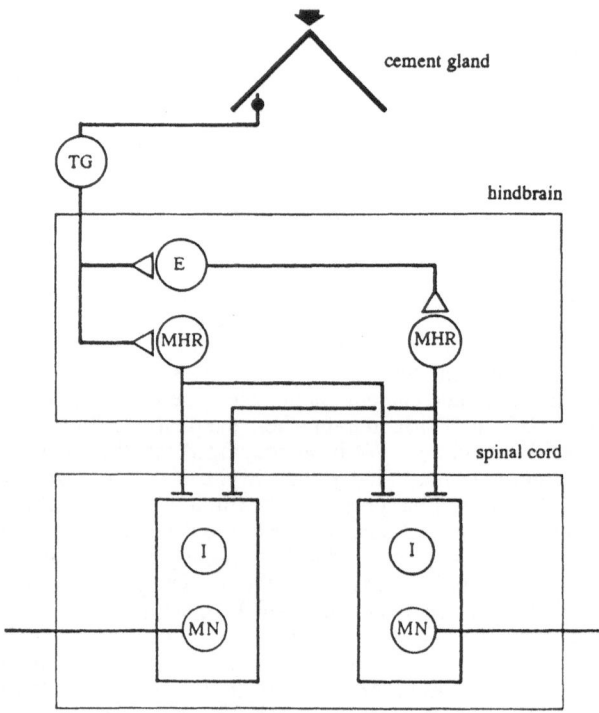

Figure 3. Proposed pathways for the stopping response. Gentle pressure to the cement gland (arrow) stimulates trigeminal "pressure" receptors (TG). TGs excite GABAergic "midhindbrain reticulospinal neurones" (MHR) and decussating excitatory interneurones (E). Es excite MHRs on the opposite side. The bilateral descending axons of MHRs produce GABAergic inhibition to turn off activity in spinal central pattern generator interneurones (I) and motoneurones (MN).Circles represent populations of neurones (Boothby and Roberts, 1992b).

ACKNOWLEDGEMENTS. The work described here was performed with Kate Boothby. I would like to thank her and my other colleagues in Bristol for their help and the SERC for support.

REFERENCES

Boothby, K.M. and Roberts, A. (1992a) The stopping response of *Xenopus laevis* embryos: behaviour, development and physiology. *Journal of Comparative Physiology* **170**, 171-180

Boothby, K.M. and Roberts, A. (1992b) The stopping response of *Xenopus laevis* embryos: Pharmacology and intracellular physiology of rhythmic spinal neurons and hindbrain neurons. *Journal of Experimental Biology* **169**, 65-86.

Clarke J.D.W. and Roberts A. (1984) Interneurones in the *Xenopus* embryo spinal cord: sensory excitation and activity during swimming. *Journal of Physiology* **354**, 345-362.

Dale, N., Ottersen, O.P., Roberts, A. and Storm-Mathisen, J. (1986) Inhibitory neurones of a motor pattern generator in *Xenopus* revealed by antibodies to glycine. *Nature* **324**, 255-257.

274

Hayes B.P. and Roberts A. (1983) The anatomy of two functional types of mechanoreceptive 'free' nerve-endings in the head skin of *Xenopus* embryos. *Proceedings of the Royal Society London (B)* **218**, 61-76.

Holstege, J.C. (1991) Ultrastructural evidence for GABAergic brainstem projections to spinal motorneurons in the rat. *Journal of Neuroscience* **11**, 159-167.

Holstege, J.C. and Bongers, C.M.H. (1990) Ultrastructural evidence that brainstem projections to spinal motorneurons contain glycine. *European Journal Neuroscience* (Suppl.) **3**, 96.

Holstege, J.C. and Kuypers, H.G.J.M. (1982) The anatomy of brainstem pathways to the spinal cord in the cat. A labelled amino acid tracing study. *Progress in Brain Research* **57**, 177-183.

Kuypers, H.G.J.M. (1981) Anatomy of descending pathways. pp 597-666, in *Handbook of Physiology*, Section 1: The nervous system, II Motor control, Part 1. ed. Brooks, V.B. American Physiological Society, Bethesda.

Llinas, R. and Terzuolo, C.A. (1964) Mechanisms of supraspinal actions upon spinal cord activities. Reticular inhibitory mechanisms on alpha-extensor motoneurons. *Journal of Neurophysiology* **27**, 579-591.

Magoun, H.W. (1963) Reticulo-spinal influences and postural regulation. pp 23-38 in *The waking Brain*. ed. Thomas. Springfield, Illinois.

Magoun, H.W. and Rhines, R. (1946) An inhibitory mechanism in the bulbar reticular formation. *Journal of Neurophysiology*, **9** 165-171.

Mori, S. (1987) Integration of posture and locomotion in acute decerebrate cats and in freely moving cats. *Progress in Neurobiology* **28**, 161-195.

Roberts, A (1980) The function and role of two types of mechanoreceptive "free" nerve endings in the head skin of amphibian embryos. *Journal of Comparative Physiology A*. **135**: 341-348.

Roberts, A. (1990) How does a nervous system produce behaviour? A case study in neurobiology. *Science Progress Oxford* **74**, 31-51.

Roberts, A. and Blight, A.R. (1975) Anatomy, Physiology and behavioural role of sensory nerve endings in the cement gland of embryonic *Xenopus*. *Proceedings of the Royal Society London B*. **296**: 195-212.

Roberts A. and Roberts B.L. (1983) *Neural origin of rhythmic movements*. SEB Symposium XXXVII. Cambridge University Press, Cambridge.

Roberts, A. and Sillar, K.T. (1990) Characterisation and function of spinal excitatory interneurons with commissural projections in *Xenopus laevis* embryos. *European Journal of Neuroscience* **2**, 1051-1062.

Roberts, A., Dale, N., Ottersen, O.P. and Storm-Mathisen, J. (1987) The early development of neurons with GABA immunoreactivity in the central nervous system of *Xenopus laevis* embryos. *Journal of Comparative Neurology*. **261**, 435-449.

Roberts, A., Dale, N., Ottersen, O.P. and Storm-Mathisen, J. (1988) Development and characterization of commissural interneurons in the spinal cord of *Xenopus laevis* embryos revealed by antibodies to glycine. *Development* **103**, 447-461.

Sillar, K.T. and Roberts, A. (1992) The role of premotor interneurones in phase-dependent modulation of a cutaneous reflex during swimming in *Xenopus laevis* embryos. *Journal of Neuroscience* **12**, 1647-1657.

NEUROMODULATION AND MOTOR PATTERN GENERATION IN THE CRUSTACEAN STOMATOGASTRIC NERVOUS SYSTEM

P. S. KATZ

Department of Neurobiology and Anatomy
University of Texas Medical School
P.O. Box 20708
Houston, Texas 77225 USA

SUMMARY

Neuromodulation is critically important to the functioning of nervous systems, yet it is not included in most descriptions of neuronal circuits. Neuromodulation has been extensively studied in the stomatogastric nervous system of decapod crustacea. Here it has been shown that neuromodulatory inputs to central pattern generator circuits play four key roles. 1) They can initiate and maintain rhythmic neuronal activity. 2) They allow a single, anatomically defined circuit to produce many different outputs. 3) They can cause cells to switch their activity from one neuronal circuit to another. 4) They can reconfigure entire networks so that previously independent circuits can function together in a coordinated fashion. Thus, the classical synaptic connections of a neuronal circuit are not enough to explain the behavioural output of that circuit; neuromodulation can alter the output by affecting cellular and synaptic properties. This imparts a greater flexibility upon nervous systems than can be attained through simple excitatory/inhibitory interactions.

INTRODUCTION

There are two types of synaptic transmission in the nervous system; synapses that mediate excitation or inhibition, and synapses that modulate the properties of other synapses or cells. Mediating (or transmitting) synapses generally utilize classical synaptic neurotransmission; a neurotransmitter activates a ligand-gated ion channel, resulting in a rapid synaptic current which causes either an excitatory post-synaptic potential (EPSP) or an inhibitory post-synaptic potential (IPSP). EPSPs and IPSPs are the basic mode of communication used in the nervous system. Modulating synapses, on the other hand, generally act via a second messenger system and, as a result, their effects have a more prolonged time course. Often, they do not evoke discrete PSPs. Furthermore, since neuromodulatory synapses alter cellular and synaptic properties, their effects can be conditional upon the state of the post-synaptic neurone.

Many neurotransmitters participate in both kinds of synaptic transmission. The small molecule transmitters such as acetylcholine, glutamate, and GABA activate receptors which are directly coupled to ion channels as well as receptors which are coupled to second messenger systems. For example, acetylcholine acts at both nicotinic and muscarinic receptors. Furthermore, there are hundreds of peptide neurotransmitters which have neuromodulatory actions. Thus, neuromodulation is extremely prevalent in the nervous system. Yet most circuit descriptions deal only with fast synaptic connections; the role of neuromodulation in the production of behaviour by the nervous system needs to be better understood.

Neural Control of Movement, Edited by W.R. Ferrell
and U. Proske, Plenum Press, New York, 1995

Modulatory Synapses Act to Change Cellular Properties

As an example of the roles of mediating and modulatory synapses, let us consider a sensory to motor reflex in the stomatogastric system of crabs. (The stomatogastric system will be explained in more detail below.) In the crab, there is a primary sensory neurone called GPR which synapses on a motoneurone called DG (Fig 1) (Katz *et al.,* 1989). The synapse from GPR to DG exhibits both mediating and modulatory components (Katz and Harris-Warrick, 1989). The mediating component is a rapid depolarization due to acetylcholine being released from GPR. The acetylcholine acts on nicotinic receptors, resulting in an EPSP with a rapid time course (Fig 1D).

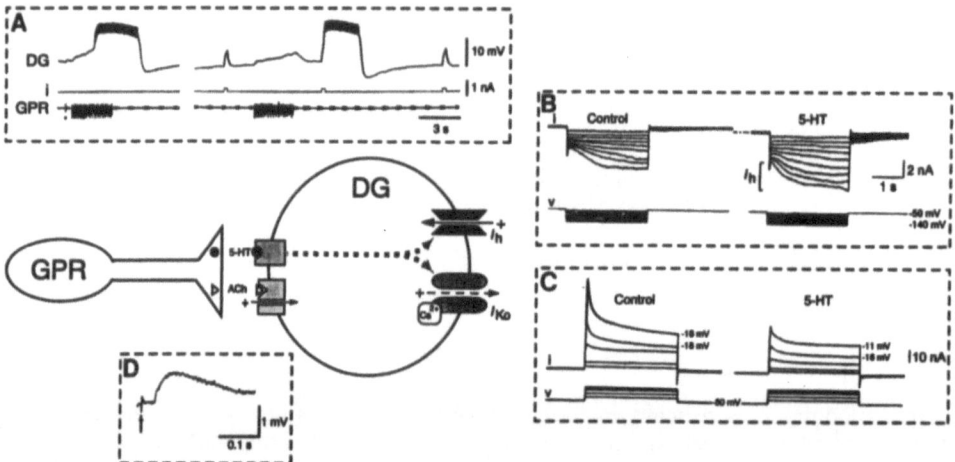

Figure 1. A primary sensory neurone called GPR synapses with a gastric mill motoneurone (DG). GPR has both mediating and modulatory effects by releasing both acetylcholine and serotonin. **A:** Stimulation of GPR triggers a plateau potential in DG. GPR induces the ability to produce a the plateau. Only immediately after GPR stimulation can a brief depolarizing current pulse evoke a prolonged plateau potential. Prior to GPR stimulation, the same current pulse is subthreshold for even an action potential. **B:** Serotonin enhances an inward current (I_h). **C:** Serotonin decreases an outward current (I_{KO}). **D:** The acetylcholine released GPR is responsible for rapid nicotinic EPSPs. (From Katz and Harris-Warrick, 1989; Kiehn and Harris-Warrick, 1992a,b)

The modulatory component of the synapse is due to the release of a cotransmitter, serotonin. Serotonin acts on one or more receptors (Zhang and Harris-Warrick, 1991) to affect at least two membrane conductances (Kiehn and Harris-Warrick, 1992a,b); serotonin increases an inward current called I_h or the "sag current", and serotonin decreases an outward calcium-dependent potassium current (I_{KO}). By altering these two conductances, serotonin changes the balance of currents in the cell, enabling DG to stay in a depolarized state, called a plateau potential, for many seconds following a brief depolarizing input. Thus, GPR, through the release of serotonin, endows DG with the ability to fire plateau potentials (Fig 1A).

The serotonin acts synergistically with the acetylcholine to allow a train of GPR spikes to trigger a plateau potential. The released serotonin induces the ability for plateaus while the acetylcholine provides the depolarizing trigger. The plateau is necessary for DG to carry out its reflex function. Thus, the induction of cellular properties by the neuromodulatory component of this synapse is an integral part of the function of the synapse.

We have seen that neuromodulation can alter the properties of neurones, endowing them with abilities that are crucial for their participation in circuits. Now, let's examine directly the role that neuromodulation plays in the control of neuronal circuits.

Four Sources of Neuromodulatory Input to Motor Circuits

One place that these questions about the role of neuromodulation in circuit function have been addressed is in the organization of pattern generating circuits. Central pattern generators (CPGs) are neuronal circuits which produce the timing cues for rhythmic behaviours such as walking, swimming, chewing, or breathing. There are four basic sources of neuromodulatory input to CPG circuits (Fig 2B) (Katz and Harris-Warrick, 1990a). 1) Other parts of the nervous system, including descending neurones from higher neuronal centres and modulatory neurones from other CPG circuits (Harris-Warrick and Marder, 1991) are well established sources of neuromodulation. 2) A less well recognized source of neuromodulatory input to a CPG is from primary sensory neurones (Katz *et al.,* 1989). 3) Circulating neurohormones affect CPG circuits. 4) Neurones intrinsic to the pattern generating circuit itself can have neuromodulatory effects within the same circuit (Katz *et al.,* 1994).

The Stomatogastric Nervous System, a Model System for Studying Neuromodulation

Invertebrate nervous systems are useful for examining the cellular rules which govern nervous system functions. In particular, the stomatogastric nervous system of decapod crustaceans has proven to be an extremely important system for understanding neuromodulation of pattern generating circuits (Harris-Warrick *et al.,* 1992). The stomatogastric ganglion consists of only 30 neurones which comprise all of the component neurones for two CPG circuits. These two circuits control the rhythmic movements of the gastric mill and pyloric regions of the foregut in animals such as lobsters and crabs. The stomatogastric nervous system can be removed from the animal, and maintained *in vitro* where it will continue to produce rhythmic motor patterns similar to those observed in the intact animal.

Thanks to many years of work, the connectivity of the pyloric and gastric mill CPG circuits has been determined and a complete wiring diagram exists (Fig 2A) (Harris-Warrick *et al.,* 1992). However, the wiring diagram does not predict the behavioural output produced by the circuit; knowing the mediating connections of a circuit is not sufficient to explain its activity pattern. This is because the properties of cells in the circuit are critically important to the production of the output pattern. In the pyloric circuit for example, it is important to know 1) that the AB cell is capable of endogenous oscillations, 2) that all of the cells exhibit post-inhibitory rebound, and 3) that some cells display more A-current than others (Harris-Warrick and Marder, 1991). These properties are not represented in the circuit diagram, but they are as important as the connections between the neurones themselves. Altering any of these properties results in a different motor pattern being produced by the same circuit. In fact, neuromodulatory inputs to the circuit are capable of doing just that.

Neuromodulators Initiate and Maintain Rhythmic Activity

Stomatogastric neurones appear to be dependent upon neuromodulatory inputs to express the very properties which are necessary for the production of any rhythmic output at all; when the stomatogastric ganglion is acutely isolated from modulatory inputs, it ceases to produce any rhythmic output. Rhythmic activity can be restored by applying any one of a number of neuromodulatory substances to the bathing medium. Without the correct expression of cellular properties, the circuit can not function as a pattern generator. Thus, one role of neuromodulatory inputs is to initiate and maintain motor patterns by endowing cells with certain properties (Fig 3A).

The ability of neuromodulators to initiate rhythmic activity is not an artifact of the *in vitro* preparation; similar phenomena are observed *in vivo*. For example, the peptide, cholecystokinin (CCK) initiates and maintains gastric mill activity in freely moving spiny lobsters (Turrigiano and Selverston, 1990). Prior to feeding, the gastric mill is silent and circulating levels of CCK are low. Upon feeding the animal, CCK levels immediately rise and the gastric mill becomes active. The level of CCK in the blood corresponds to the level of gastric mill activity. If the CCK antagonist, proglumide, is injected into the animal prior to feeding, then gastric mill activity is suppressed. Furthermore, if CCK itself is injected in the animal the gastric mill is activated, even without feeding the animal. Bath application of CCK onto an isolated stomatogastric ganglion will also initiate gastric mill activity (Turrigiano and Selverston, 1989). Thus CCK seems to cause the gastric mill circuit to become active and is involved in the maintenance of that rhythmic activity.

Neuromodulators Alter Production of Ongoing Motor Patterns

Neuromodulators not only initiate and maintain rhythmic activity, but can qualitatively alter the motor pattern that is produced by a CPG circuit, causing a single circuit to produce

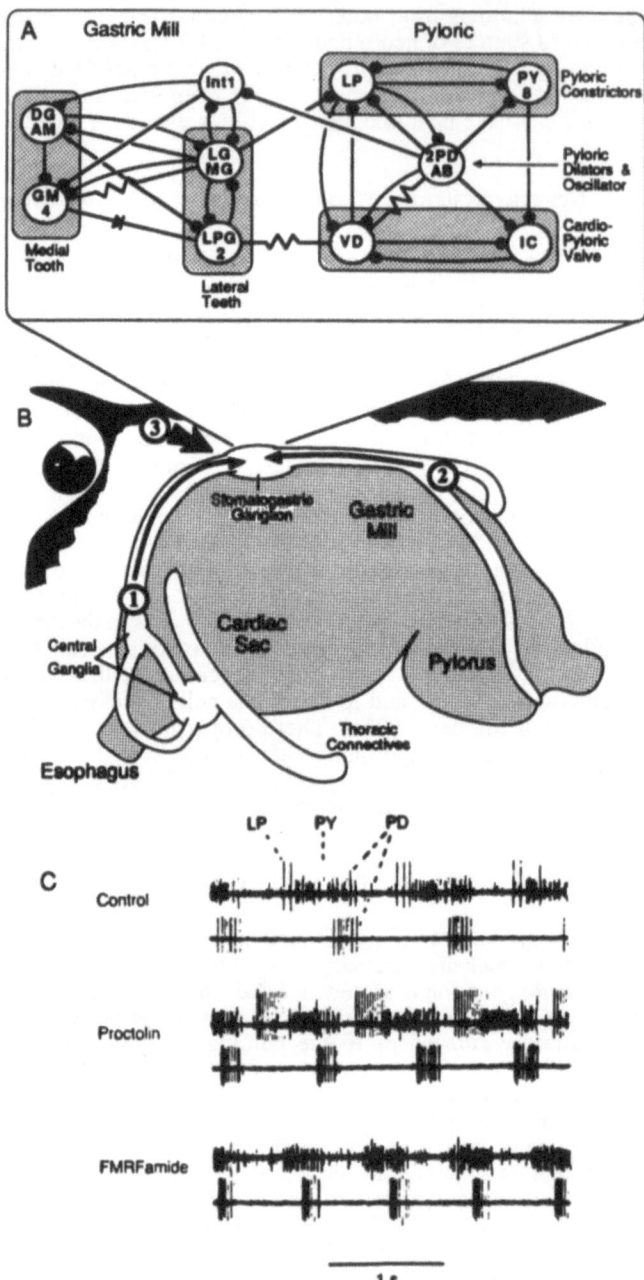

Figure 2. A: The circuit diagrams of the gastric mill and pyloric CPGs. **B:** Schematic diagram showing the locations of the regions of the foregut and the stomatogastric nervous system. Neuromodulatory inputs to the stomatogastric ganglion arise from three sources: 1) neurones in other parts of the nervous system, 2) primary sensory neurones such as GPR, and 3) circulating neurohormones. As yet, none of the neurones intrinsic to the stomatogastric ganglion have been shown to have modulatory effects within the ganglion. **C:** Extracellular recordings of the pyloric motor pattern. Addition of the peptides: Proctolin and FMRFamide cause different patterns to be produced by a single circuit. (from Katz and Harris-Warrick, 1990a and Marder, 1984).

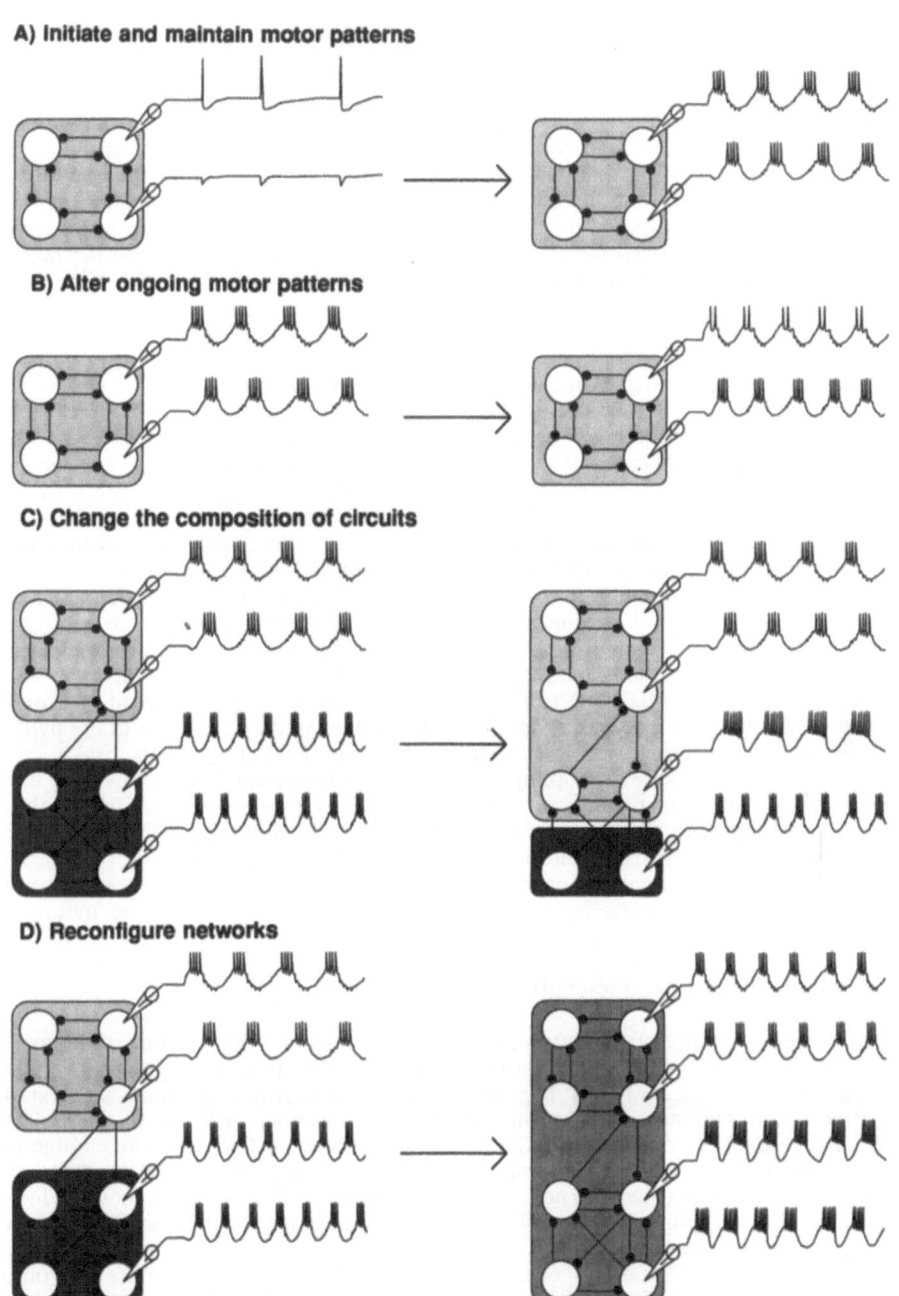

A) Initiate and maintain motor patterns

B) Alter ongoing motor patterns

C) Change the composition of circuits

D) Reconfigure networks

Figure 3. The functions of neuromodulation.

multiple outputs. For example, application of different peptide neuromodulators can alter the output of the pyloric motor pattern (Fig 2C).

Neuromodulatory changes can also be observed by stimulating neurones that release neuromodulatory substances. For instance, stimulation of the serotonergic GPR cells alters the output of the pyloric pattern (Katz and Harris-Warrick, 1990b). These neuromodulatory neurones have different effects on each of the cells in the pyloric circuit, lasting for up to a minute. The effects include an enhancement of oscillation frequency, prolonged excitation of some cells, and prolonged inhibition of other cells. The result is that the level of activity of the neurones changes and the phase relations between the neurones changes. Thus the neuromodulatory abilities of GPR allow this cell to have both short and long-lasting effects on the behavioural output of the circuit.

The use of neuromodulatory synapses by GPR also allows this sensory neurone to have phase-independent effects on the pyloric circuit. GPR responds to movements of the gastric mill, which has a period of oscillation between 5 and 15 seconds. Yet it affects the pyloric circuit, which produces a motor pattern with an oscillation period of 1-2 sec. Since the GPR synaptic effects are not mediated by synaptic potentials, they do not have to be precisely timed to the ongoing pyloric pattern. As a result, GPR can fire at any time in the pyloric pattern and, by biasing the behaviour of cells in the circuit, its effects will be integrated automatically into the rhythmic pattern. Thus, periodic neuromodulatory input acts as a phase-independent mechanism to alter the ongoing output of a CPG circuit.

Neuromodulatory Inputs Alter the Neuronal Circuit Composition

By changing the properties of cells, or altering the strengths of synapses, neuromodulatory inputs can cause neurones to switch allegiance from one pattern generating circuit to another (Fig. 3 C), thereby changing the composition of the circuits. There are two good examples of this in the stomatogastric system. The first is that when the gastric mill is silent, two gastric mill neurones (MG and LG) can be recruited into the pyloric motor pattern by brief stimulation of the GPR cells (Katz and Harris-Warrick, 1991).

The second example is that when the cardiac sac CPG (which controls another portion of the foregut) is quiescent, the VD neurone acts as a member of the pyloric circuit. However, when the cardiac sac is active, VD switches allegiance and fires in phase with the slower cardiac sac rhythm (Hooper and Moulins, 1989). The mechanism underlying this switch has been determined. While VD is part of the pyloric circuit, it exhibits plateau potentials. These plateau properties are necessary for VD to fire at the rapid cycle period of the pyloric pattern. However, when the cardiac sac is activated, neuromodulatory inputs disable these plateau properties, effectively decoupling VD from the pyloric circuit and allowing it to be driven by the cardiac sac pattern generator. Thus, by enabling or disabling cellular properties, neuromodulatory inputs can cause cells to switch from one circuit to another.

This is a rather surprising result because it means that the composition of neuronal circuits is not uniquely determined by anatomical connections. Rather, neurones can be put into and taken out of circuits by altering their ability to respond to the mediating connections of that circuit.

Neuromodulatory Inputs can Reconfigure Entire Networks

Finally, neuromodulatory inputs can cause an entire reorganization of neuronal networks, forcing circuits that operate independently to instead operate conjointly (Fig 3D). There are two examples of this happening in the stomatogastric nervous system. Application of the peptide, red pigment concentrating hormone (RPCH) to the isolated stomatogastric nervous system causes the gastric mill circuit to fuse with the cardiac sac circuit (Dickinson et al., 1990). This results in an entirely new pattern being produced at a frequency intermediate between the normal gastric mill and cardiac sac frequencies. One of the mechanisms underlying the effects of RPCH is a large change in the synaptic efficacy of synapses coordinating the two circuits.

A second example of the ability of neuromodulatory inputs to reconfigure a network is seen with the PS neurone (Meyrand et al., 1991, 1994). When PS is activated, it imposes its own rhythm on the gastric mill, pyloric, and oesophageal networks, causing these three pattern generating circuits to act as one. It does this by the same mechanisms that have already been mentioned: alteration of cellular properties, prolonged biasing of membrane potentials, and modulation of synaptic efficacy.

Neuromodulation Endows Circuits with Enormous Flexibility

It is clear why neuromodulation is critically important for the ability of neuronal circuits to function. Neuromodulators allow circuits to be activated when needed (Fig 3A). Neuromodulatory inputs also allow a single circuit to produce a multitude of different outputs (Fig 3B). Neuromodulation allows inputs to act in a phase-independent fashion, freeing them from the necessity of being precisely timed to a rhythmic pattern.

Neuromodulatory inputs can alter the composition of circuits (Fig 3C); thus, the neuronal membership of a circuit can vary depending on the circumstances. Finally, neuromodulatory inputs can completely reconfigure a network, allowing circuits to operate in different modes: either jointly or independently (Fig 3D).

These abilities of neuromodulation arise because cellular properties are crucial for the determination of the output of a circuit; the anatomical connectivity of the circuit does not uniquely determine its output. By changing the properties of individual neurones, neuromodulatory inputs endow motor circuits flexibility that cannot be attained through mediating synapses alone.

REFERENCES

Dickinson P.S., Mecsas C. and Marder E. (1990) Neuropeptide fusion of two motor pattern generator circuits *Nature* **344**, 155-158.

Harris-Warrick R.M. and Marder E. (1991) Modulation of neural networks for behavior. *Annual Review of Neuroscience* **14**, 39-57.

Harris-Warrick R.M., Marder E., Selverston A.I. and Moulins M. (eds) (1992) *Dynamic Biological Networks, The stomatogastric nervous system.* MIT Press, Cambridge, MA.

Hooper S.L. and Moulins M. (1989) Switching of a neuron from one network to another by sensory induced changes in its membrane properties. *Science* **244**, 1587-1589.

Katz P.S., Eigg M.H. and Harris-Warrick R.M. (1989) Serotonergic/cholinergic muscle receptor cells in the crab stomatogastric nervous system. I. Identification and characterization of the gastropyloric receptor cells. *Journal of Neurophysiology* **62**, 558-570.

Katz P.S., Getting P.A. and Frost W.N. (1994) Dynamic Neuromodulation of synaptic strength intrinsic to a central pattern generator circuit. *Nature* **367**, 729-731.

Katz P.S. and Harris-Warrick R.M. (1989) Serotonergic/cholinergic muscle receptor cells in the crab stomatogastric nervous system. II. Rapid nicotinic and prolonged modulatory effects on neurons in the stomatogastric ganglion. *Journal of Neurophysiology* **62**, 571-581.

Katz P.S. and Harris-Warrick R.M. (1990a) Actions of identified neuromodulatory neurons in a simple motor system. *Trends in Neuroscience* **13**, 367-373.

Katz P.S. and Harris-Warrick R.M. (1990b) Neuromodulation of the crab pyloric central pattern generator by serotonergic/cholinergic proprioceptive afferents. *Journal of Neuroscience* **10**, 1495-1512.

Katz P.S. and Harris-Warrick R.M. (1991) Recruitment of crab gastric mill neurons into the pyloric motor pattern by mechanosensory afferent stimulation. *Journal of Neurophysiology* **65**, 1442-1451.

Kiehn O. and Harris-Warrick R.M. (1992a) Serotonergic stretch receptors induce plateau properties in a crustacean motor neuron by a dual-conductance mechanism. *Journal of Neurophysiology* **68**, 484-495.

Kiehn O. and Harris-Warrick R.M. (1992b) 5-HT modulation of hyperpolarization-activated inward current and calcium-dependent outward current in a crustacean motor neuron. *Journal of Neurophysiology* **68**, 496-508.

Marder E. (1984) Mechanisms underlying neurotransmitter modulation of neuronal circuits. *Trends in Neuroscience* **7**, 48-53.

Meyrand P., Simmers J. and Moulins M. (1991) Construction of a pattern-generating circuit with neurons of different networks. *Nature* **351**, 60-63.

Meyrand P., Simmers J. and Moulins M. (1994) Dynamic construction of a neural network from multiple pattern generators in the lobster stomatogastric nervous system. *Journal of Neuroscience* **14**, 630-644.

Turrigiano G.G. and Selverston A.I. (1989) Cholecystokinin-like peptide is a modulator of a crustacean central pattern generator. *Journal of Neuroscience* **9**, 2486-2501.

Turrigiano G.G. and Selverston A.I. (1990) A cholecystokinin-like hormone activates a feeding-related neural circuit in lobster. *Nature* **344**, 866-868.

Zhang B. and Harris-Warrick R.M. (1994) Multiple receptors mediate the modulatory effects of serotonergic neurons in a small neural network. *Journal of Experimental Biology* **190**, 55-77.

MECHANOSENSORY SIGNAL PROCESSING: IMPACT ON AND MODULATION BY PATTERN-GENERATING NETWORKS, EXEMPLIFIED IN LOCUST FLIGHT AND WALKING

H. WOLF

Fakultät für Biologie
Universität Konstanz
Postfach 5560-M624
D-78434 Konstanz, Germany

SUMMARY

In the control of locomotor movements, central pattern generating networks and sensory input interact in different ways. Sensory feedback may set the timing for the generation of central pattern elements, and in this way serve to trigger critical transitions between (centrally programmed) phases of a movement. Conversely, central network action may modulate sensory signal processing in a phase-dependent manner, and thus guarantee appropriate reflex responses at all phases of a cyclic movement. The former aspect is exemplified here by the reset of the wingbeat rhythm in the locust by input from a wing proprioceptor, the tegula. The latter aspect is illustrated by the modulatory effects a rhythmic central drive has on the processing of leg mechanosensory signals by spiking local interneurones in the locust. Both mechanisms will usually coexist and interact in the generation of the motor output, requiring close interaction of central and peripheral elements.

Certain traits of these insect motor control networks, and possibly of neural circuits in general, appear to reflect developmental and evolutionary cues, rather than genuine computational requirements. This concerns in particular the frequent observation of "weak" or "redundant" synaptic contacts in neural networks and the high degree of interconnectivity.

INTRODUCTION

The role of sensory feedback in the control of rhythmic locomotor activity varies depending on the behaviour being performed. Some motor programmes appear to be of purely central origin, as those that control swimming in the pelagic pteropod *Clione limacina* (Arshavsky *et al.,* 1985). More commonly, central pattern-generating networks interact with sensory signals, usually from mechanoreceptors that monitor certain aspects of the movement, in the production of locomotor movements. In some cases, sensory input appears to be required for the mere generation of rhythmicity, as in stick insect walking (Bässler & Wegner 1983, Bässler 1993b). According to their various roles in motor control, the processing of mechanosensory signals and their interaction with central pattern-generating circuits may be controlled by different mechanisms.

The following examples illustrate two characteristic but not mutually exclusive ways in which central network action and sensory feedback may be integrated in the control of rhythmic movement. In locust flight, input from the tegula, a complex receptor organ at the wing base, overrides timing cues of a central oscillator network in the generation of the flight motor command. The tegula apparently produces this effect by contacting most, if not all, interneurones of the flight oscillator. By means of this immediate access to the central network, the tegula resets the oscillator in each wingbeat cycle when it is stimulated by the downstroke movement of the wing. This mechanism guarantees rapid transition through an aerodynamically unfavourable downstroke attitude of the wing. In contrast, during walking signals from leg mechanoreceptors are modulated, gated, and apparently distributed into different sensory-motor channels, depending on the phase of the leg movement. All classes of neurones involved in the control of leg position and local postural reflexes (Burrows 1989) appear to receive rhythmic drive from the walking-pattern generator and, thus, to participate in this phase-dependent modulation of sensory signal flow. During walking, these mechanisms serve to assure e.g. appropriate reflex responses during the different phases of the leg movement.

Motor control networks exhibit a high degree of connectivity and a partially parallel layout which often appears redundant in the context of the considered function. Besides commonly perceived merits of parallel processing and interconnectivity, these character- istics may reflect developmental and evolutionary strategies.

The Role of the Tegula in Locust Wing Movement Control

In locust flight, central rhythm generating networks and numerous sense organs, in particular proprioceptors associated with the wing hinge, contribute to the generation of the flight motor pattern (e.g. Wendler 1985). On the one hand, significant aspects of the flight motor pattern are generated by the central oscillator and, indeed, an isolated ventral cord preparation will generate a flight pattern with reduced cycle frequency under the application of octopamine (Stevenson & Kutsch 1987). On the other hand, modification of the centrally generated pattern by sensory feedback is essential for the production of functional wing movements.

Among the wing receptors the tegula, a complex receptor organ of the anterior wing base (Kutsch et al., 1980), proved to be of particular importance for flight pattern generation (Wolf & Pearson 1988, Wolf 1991). During flight, the tegula is excited by the downstroke movement of the wing. The discharge of the receptor axons begins a few ms after the wing has passed the upper reversal point (Neumann 1985). A discharge of the tegula afferents, in turn, initiates a burst of action potentials in the wing elevator motoneurones. This activation of elevator motoneurones (and the concomitant suppression of depressors) by tegula input occurs significantly earlier than the discharge of elevator motoneurones driven by the central oscillator of a dissected ventral cord preparation - or by the otherwise intact flight system of a locust with excised tegulae (middle trace in Fig. 1B).

Tegula input thus overrides timing cues of the central oscillator. At similar wingbeat frequencies, the upstroke movement of the wings is delayed after tegula removal by an average of 13ms, compared to the intact situation (top trace in Fig. 1B).

If the tegula afferents of an intact locust are stimulated randomly during tethered flight by means of implanted hook electrodes, each stimulus effects a reset of the wingbeat rhythm to the phase of wing elevation (Fig. 1A). This reset is independent of the phase of stimulus presentation and demonstrates more clearly than tegula removal (Fig. 1B) that input from the proprioceptor functions to initiate elevator motoneurone activity, thereby overriding central timing cues.

What is the functional significance of this dramatic effect of tegula input? To address this question, I have examined the function of the locust tegula in the control of wing movement and the generation of aerodynamic force (Wolf 1993). As demonstrated in Fig. 1B, tegula removal delays the upstroke movement of the wing, that is, it extends the time period during which the wings stay near the lower reversal point of the stroke trajectory. The wings remain near this position in an aerodynamically unfavourable attitude, in almost horizontal or even slightly pronated (anterior wing margin pointing downward) angular settings. The flip from the pronated downstroke attitude into the supinated (anterior wing margin pointing upward) upstroke attitude is brought about only by the (delayed) activation of elevator muscles. The delayed upstroke movement observed as a consequence of tegula removal causes prolonged and increased generation of the negative lift forces because the flip into an aerodynamically appropriate wing profile (Nachtigall 1981) is also delayed. In particular, the peak of negative lift production associated with the beginning of the upstroke movement is enhanced and extended (lower trace in Fig. 1C, dotted area).

286

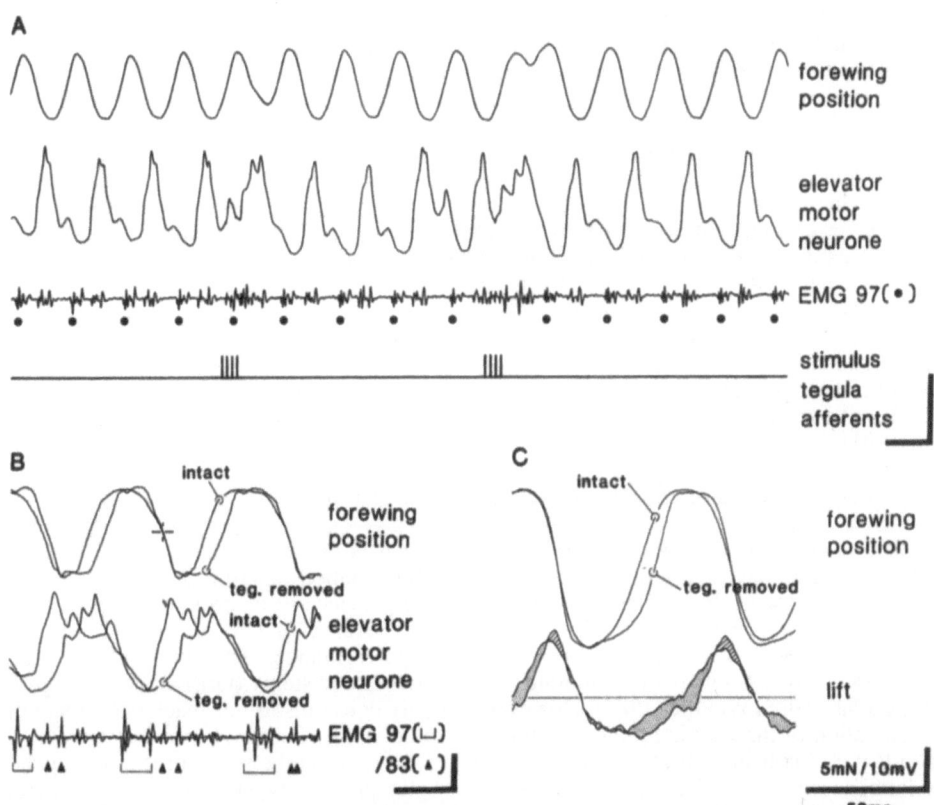

Figure 1. The tegula is an important element of the locust flight pattern generator. (A) Stimulation of tegula afferents (electric stimulus trains in bottom trace) in an intact, tethered flying locust resets the wing movement to the elevation phase by initiation of elevator motoneurone activity. Depressor motoneurone activity is suppressed by the stimulus; note missing depressor discharge (dots) in the electromyogram (EMG) after the 2nd stimulus train. (B) Two recordings with incidentally matching wingbeat frequencies, made before and after tegula removal, were superimposed taking the middle of the downstroke as reference (cross lines in top trace). Arrowheads mark the 1st spikes of elevator motoneurone discharges in the EMG, illustrating the delayed onset of elevator muscle activity after tegula removal (see also intracellular motoneurone recording, middle trace). (C) Superposition as in (B) but each trace represents the average of 32 recordings. The horizontal line in the lower trace marks zero lift. Areas are dotted where lift generation was smaller after tegula removal and hatched where it was larger. Slightly higher peak lift was associated with the downstroke movement after tegula removal, probably due to the more rapid movement (note that inertial components make a significant contribution to the force measurements particularly near the reversal points). The extended generation of negative lift observed near the downstroke position and during the upstroke is much more pronounced. Net lift was about 0.5mN before tegula removal and just negative afterwards. Each wing of the locust is equipped with one tegula organ. The properties described here concern the hindwing tegulae. The organs of the forewings are of minor, if any, importance in this context (Büschges*et al.*, 1992).

These data suggest that it is the function of the tegula to signal to the central nervous oscillator the impending completion of the wing downstroke, thereby allowing initiation of an immediately subsequent upstroke movement. In this way, prolonging of an aerodynamically unfavourable wing position near the lower reversal point is avoided (the upper reversal point, by contrast, is much less critical because here the wings will remain in a supinated attitude and thus generate positive lift with front wind present (Nachtigall 1981) - and actually do so in gliding locusts). The inherent variability of any motor output produces a need for sensory feedback control of critical transitions between phases of a movement, in the case of locust flight between the downstroke and the upstroke movement of the wings. This explains the powerful effect tegula input has on the central aspects of flight pattern generation, exemplified by the reset properties shown in Fig. 1A.

But how is this reset brought about on a cellular level? How do sensory feedback from the tegula and activity of the central pattern generator interact? The tegula makes extensive contact with all flight interneurones characterized to date, and in particular with the interneurones of the central oscillator. These connections follow a distinct pattern and consistently produce excitation of wing elevator motoneurones and inhibition of wing depressor motoneurones (Fig. 2B; Pearson & Wolf 1988): interneurones which excite elevator motoneurones receive monosynaptic excitatory input from the tegula, while inhibitory interneurones connected to the elevator motoneurones receive inhibitory input *(via* an layer of inhibitory neurones which invert the sign of the afferent synapses). The reverse scheme of connectivity is true for depressor motoneurones. The tegulae also make excitatory contacts directly with elevator motoneurones of the ipsilateral hemisegment.

Moreover, tegula input elicits plateau potentials and concomitant bursts of action potentials in elevator interneurones, in this way triggering an elevator motoneurone discharge (Ramirez & Pearson 1991). It is apparently by virtue of this immediate access to central oscillator components that the tegula resets the oscillator in every wingbeat cycle if and when it is stimulated, restarting the rhythm with an elevator discharge (Fig.1; review in Wolf 1991).

Modulation of Mechanoreceptive Signal Flow in Locust Walking

For an animal to walk in a natural environment it is essential that its motor control mechanisms integrate a variety of sensory signals. Information regarding, for instance, leg position, contact with the supporting substrate, or the presence of obstacles, is provided by mechanoreceptors on legs and associated structures. Hence one would expect sensory feedback to have pronounced effects on motor performance. This has indeed been demonstrated for a number of vertebrate and invertebrate systems (Grillner & Rossignol 1978, Bässler 1983, Wendler 1985). However, not only does sensory feedback participate in the generation of the walking motor program, the processing of sensory information is also dependent on the behavioural situation and, during walking, on the phase of the step cycle. The latter aspect, phase-dependent modulation of signal processing, shall be addressed here. Modulation of reflex pathways during walking is essential to guarantee appropriate motor responses during each phase of the step cycle. For instance, stimulation of tarsal hair sensilla on the middle leg of the locust will produce a noticeable response only toward the end of the stance phase and during the swing phase of the leg movement (Wolf 1992). During the stance phase a response of the stimulated leg would jeopardize mechanical stability of the animal and is not observed.

Mechanosensory signals must thus be routed and processed differently during different phases of the leg movement. The neural network which transmits mechanosensory information to the appropriate set of motoneurones in the context of postural reflexes has been investigated in considerable detail (review by Burrows 1989). Although it is a complex and highly interconnected network, a distinct structure is apparent (Fig. 2A). Given the physiological and structural properties of that network - for example, current injection into a single interneurone may produce a coordinated leg movement involving several joints - there is little doubt that this circuit is also involved in the control of walking. Indeed, in the stick insect, the contribution of several elements of this network in the control of leg movements and walking behaviour has been described (Schmitz *et al.,* 1991, Bässler 1983, 1993b).

At least five populations of neurones are involved in the transmission and processing of mechanosensory signals. First, there are the afferent neurones arising from the leg mechanoreceptors. These afferents project onto a first layer of interneurones, the spiking local interneurones ("local" refers to neurones restricted to the segmental ganglion). The afferents also project, though less frequently, onto nonspiking and intersegmental interneurones and onto motoneurones. The spiking local interneurones collate input from arrays of mechanoreceptors and organize them into functionally relevant receptive fields. They transmit this information primarily to nonspiking local interneurones but also to motoneurones and intersegmental (spiking) interneurones, the latter subserving intersegmental leg coordination. The nonspiking interneurones make lateral interactions and drive the leg motoneurones. A given nonspiking neurone contacts a subset of motoneurones, which is usually responsible for the control of a particular leg movement. By means of their tonic, graded release of transmitter, nonspiking interneurones subserve the delicate adjustment of the motoneurones' membrane potentials and discharge rates and thus allow subtle control of muscle contraction.

The nonspiking interneurones would appear as the appropriate site for a modulation of mechanosensory signal flow. They are the primary premotor elements and the point of convergence of sensory information from local sources and adjacent segments. Indeed, nonspiking interneurones of the locust have been implicated in the adjustment of the gain of

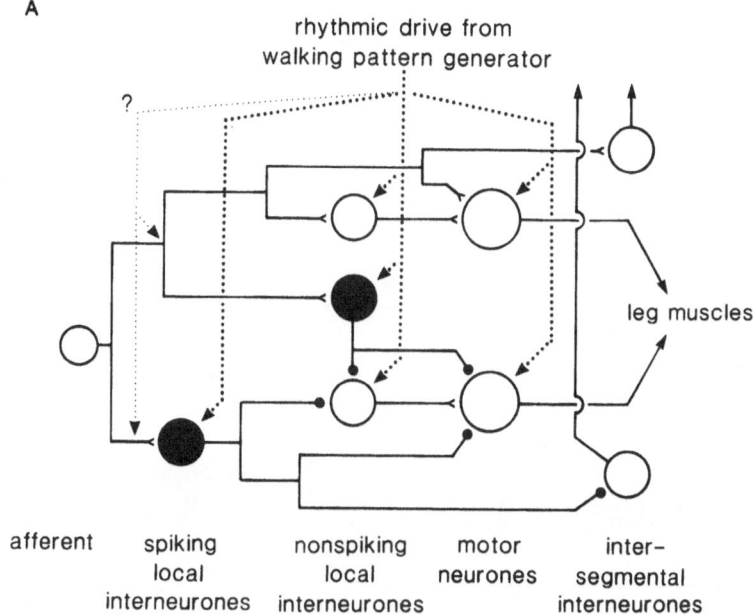

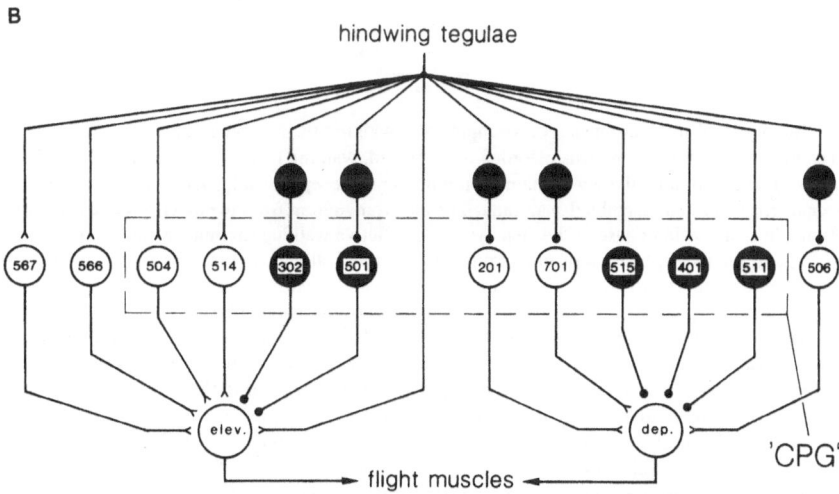

Figure 2. Circuit diagrams of, **A**, the local network mediating postural reflexes in the locust and, **B**, the connectivity pattern of the tegula afferents with interneurones of the flight oscillator. In (A) the dotted lines indicate synaptic drive the respective populations of neurones receive from the walking-pattern generator. This drive may be excitatory, inhibitory, or both. In **B** only those interneurones are shown which are know to make monosynaptic connections with flight motoneurones. "CPG" marks interneurones which are members of the oscillator network proposed by Robertson and Pearson (1983, 1985), although not all of them are able to reset the flight rhythm and some, for instance neurone 201, appear to be primarily involved in flight steering rather than pattern generation. Inhibitory neurones are represented by filled circles.

postural reflexes in response to segmental and intersegmental mechanosensory stimuli (Burrows 1989, Laurent & Burrows 1989) and modulation of their membrane potential in the rhythm of the step cycle has been observed in tethered walking stick insects (Schmitz *et al.*, 1991). Such (rhythmic) changes in membrane potential would affect the properties of nonspiking interneurones with regard to signal processing and propagation, primarily because of pronounced nonlinearities in the neurones' membrane properties (Laurent 1990, 1993). However, during walking modulation of reflex pathways by nonspiking interneurones has not yet been examined (see however Wolf and Büschges 1994).

The presence of rhythmic input from walking pattern-generating networks and their actual interaction with mechanosensory signal processing were recently demonstrated for spiking local interneurones by Gilles Laurent and myself (Wolf & Laurent 1993). Spiking local interneurones were recorded intracellularly in tethered walking animals (Fig. 3A).

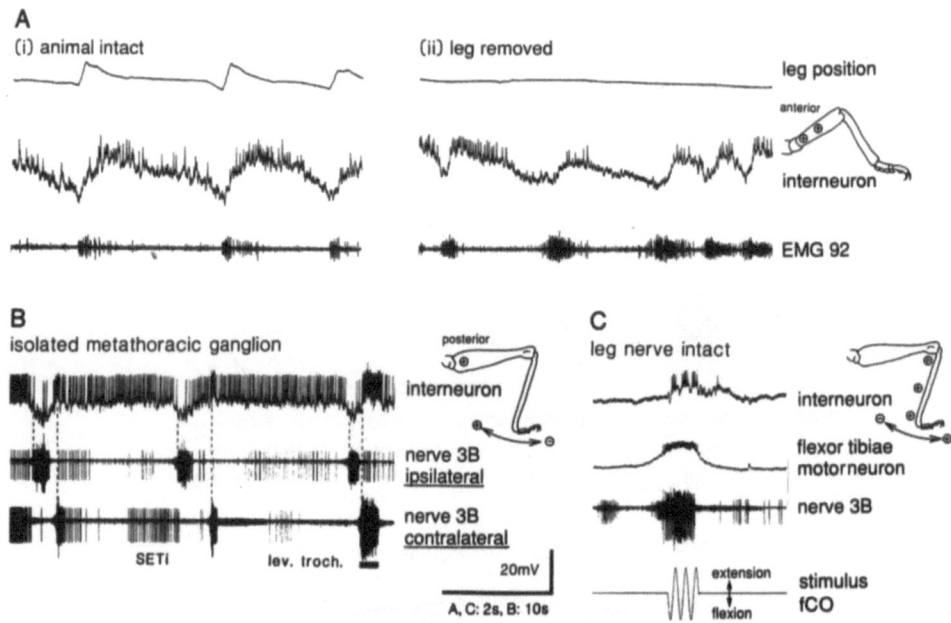

Figure 3. Spiking local interneurones receive input from walking pattern-generating networks. **A:** The cell body of a spiking local interneurone was recorded in a tethered, walking locust (method in Wolf 1990), before (i) and after (ii) amputation of the leg containing the neurone's receptive field. After amputation, the motor pattern became irregular and resembled searching. The electromyogram from the anterior coxa rotator muscle (EMG 92) monitors the swing phase of the step cycle. **B:** A fictive walking rhythm was induced in an isolated metathoracic ganglion by application of pilocarpine. The fictive walking rhythm was monitored by bilateral suction electrode recordings from nerves 3B. Activity of the slow extensor tibiae motoneurone (SETi) marks the stance phase, discharge of the levator trochanteris motoneurones marks the swing phase of the fictive leg movement. The interneurone is hyperpolarized during the ipsilateral and depolarized during the contralateral fictive swing phase. **C:** The femoral chordotonal organ of a hindleg was left attached to the ganglion in a fictive walking preparation and stimulated sinusoidally (fCO, bottom trace). In the recording segment shown here the stimulus was delivered during the swing phase, marked by a levator trochanteris discharge and depolarization of the flexor tibiae motoneurone. Interaction of sensory input and central drive is apparent in the interneurone recording. Stimulus frequency was too high in this case to affect the flexor motoneurone discharge. The inset diagrams indicate the receptive field of the interneurone. Excitatory (+) and inhibitory (-) regions of touch sensitivity and direction of movement sensitivity are indicated.

This revealed rhythmic changes in the neurones' membrane potentials and spike discharges which were, as a rule, closely related to leg movement (Fig. 3Ai). These rhythmic modulations of the activity pattern changed after leg amputation but never disappeared (Fig. 3Aii), indicating that sensory input from an interneurone's receptive field contributed to but was not wholly responsible for the rhythmic activity observed in intact animals.

Application of the muscarinic agonist pilocarpine can elicit a fictive walking pattern in isolated ventral cord preparations of the locust (Ryckebusch & Laurent 1993). Recordings from local spiking interneurones in such a preparation - the data in Fig. 3B are from an isolated metathoracic ganglion - are made in the complete absence of sensory feedback. In this way, rhythmic input from (central) walking-pattern generator networks could be demonstrated convincingly. A distinct rhythmic drive was observed in about 75% of the recorded spiking local interneurones and was usually related to the fictive walking rhythm of the ipsilateral leg. Bilateral (Fig. 3B) or purely contralateral input was also observed, however. This rhythmic drive interacted effectively with sensory input from, for instance, a leg chordotonal organ left attached to the nerve cord preparation (Fig. 3C). These results indicate that local reflexes are modulated by the action of circuits generating the walking pattern on the spiking local interneurones and that these interneurones contribute to the phase-dependent processing of sensory information during walking. Phase-dependent modulation thus already occurs at the first level of central neurones.

Burrows and Laurent (1993) have provided evidence that even at the afferents' terminals presynaptic inhibition occurs and may influence signal transmission onto central neurones. Their experiments were carried out in immobilized, dissected preparations and thus a possible rhythmic activation of the inhibitory (and apparently GABAergic) presynaptic terminals during walking could not be observed. Such rhythmic activation appears to be a distinct possibility, however, and has actually been reported for crustacean presynaptic inhibition of afferent fibres during walking (review in Clarac et al., 1992).

It is a matter of course that the remaining group of neurones, the motoneurones, whose membrane potentials oscillate in the cycle of the leg movement, will gate sensory input (e.g. Wolf 1992) received either directly from sensory afferents or from the spiking local interneurones. The rhythmic drive the motoneurones receive is inseparably linked to their function as motoneurones and thus it is questionable whether this gating is of actual functional significance or rather an epiphenomenon. This contrasts with the situation in the spiking local interneurones or the afferent terminals where a priori rhythmic modulatory input would not be expected to occur.

In summary, modulation of the transmission and the processing of mechanosensory signals appears to take place on all levels of the network outlined in Fig. 2A. Synaptic drive from the walking-pattern generator network impinges on all groups of interneurones and possibly on the afferent terminals, thereby allowing for numerous and intricate modulatory effects yet to be understood in detail.

Significance of the Properties of Mechanosensory Signal Processing

In any given motor control system, both aspects of sensory signal processing - modulation of pattern generator function by sensory input and modulation of sensory signal transmission by pattern generator networks - may, and usually will, coexist. In general, functionally critical transitions between phases of a cyclic movement will have to be under strict sensory control because central programming of subtleties in a movement trajectory is impossible, due to the omnipresence of external perturbations and inherent variability of the motor output. The function of the locust tegula in flight pattern generation is an example of such a control mechanism. Sensory-motor pathways will be subject to gating and phase-dependent re-routing if phase-dependent control of the resulting motor output is required. Aspects of the modulation of sensory signal processing during walking in the locust were given as examples here.

Apart from these rather trivial principles based on functional requirements, there is considerable diversity and an apparent lack of a common layout of motor control circuits (e.g. Getting 1988) in different groups of animals. Apparently evolution, when acting on the substrates provided by the nervous systems of different organisms, produced a large number of solutions for similar problems. However, several intriguing aspects reappear. Some are of obvious functional significance, like mutually inhibitory connections between halfcentres, the significance of others is less evident. For example, there is usually a high degree of connectivity and, as a result, weak synaptic connections are often observed which appear functionally redundant in the examined context. With regard to the two examples given above one may ask: Why does the walking-pattern generator contact all groups of neurones involved in the processing of mechanosensory signals from the leg - and not just impinge on the nonspiking interneurones? And why does tegula input project onto all known interneurones of the flight oscillator, while contact with the few interneurones that are able to reset the flight rhythm (Robertson & Pearson 1983, 1985) would appear to be sufficient? One would expect these traits to have functional implications since they are evidently not caused by a common origin of the two networks.

Apart from the more frequently discussed merits of parallel processing - applicability to complex problems and rapid computation, robustness against malfunction of individual elements, possibility of implementing learning properties, etc. - there might be other underlying principles, both evolutionary and developmental in nature. The high degree of interconnectivity observed in nervous systems - even if they are not used as associative memory - may reflect a stable strategy in terms of evolutionary change. With changing requirements posed on the nervous system it may be easier to change the efficacy of already existing synapses, decreasing their gain to insignificance or activating previously "dormant" contacts, rather than to establish completely new axonal pathways to acquire novel functional properties. The latter would imply comparatively elaborate - and hence expensive in terms of genetic control - changes in axonal pathfinding during development. It might be advantageous, therefore, to maintain a high level of connectivity among central neurones even if only a fraction of these contacts are essential for the control of behaviour of the present species. This does not mean that the "redundant" connections are non-functional in the contexts mentioned above, for example, parallel signal processing. Instead, they appear to be of minor importance and in several cases it is obvious that the circuit can function without them (e.g. Ronacher *et al.*, 1988; Wolf*et al.*, 1988; see also Horsmann 1985). The fact that neural structures are inexpensive in terms of energy and material demands compared to, for example, muscle tissue, lends support to such hypotheses.

A "conservative" structure of the nervous system has often been reported, for instance, with regard to motoneurones supplying remnant flight muscles of flightless grasshoppers (Arbas 1986), or locust leg motoneurones which have acquired different functions in adjacent body segments (Wilson 1979). With immediate significance for the processing of leg mechanosensory signals discussed above, this appears to be true also for the basic organization of the network controlling movement of the femur-tibia joint in locusts and stick insects (Bässler 1993a).

ACKNOWLEDGEMENTS. I thank K.G. Pearson and G.L. Laurent for their cooperation and generous support in the course of several investigations mentioned above. R. Kittmann and A. Büschges provided important comments on initial drafts of the manuscript and Mary Anne Cahill corrected the English text. Financial support was provided by the Medical Research Council of Canada, the Alberta Heritage Foundation, and the Deutsche Forschungsgemeinschaft (SFB 156 and Heisenberg Fellowship).

REFERENCES

Arbas E.A. and Tolbert L.T. (1986). Presynaptic terminals persist following degeneration of "flight" muscle during development of a flightless grasshopper. *Journal of Neurobiology* **17**, 627-636.

Arshavsky Y.I., Beloozerova I.N., Orlovsky G.N., Panchin Y.V. and Pavlova G.A. (1985). Control of locomotion in marine mollusc *Clione limacina*. III. On the origin of locomotory rhythm. *Experimental Brain Research* **58**, 273-284.

Bässler U. (1983). Neural basis of elementary behaviour in stick insects. ed Braitenberg V. 1-169. Springer Verlag, Berlin, Heidelberg, New York.

Bässler U. and Wegner U. (1983). Motor output of the denervated thoracic ventral nerve cord in the stick insect *Carausius morosus. Journal of Experimental Biology* **105**, 127-145.

Bässler U. (1993a). The femur-tibia control system of stick insects - a model system for the study of the neural basis of joint control. *Brain Research Reviews* **18**, 207-226.

Bässler U. (1993b). The walking (and searching-) pattern generator of stick insects, a modular system composed of reflex chains and endogenous oscillators. *Biological Cybernetics* **69**, 305-317.

Burrows M. (1989). Processing of mechanosensory signals in local reflex pathways of the locust. *Journal of Experimental Biology* **146**, 209-227.

Burrows M, and Laurent G. (1993). Synaptic potentials in the central terminals of locust proprioceptive afferents generated by other afferents from the same sense organ. *Journal of Neuroscience* **13**, 808-819.

Büschges A., Ramirez J-M., Driesang R. and Pearson K.G. (1992). Connections of the forewing tegulae in the locust flight system and their modification following partial deafferentation. *Journal of Neurobiology* **23**, 44-60.

Clarac F., El Manira A. and Cattaert D. (1992). Presynaptic control as a mechanism of sensory-motor integration. *Current Opinion in Neurobiology* **2**, 764-769.

Getting P.A. (1988). Comparative analysis of invertebrate central pattern generators. In: *Neural control of rhythmic movements* eds Cohen A.H., Rossignol S. and Grillner S. 101-127. Wiley & Sons, New York.

Grillner S. and Rossignol S. (1978). On the initiation of the swing phase of locomotion in chronic spinal cats. *Brain Research* **146**, 269-277.

Horsmann U. (1985). Der Einfluß propriozeptiver Windmessung auf den Flug der Wanderheuschrecke und die Bedeutung descendierender Neuronen der Tritocerebralcommissur. *Ph.D thesis university of Cologne.*

Kutsch W., Hanloser H. and Reinecke M. (1980). Light- and electron-microscopic analysis of a complex sensory organ: the tegula of *Locusta migratoria. Cell and Tissue Research* **210**, 461-478.

Laurent G. (1990). Voltage-dependent nonlinearities in the membrane of locust nonspiking local interneurons, and their significance for synaptic integration. *Journal of Neuroscience* **10**, 2268-2280.

Laurent G. (1993). A dendritic control mechanism in axonless neurons of the locust, *Schistocerca gregaria. Journal of Physiology (Lond)* **470**, 45-54.

Laurent G. and Burrows M. (1989). Intersegmental interneurons can control the gain of reflexes in adjacent segments of the locust by their action on nonspiking local interneurons. *Journal of Neuroscience* **9**, 3030-3039.

Nachtigall W. (1981). Der Vorderflügel großer Heuschrecken als Luftkrafterzeuger. I. Modellmessungen zur aerodynamischen Wirkung unterschiedlicher Flügelprofile. *Journal of Comparative Physiology* **142**, 127-134.

Neumann L. (1985). Experiments on tegula function for flight coordination in the locust. In: *Insect locomotion* eds Gewecke M. ans Wendler G. 149-156. Parey Verlag, Berlin, Hamburg.

Pearson K.G. and Wolf H. (1988). Connections of hindwing tegulae with flight neurones in the locust, *Locusta migratoria. Journal of Experimental Biology* **135**, 381-409.

Ramirez J-M. and Pearson K.G. (1991). Octopaminergic modulation in the flight system of the locust. *Journal of Neurophysiology* **66**, 1522-1537.

Robertson R.M. and Pearson K.G. (1983). Interneurones in the flight system of the locust: distribution, connections and resetting properties. *Journal of Comparative Neurology* **215**, 33-50.

Robertson R.M. and Pearson K.G. (1985). Neural networks controlling locomotion in locusts. In: *Model neuronal networks and behaviour* ed Selverston A.I. 21-35. Plenum Press, New York, London.

Ronacher B., Wolf H. and Reichert H. (1988). Locust flight behavior after hemisection of individual thoracic ganglia: evidence for hemiganglionic premotor centers. *Journal of Comparative Physiology A* **163**: 749-759.

Ryckebusch S. and Laurent G. (1993). Rhythmic patterns evoked in locust leg motor neurons by the muscarinic agonist pilocarpine. *Journal of Neurophysiology* **69**, 1583-1595.

Schmitz J., Büschges A. and Kittmann R. (1991). Intracellular recordings from nonspiking interneurons in a semi-intact, tethered walking insect. *Journal of Neurobiology* **22**, 907-921.

Stevenson P.A. and Kutsch W. (1987). A reconsideration of the central pattern generator concept for locust flight. *Journal of Comparative Physiology A* **161**, 115-129.

Wendler G (1985). Insect locomotory systems: control by proprioceptive and exteroceptive inputs. In: *Insect locomotion* eds Gewecke M. and Wendler G. 245-254. Parey Verlag, Berlin, Hamburg.

Wilson J.A. (1979). The structure and function of serially homologous leg motor neurons in the locust. I. Anatomy. *Journal of Neurobiology* **10**, 41-65.

Wolf H. (1990). Activity patterns of inhibitory motoneurons and their impact on leg movement in tethered walking locusts. *Journal of Experimental Biology* **152**, 281-304.

Wolf H. (1991). Sensory feedback in locust flight patterning. In: *Locomotor neural mechanisms in arthropods and vertebrates* eds Armstrong D.M. and Bush B.M.H. 134-148. Manchester University Press, Manchester, New York.

Wolf H. (1992). Reflex modulation in locusts walking on a treadwheel - intracellular recordings from motoneurons. *Journal of Comparative Physiology A* **170**, 443-462.

Wolf H. (1993). The locust tegula: significance for flight rhythm generation, wing movement control, and aerodynamic force production. *Journal of Experimental Biology* **182**, 229-253.

Wolf H. and Pearson K.G. (1988). Proprioceptive input patterns elevator activity in the locust flight system. *Journal of Neurophysiology* **59**, 1831-1853.

Wolf H., Ronacher B. and Reichert H. (1988). Patterned synaptic drive to locust flight motoneurons after hemisection of thoracic ganglia. *Journal of Comparative Physiology A* **163**, 761-769.

Wolf H. and Laurent G.L. (1993). Rhythmic modulation of the responsiveness of locust sensory local interneurons by walking pattern generating networks. *Journal of Neurophysiology* **71**, 110-118.

Wolf H. and Büschges A. (1994). Insect nonspiking local interneurons and the control of leg swing in walking. In: *Sensory transduction (proceedings of the 22th Göttingen neurobiology conference)* eds Elsner N. and Breer H.   Poster 282, Thieme Verlag, Stuttgart, New York.

ESCAPE AND SWIMMING IN GOLDFISH: A MODEL SYSTEM FOR STUDYING INTERACTIONS BETWEEN MOTOR NETWORKS.

J. R. FETCHO

Department of Neurobiology and Behavior
State University of New York at Stony Brook
Stony Brook, NY 11794-5230

SUMMARY

Interactions between the motor networks for different behaviours occur in all vertebrates, but little is know about how these are accomplished at a cellular level. A useful model for studying these interactions would be one in which two well defined, interacting motor behaviours can be elicited in a fictive preparation in which a cellular analysis is possible. One such model system is the interactions between escape and swimming networks in goldfish. Fish can perform escapes in the midst of swimming, using some of the same axial muscles for both. Fictive versions of both swimming and escape can also be elicited in paralyzed fish in which intracellular recordings are possible. Studies of the interactions between escape and swimming indicate a powerful ability of escape circuits to override the swimming motor pattern in both freely swimming fish and in fictive preparations. A single spike in the Mauthner cell that initiates the escape can produce an output appropriate for escape regardless of the phase of the swimming rhythm at which the cell is fired. The Mauthner cell can not only override the swimming output, it can also reset the swimming motor rhythm in a way that may allow a smooth transition from escape into subsequent swimming. Similarities between the cellular organization of escape and swimming networks suggest neurones at which the interactions between the two circuits might occur. The goldfish preparation therefore offers a unique opportunity to study the cellular basis of interactions between the output of a single cell (the Mauthner cell involved in escape) and a rhythm generating network (swimming), in a situation that mimics an interaction that occurs during behaviour. The similarities between these interactions and those observed between the reticulospinal system and spinal rhythm generating networks in mammals indicate that the goldfish model may provide broad insights into interactions between the motor networks of vertebrates.

INTRODUCTION

Interactions between the circuits for different motor behaviours occur in all vertebrates. Most humans, for example, find it relatively easy to run across a soccer field and kick a ball in the midst of running, even though the running and kicking use some of the same muscles, but activate them in different ways. Even though interactions between the networks for different motor behaviours are fundamental to motor control we do not know much about how they occur at a cellular level in vertebrates.

Some of the most important interactions among motor networks occur between descending pathways and local spinal networks. This is probably a consequence of the separation of the special sensory organs (eyes, ears, etc.) and the decision making apparatus (the brain) in the head from much of the motor apparatus, including locomotor rhythm generating circuits, in the spinal cord. In this paper, I present a brief overview of our attempts to develop a model system for studying the interactions between two motor behaviours - one of which involves powerful descending pathways, and another in which spinal circuits play a predominant role.

Requirements of a Model

There are several important requirements of a model system for understanding the cellular basis of interactions between motor networks. First, at least two interacting motor behaviours having a clear functional role must be identified. Second, there should ideally be studies of the interactions between these motor behaviours in the freely moving animal, to allow a comparison between cellular and behavioural data. Third, intracellular studies require that the behaviours can be elicited in a paralyzed animal; that is there must be "fictive" versions of both behaviours and it must be possible to activate the fictive versions of both behaviours in the same preparation to study their interactions. Finally, the interactions studied should have a broad relevance among vertebrates. They should involve behaviours or pathways that are common among vertebrates and therefore are likely to provide data that will apply to many species. If the pathways chosen are primitive ones for vertebrates that are also found in mammals, such studies are likely to provide insight into mammalian organization as well.

Overview of the Goldfish System

The model animal we have chosen based on the above criterion is fish. The behaviours we study are escape and swimming in goldfish. Both behaviours are produced by the body or axial muscles of the fish. There are substantial behavioural data characterizing both escape and swimming behaviours in fish, with especially good behavioural data dealing with escape in goldfish (Grillner and Kashin, 1976; Roberts, 1981; Wallen and Williams, 1984; Eaton et al., 1988; Williams et al., 1989; Foreman and Eaton, 1993). Recently Jayne and Lauder have used electromyography to provide the first data addressing the functional interactions between escape and swimming behaviours in freely swimming fish. Our approach has been a somewhat more reductionist one. We have focused on developing a goldfish preparation in which we can elicit both escape and swimming behaviours in a paralyzed animal in which we can use intracellular recording to study the cellular basis of the interactions observed in behaving fish(Fetcho and Faber, 1988; Fetcho, 1990, 1991a, 1992a, 1992b; Fetcho and Svoboda, 1993). In the following sections I deal first with the characterizations of the fictive escape and swimming preparations. These separate observations of the fictive escape and swimming behaviours and their networks led to some predictions about the interactions between the two. These have been tested by both our recent experiments in the fictive preparation as well as studies of the interactions in freely swimming sunfish. I deal with these later in the paper.

Spinal Escape Pathways

The behavioural features of the escape in goldfish are well understood, largely as the result of studies by Robert Eaton and colleagues (Eaton et al., 1988; Foreman and Eaton, 1993). During an escape, the fish bends very rapidly (in roughly 20 msec) to one side so that the body briefly forms a C-shaped bend. This initially very fast and forceful bend swivels the fish about its centre of mass so that it points away from a potential threat (e.g. a predator in a natural situation). The fish then bends to the opposite side, propelling it away from the threat. This behaviour is initiated by the firing of one of a pair of reticulospinal neurones, the Mauthner cells (M-cells). The cell bodies of the M-cells in the hindbrain receive sensory input from many modalities including the auditory system, the visual system and the lateral line (Faber et al., 1989). The axons of the M-cells cross in the brain and extend down the entire length of the spinal cord. Recordings of the extracellular firing of the M-cell in freely swimming fish demonstrate that one and usually only one of the two Mauthner cells fires a single action potential in conjunction with every escape; the left Mauthner cell firing in conjunction with C-bends to the right and the right in conjunction with leftward C-bends (Zottoli, 1977). Selective activation of a single Mauthner axon (M-axon) in a freely swimming fish leads to a powerful C-bend similar to the early stages of more naturally elicited escapes (Nissanov *et al.*, 1990). Thus, the Mauthner cell has provided a very fruitful model system in which a single spike in one reticulospinal neurone produces a major portion of a behaviour that can make the difference between life or death for the fish.

The large size of the Mauthner cell has made it especially suited for studies of its input and output connections. Numerous physiological and morphological studies documenting the circuitry of the M-cell make the network of the M-cell and the behaviour it produces one of the best understood motor behaviours in any vertebrate (Faber *et al.*, 1989). Here I focus on some of the cellular details of the spinal network of the Mauthner cell which provided much of the impetus for studies of the interactions between escape and swimming networks. Simultaneous intracellular recordings from the M-axon and spinal interneurones and motoneurones, in conjunction with dye labelling of spinal cells, have led to the identification of some of the major neurones in the spinal output network of the M-cell (Fetcho and Faber, 1988; Fetcho, 1990; Fetcho, 1992a). In most cases, the synaptic connections in the network have been studied with pairwise recordings from pre and postsynaptic cells as well as light and electron microscopically. Our work has focused primarily on the physiology and light microscopy and Yasargil's group has studied the connections with the electron microscope(Celio *et al.*, 1979; Yasargil and Sandri, 1990). The microscopy and physiology from different laboratories have led to the same conclusions concerning the connectivity in the network; therefore, the evidence for many of the connections is very strong.

The right side of figure one shows a diagram of the spinal network of the Mauthner cell as we currently understand it. All of the diagrammed **monosynaptic** output connections of the M-cell have been studied both physiologically and by light and electron microscopy. This is also true for outputs of the commissural inhibitory interneurones, but the outputs of the descending interneurones have been only studied physiologically. The Mauthner axon monosynaptically, chemically excites large ipsilateral motoneurones. These large motoneurones are the largest in the goldfish motor column and they innervate exclusively the faster muscle fibre types in the myomeres (Fetcho, 1986; Fetcho, 1987). The M-axon also disynaptically excites the same large motoneurones via interneurones that are chemically excited by the M-cell and are in turn electrotonically coupled to the motoneurones (Fetcho, 1992a). These interneurones have axons that descend in the spinal cord (hence the name descending interneurone) and branch extensively in the motor column where they apparently contact many motoneurones. The combination of a monosynaptic and disynaptic input to the motoneurones leads to a powerful excitation of the large ipsilateral motoneurones and an associated powerful C-bend when the M-axon is fired. The excitation of motoneurones on the side ipsilateral to the M-axon is accompanied by an inhibition of motoneurones on the opposite side. The identified interneurones mediating this are commissural interneurones that are electrotonically excited by the Mauthner axon and cross to the opposite side of the cord to inhibit contralateral motoneurones (Fetcho, 1990). These cells inhibit not only the contralateral motoneurones, but also the contralateral descending and commissural interneurones. The inhibition of the contralateral motoneurones and descending interneurones serves to block excitation of motoneurones on the side opposite the C-bend. The inhibition of the contralateral commissural cells prevents them from interfering, via their crossed inhibition, with the C-bend being produced on the opposite side. The end result is the very powerful C-bend to the side of the active M-axon.

One of the obvious outcomes of the studies of the spinal network for escape was the similarity between the escape network and the spinal networks for swimming in other species (Fetcho, 1992b). The best understood rhythmic motor circuits in any vertebrate are those responsible for swimming in frog embryos and lampreys (Cohen and Wallen, 1980; Buchanan and Cohen, 1982; Roberts, 1990; Grillner and Matsushima, 1991). A comparison between the escape network in goldfish and swimming networks in other vertebrates (figure 1) reveals a similar circuit organization in each. The two classes of interneurones in the escape network are similar to interneurones in swimming networks. Descending interneurones in both networks monosynaptically excite motoneurones and are activated in conjunction with ipsilateral motoneurones during bending to one side. Commissural inhibitory interneurones in both cases are active on the side of the bend and cross the spinal cord to inhibit motoneurones, descending interneurones and commissural interneurones on the opposite side. These similarities raised the possibility that the M-axon was simply tapping into some of the local spinal networks involved in swimming to produce

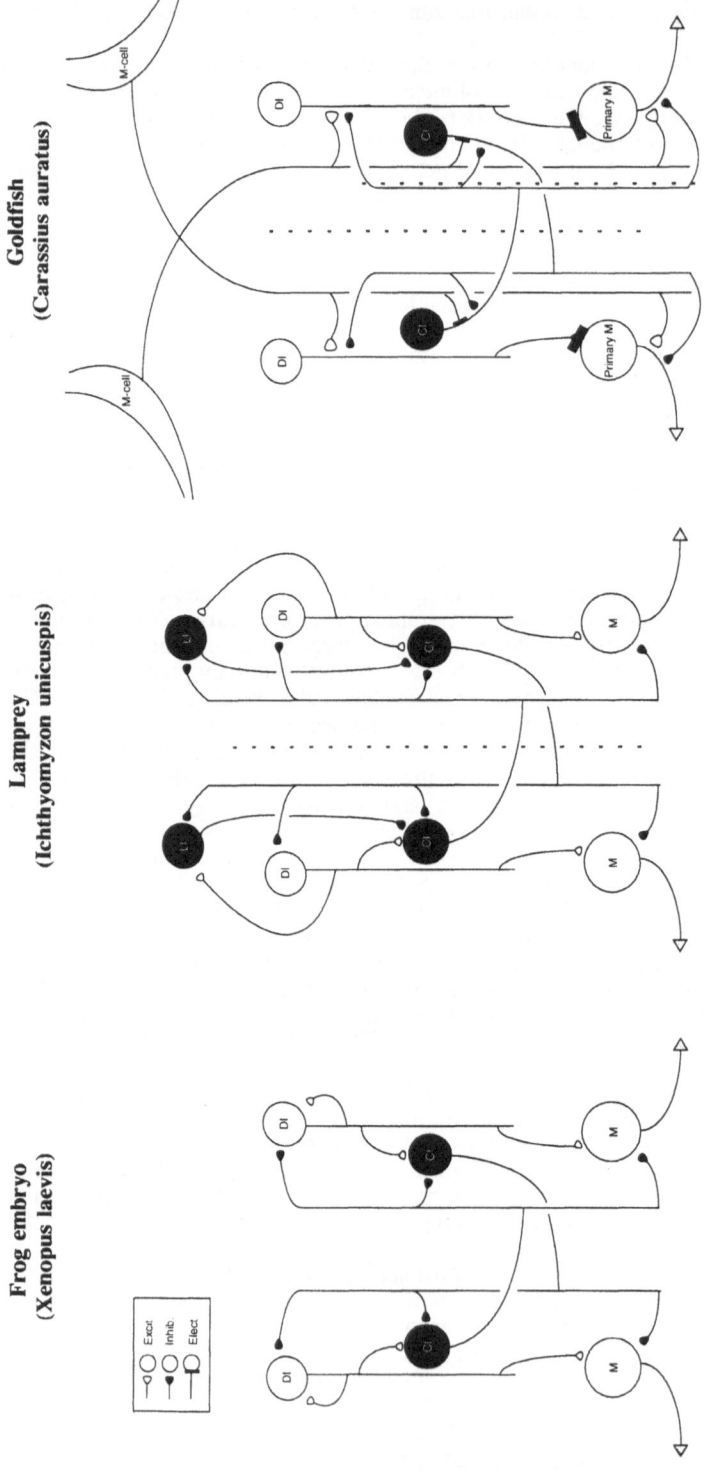

Figure 1. Diagrams of the premotor networks important for swimming in frog embryos and lampreys and for escape in goldfish. Excitatory cells are shown as open circles, inhibitory cells are filled. DI, descending interneurone; CI, commissural interneurone; M, motoneurone; Primary M, primary motoneurone. Inset shows the types of the synaptic connections: Excit., chemical excitatory synapse; Inhib., chemical inhibitory synapse; Elect., electrotonic synapse. Further details and references to the frog and lamprey work are in the text. From Fetcho, 1992b.

an escape. All of the output synapses of the M-axon are located on the initial segment or axon hillock region of the postsynaptic cells where they might be especially effective in overriding other inputs to the cells. These observations suggested that the M-cell might have the ability to override the activity from swimming motor networks and perhaps might also reset the swimming rhythm if the M-axon tapped into some of the interneurones in the swimming rhythm generating circuits. In order to study the influence of the M-axon on swimming motor patterns and examine what neurones might be shared by escape and swimming networks, we developed a fictive swimming preparation of goldfish.

Fictive Swimming of Goldfish

Swimming involves a rhythmic bending of the fish from side to side. Studies of swimming circuits require a reliable method of eliciting the behaviour. In frog embryos, a tap to the head activates the swimming motor pattern (Roberts, 1990). However, in other species a sensory stimulus is not usually sufficient to produce a sustained swimming motor output. Instead, two major approaches have been adopted. In the first, excitatory amino acids or their analogs are applied to a solution bathing the spinal cord (Cohen and Wallen, 1980). This apparently raises excitability in the cord and leads to the activation of the swimming motor circuits. In the second approach, a constant series of stimuli are applied to an area in the midbrain, the midbrain locomotor region (Kashin et al., 1974; McClellan and Grillner, 1984; McClellan, 1986). The midbrain locomotor region is an evolutionarily conserved area among vertebrates that when stimulated leads to activation of the rhythm generating networks in spinal cord. Thus, in a cat, stimulation of the midbrain locomotor region leads to walking or other rhythmic limb movements whereas in a fish stimulation of the comparable region leads to swimming (Shik et al., 1966; McClellan, 1986). We chose to elicit swimming in goldfish via stimulation of the midbrain locomotor region because we could do this in vivo (in which bathing with amino acids is difficult) and because the brain stimulus offered good control over when the swimming occurred, because swimming was largely confined to the times during which the brain stimulus was applied.

Our approach to eliciting swimming was similar to that used first by Kashin et al. (1974) who showed that midbrain stimulation lead to alternating movements of the tail of carp. Similar stimulation (monopolar, cathodal, 100Hz., 0.2msec duration, 25-200μA) in goldfish leads to alternating left/right bending movements of the tail, with the frequency of tail movements related linearly to the strength of the brain stimulation over a large range of stimulus strengths (Fetcho and Svoboda, 1993). The frequency of the tail beats in response to brain stimulation was from 2-16Hz, corresponding very well to the range of tail beat frequencies observed in swimming goldfish (Bainbridge, 1958). Electromyograms (EMGs) recorded on opposite sides of the body during stimulation of the midbrain alternated, with bursts of activity on first one side of the body, then the other, in accord with the left and rightward bending. The EMGs also showed a rostrocaudal delay, with bursts from muscle in more rostral body segments beginning prior to those in caudal segments.

The major features of the EMG recordings were also seen in the pattern of activity recorded from spinal motor nerves after paralyzing the fish to produce a "fictive" preparation (Fig. 2; Fetcho and Svoboda, 1993). This fictive preparation was studied in detail to determine whether it showed all of the major features of the swimming motor pattern recorded from freely swimming fish (Grillner and Kashin, 1976). These features include: 1) an alternation of activity on opposite sides of the body, with the burst on one side occupying roughly half of the time between successive bursts (cycle time) on the same side ; 2) As fish swim faster, the cycle time decreases and this is accompanied by a proportional decrease in the burst duration, such that the burst duration remains a constant fraction of cycle time; 3). There is a rostrocaudal wave of activity along each side of the body during swimming, that leads to the delay between the activation of rostral and caudal segments; 4) This delay remains a constant fraction of cycle time as swimming speed increases because the delay decreases in proportion to decreases in cycle time.

We would expect to find all of these features in a fictive swimming model. In the fictive experiments using goldfish, we varied the cycle time ("swimming speed") by varying the current intensity of the brain stimulus. We observed that the bursts alternated on opposite sides of the body (Fig. 2), with the burst duration varying linearly with cycle time, occupying, on average, 50.6% of the cycle time (Fetcho and Svoboda, 1993). There was also a rostrocaudal delay along the body (Fig. 2, Fetcho and Svoboda, 1993). This delay as a fraction of the cycle time remained roughly constant in each fish, averaging roughly 2.1%

of the cycle time per segment of fish. This number implies that there is roughly 60% of a wave of activity along the nearly 30 segment goldfish at any point in time. This value is comparable to other short bodied fish, but is less than the entire cycle of activity observed along the body during swimming of the more elongate lampreys (Williams *et al.*, 1989). Thus, all of the major features of the swimming motor pattern were observed in the fictive swimming goldfish, making it a favourable model for studying the interactions between swimming and escape networks (see Fetcho and Svoboda, 1993 for a discussion of some subtle differences between the fictive and freely swimming fish).

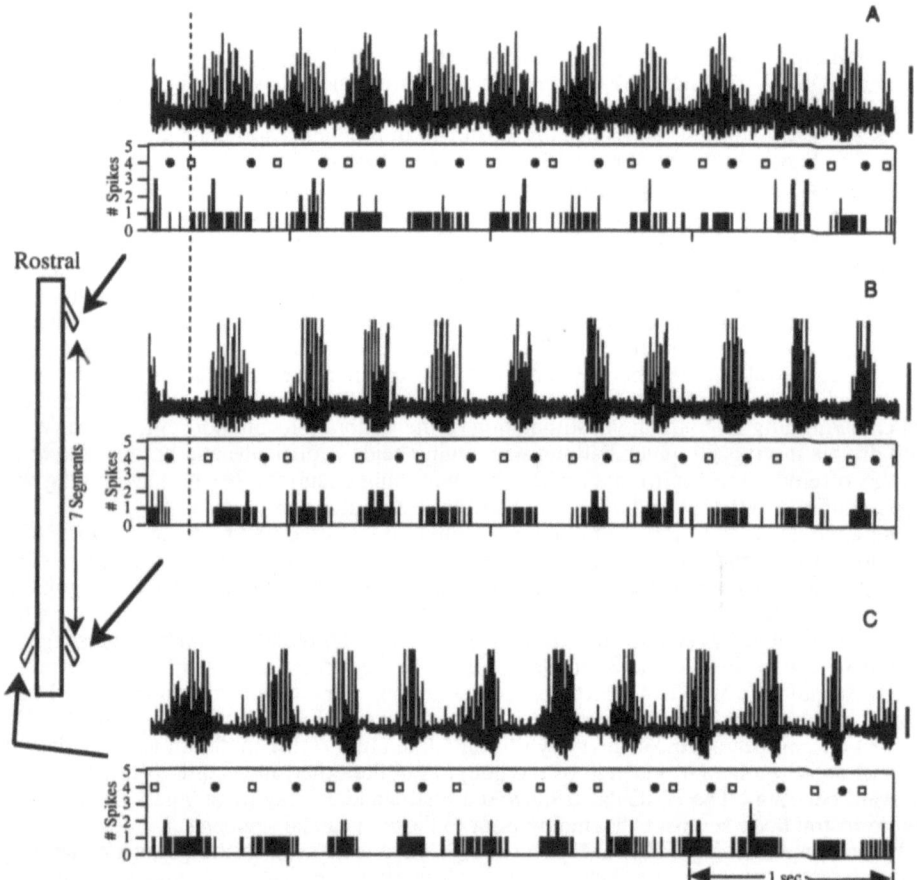

Figure 2. Motor pattern produced by stimulation of the midbrain in goldfish. Recordings from three different dorsal rami of the ventral root are shown. The locations of the recording sites are shown at left. Two of the nerves were on opposite sides of the same segment. The other was seven segments rostral to those. For each nerve, the raw data recording is shown along with corresponding spike   histograms that show computer identified starts (open squares) and ends of bursts (filled circles). Recordings shown in A and B are from nerves on the same side of the body and show that activity at the rostral recording site begins before the activity at the caudal site on the same side. The dotted line marks the start of activity in nerve A to allow easy visualization of the delay between A and B. Activity in nerves on opposite sides of the same body segment(B and C), alternates. Recordings are from segments 17 and 24. Stimulus parameters are 91Hz, 100μAmp. Scale bar at right in A-C = 200μV. From Fetcho and Svoboda, 1993.

Interactions Between Escape and Swimming

Swimming and escape both involve bending of the body and tail, but in other respects they are very different. Escapes are very rapid and forceful, "one shot" responses whereas swimming is usually a slower, sustained rhythmic bending from side to side. The two behaviours can occur independently of one another, with swimming proceeding in the absence of escapes and escapes occurring from rest in a non-swimming fish. However, we can imagine a situation in which a fish is attacked by a predator in the midst of swimming - a situation in which we might expect an escape turn to occur during swimming. Indeed there is now strong evidence from behavioural and electromyographic data that escape C-bends can occur in the midst of swimming (Jayne and Lauder, 1993).

The similarities between escape networks in goldfish and swimming networks in other species led to two major predictions about the interactions between the two behaviours when an escape occurs in the midst of swimming. Both of the predictions have been tested and substantiated in recent experiments. First, the connections of the Mauthner cell and some its output neurones onto the initial segment or axon hillock of spinal motoneurones and interneurones suggested a potentially powerful ability of the Mauthner cell to override other inputs to spinal cells during escapes. Recently, Jayne and Lauder (1993) have demonstrated how powerful the escape circuits really are in a freely swimming fish. They elicited escapes in sunfish swimming in flow tanks and recorded the electromyographic activity in the axial muscles during the escape. They found that fish could generate escapes in the midst of swimming at any phase of the swimming cycle. Furthermore, a detailed kinematic analysis revealed that only one of twelve kinematic variables differed significantly between escape responses elicited from swimming fish versus those elicited from a standstill. Thus, escape circuits can not only override swimming circuits, but they can do so and still produce an output that is very similar to that produced in the absence of swimming. This indicates a very powerful override ability.

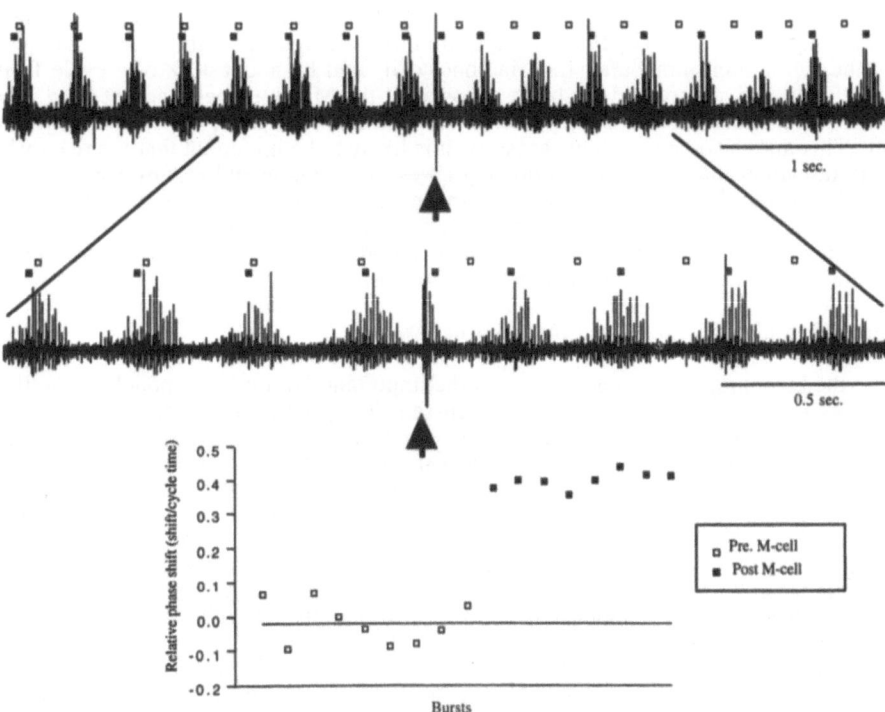

Figure 3. Example of the effect of the Mauthner cell on the swimming rhythm. The traces show extracellular recordings of the swimming rhythm produced by midbrain stimulation. The second trace is an enlargement of a portion of the first. The M-axon on the same side of the body as the motor nerve was fired at the arrow. It produced an override of the swimming rhythm leading to a burst of activity in the nerve when there would not normally be one during swimming. It also reset the swimming rhythm, as shown by the difference between the actual locations of bursts (solid squares) and the locations (open squares) where bursts would be predicted to occur if the M-axon had no effect on the rhythm. The magnitude of the reset is quantified in the graph at the bottom as the difference between the actual and predicted location of bursts divided by the average interval between bursts. In this case the reset was roughly 40% of the mean time between bursts.

We have observed a similar ability of escape pathways - in our case specifically the Mauthner cell- to override swimming circuits. In our fictive preparations firing **a single spike** in the Mauthner cell in the midst of a swimming motor pattern produced by midbrain stimulation leads to an override of the swimming motor pattern to produce an output appropriate for an escape (Fetcho, 1992a). Figure three shows an example experiment in which the M-axon was fired during fictive swimming. The motor output was monitored by an extracellular recording from a motor nerve to white muscle. In this case, the M-axon on the side of the motor nerve (and contralateral to the M-cell soma) was fired at a time when the motor nerve was silent - that is, at a time when the opposite side was active during swimming. This mimics a situation in which the fish is bending to one side during swimming and a Mauthner initiated escape bend is elicited to the opposite side. In this example, we observed that the M-axon can produce its normal, brief motor output in the nerve even at a time when the nerve would normally be silent during swimming. Thus, the M-axon could override the silent period during swimming to produce a motor output consistent with escape. The M-axon could do this in the fictive preparations regardless of the time at which in fired during swimming. Its activation always produced a strong brief ipsilateral excitation and a brief shut down of contralateral activity. In summary, our fictive preparations show an ability of a single spike in the M-axon to override the swimming motor output. This is consistent with both the data from swimming fish and our predictions based upon the organization of the M-cell network.

The similarities of the escape and swimming networks among goldfish, frog embryos and lampreys also raised the possibility that the escape networks might be tapping into some of the neurones that form part of the central pattern generating circuits for swimming. If so, then one might predict that activation of escape pathways could produce a resetting of the swimming rhythm. We have observed a very powerful ability of the Mauthner cell to reset the swimming rhythm (Fetcho, 1992c). In figure three, a single spike in the Mauthner cell shifted the swimming bursts after the Mauthner cell fired by about 40% of a cycle from where we would have expected the bursts to occur if the Mauthner cell had not fired. The Mauthner cell can therefore not simply override the swimming rhythm, but can also reset the rhythm. This makes some functional sense, as it is likely to be important that the fish make a smooth transition into swimming following an escape. This would require some sort of mechanism for linking escape and swimming movements. Indeed, we consistently find that the first swimming burst following a M-axon initiated escape burst occurs to the side opposite the escape burst, which would lead to a smooth transition from the escape bend to one side into a subsequent swimming bend to the opposite side.

An Evaluation of the Model and Future Directions

Near the beginning of this paper some of the important features of a model preparation for studying behavioural interactions were discussed. The goldfish preparation has many of the important features of a useful model. The two behaviours, swimming and escape, can be elicited fictively and the interactions observed in the fictive preparation mimic the very powerful interactions observed in freely swimming fish. The results from the goldfish preparation are likely to have broad implications for vertebrate motor systems. The interactions observed in goldfish are not unique to fish. Most swimming anamniotic vertebrates (that is, most vertebrates, because most vertebrates are fish and amphibians) have Mauthner cells and are likely to have similar interactions between swimming and escape (Zottoli, 1978). For example, very similar effects of the Mauthner cell on swimming have been observed in frog tadpoles (Lee and Olin, 1992).

However, the data from goldfish are likely to be applicable not just to other swimming vertebrates, but to mammals as well. The effects of the M-cell are strikingly analogous to the effects of stimulation of reticulospinal pathways in mammals. Activation of reticulospinal pathways during walking in cats has powerful effects on the walking motor rhythm including an interruption and a resetting of the rhythm(Drew and Rossignol, 1984). The similarities between fish and mammals are not too surprising because the reticulospinal pathway is one of the most conservative pathways among vertebrates, with both the Mauthner cell and reticulospinal neurones in mammals forming monosynaptic connections with motoneurones and premotor interneurones in spinal cord (McClellan, 1986; Ohta and Grillner, 1989; Fetcho, 1992b). Some of the effects produced by reticulospinal stimulation in mammals may underlie an interaction between a mammalian startle behaviour produced by short latency reticulospinal pathways and rhythmic spinal networks. Thus, there are both pathways and behavioural interactions in mammals that are similar to escape/swimming interactions in goldfish. The differences between goldfish and mammals are 1) that in goldfish the effects are produced by one spike in a single reticulospinal cell rather than a more sustained activation of many cells, and 2) in goldfish we can record intracellularly relatively easily from interneurones (and motoneurones) in the motor networks. The goldfish preparation therefore offers a unique opportunity to study the interactions between the output of a single cell (that has important behavioural consequences) and a rhythm generating network, in a situation that mimics an interaction that occurs during behaviour. The challenge of future studies of the interactions between escape and swimming in goldfish will be to understand the cellular basis of the ability of the M-cell to override and reset the swimming rhythm. In particular, we need to know which cells are shared by swimming and escape networks and what sort of interactions at the level of single cells might account for the interactions between the networks.

ACKNOWLEDGEMENTS. Supported by NIH grant NS 26539 and an Alfred P. Sloan Research Fellowship.

REFERENCES

Bainbridge, R. (1958) The speed of swimming of fish as related to size and to the frequency and amplitude of the tail beat. *Journal of Experimental Biology* **35**: 109-133.

Buchanan, J. T. and Cohen, A. H. (1982) Activities of identified interneurons, motoneurones, and muscle fibers during fictive swimming in the lamprey and effects of reticulospinal and dorsal cell stimulation. *Journal of Neurophysiology* **47**: 948-960.

Celio, M.R., Gray, E.G., and Yasargil, G.M. (1979) Ultrastructure of the Mauthner axon collateral and its synapses in the goldfish spinal cord. *Journal of Neurocytology* **8**: 19-29.

Cohen, A. and Wallen, P. (1980) The neuronal correlate of locomotion in fish. *Experimental Brain Research* **41**: 11-18.

Drew, T and Rossignol, S. (1984) Phase-dependent responses evoked in limb muscles by stimulation of medullary reticular formation during locomotion in thalamic cats. *Journal of Neurophysiology* **52**: 653-675.

Eaton, R. C., DiDomenico, R. and Nissanov, J. (1988) Flexible body dynamics of the goldfish C-start: Implications for reticulospinal command mechanisms. *Journal of Neuroscience* **8**: 2758-2768.

Faber, D. S., Fetcho, J. R. and Korn, H. (1989) Neuronal networks underlying the escape response in goldfish: General implications for motor control. *Annals of the New York Academy of Science* **563**: 11-33.

Fetcho, J. R. (1986) The organization of the motoneurones innervating the axial musculature of   vertebrates. I. Goldfish (Carassius auratus) and mudpuppies (Necturus maculosus). *Journal of Comparative Neurology* **249**: 521-550.

Fetcho, J. R. (1987) A review of the organization and evolution of motoneurones innervating the axial musculature of vertebrates. *Brain Research Reviews* **12**: 143-280.

Fetcho, J. R. (1990) Morphological variability, segmental relationships, and functional role of a class of commissural interneurons in the spinal cord of goldfish. *Journal of Comparative Neurology* **299**: 283-298.

Fetcho, J. R. (1991a) Spinal network of the Mauthner cell. *Brain Behavior and Evolution* **37**: 298- 316.

Fetcho, J.R. (1991b) Fictive spinal motor output elicited by stimulation of the midbrain in goldfish. *Society for Neuroscience Abstracts* **17**: 121.

Fetcho, J. R. (1992a) Excitation of motoneurones by the Mauthner axon in goldfish: Complexities in a "simple" reticulospinal pathway. *Journal of Neurophysiology* **67**: 1574- 1586.

Fetcho, J. R. (1992b) The spinal motor system in early vertebrates and some of its evolutionary changes. *Brain Behavior and Evolution* **40**: 82-97.

Fetcho, J.R. (1992c) A single action potential in a reticulospinal neuron can reset the fictive swimming rhythm in goldfish. *Society for Neuroscience Abstracts* **18**: 316.

Fetcho, J. R. and Faber, D. S. (1988) Identification of motoneurones and interneurons in the spinal network for escapes initiated by the Mauthner cell in goldfish. *Journal of Neuroscience* **8**: 4192-4213.

Fetcho, J.R. and Svoboda, K.R. (1993) Fictive swimming elicited by electrical stimulation of the midbrain in goldfish. *Journal of Neurophysiology* **70**: 765-780.

Foreman, M.B. and Eaton, R.C. (1993) The direction change concept for reticulospinal control of goldfish escape. *Journal of Neuroscience* **13**: 4101-4113.

Grillner, S. (1974) On the generation of locomotion in the spinal dogfish. *Experimental Brain Research* **20**: 459-470.

Grillner, S. and Kashin, S. (1976) On the generation and performance of swimming in fish. In: *Neural control of locomotion*. Advances in Behavioral Biology Vol. 18, edited by R. M. Hermann, S. Grillner, P. S. G. Stein, and D. G. Stuart. New York: Plenum Press, pp. 181-201.

Grillner, S. and Matsushima, T. (1991) The neural network underlying locomotion in lamprey - synaptic and cellular mechanisms. *Neuron* **7**: 1-15.

Jayne, B.C. and Lauder, G.V. (1992) An electromyographic study of the startle response during steady speed swimming of fish. *Society for Neuroscience Abstracts* **18**:1405.

Jayne, B.C. and Lauder, G.V. (1993) Red and white muscle activity and kinematics of the escape response of the bluegill sunfish during swimming. *Journal of Comparative Physiology A* **173**: 495-508.

Jordan, L. M. (1986) Initiation of locomotion from the mammalian brainstem. In: *Neurobiology of Vertebrate Locomotion,* edited by S. Grillner, P. S. G. Stein, D.G. Stuart, H. Forssberg, and R. M. Herman. London: MacMillan, pp. 21-37.

Kashin, S. M., Feldman, A. G. and Orlovsky, G. N. (1974) Locomotion of fish evoked by electrical stimulation of the brain. *Brain Research* **82**: 41-47.

Lee, M.T. and Olin, A.M. (1992) Interaction between Mauthner neurons and the swimming pattern generator in Xenopus tadpoles. *Society for Neuroscience Abstracts* **18**: 315.

McClellan, A. D. (1986) Command systems for initiating locomotion in fish and amphibians: parallels to initiation systems in mammals. In: *Neurobiology of Vertebrate Locomotion,* edited by S. Grillner, P. S. G. Stein, D.G. Stuart, H. Forssberg, and R. M. Herman. London: MacMillian, pp. 3-20.

McClellan, A. D. and Grillner, S. (1984.) Activation of `fictive swimming' by electrical microstimulation of brainstem locomotor regions in an in vitro preparation of the lamprey central nervous system. *Brain Research* **300**: 357-361

Nissanov, J., Eaton, R.C. and DiDomenico, R. (1990) The motor output of the Mauthner cell, a reticulospinal command neuron. *Brain Research* **517**: 88-98.

Ohta, Y. and Grillner, S. (1989) Monosynaptic excitatory amino acid transmission from the posterior rhombencephalic reticular nucleus to spinal neurons involved in the control of locomotion in lamprey. *Journal of Neurophysiology* **62**: 1097-1089.

Roberts, A. (1990) How does a nervous system produce behaviour? A case study in neurobiology. *Science Progress* **74**: 31-51.

Roberts, B. L. (1981) The organization of the nervous system of fishes in relation to locomotion. *Symposium of the Zoological Society of London* **48**: 115-136.

Shik, M. L., Severin, F. V. and Orlovskii, G. N. (1966) Control of walking and running by means of electrical stimulation of the mid-brain. *Biofizyka* **11**: 659-666.

Sokal, R. R. and Rohlf, F. J. (1981) *Biometry.* :H.W. Freeman and Co: New York.

Wallen, P. and Williams, T. L. (1984) Fictive locomotion in the lamprey spinal cord in vitro compared with swimming in the intact and spinal animal. *Journal of Physiology* **347**: 225-239.

Williams, T. L., Grillner, S., Smoljaninov, V. V., Wallen, P., Kashin, S. and Rossignol, S. (1989) Locomotion in lamprey and trout: The relative timing of activation and movement. *Journal of Experimental Biology* **143**: 559-566.

Yasargil, G.M., and Sandri, C. (1990) Topography and ultrastructure of commissural interneurons that may establish reciprocal inhibitory connections of the Mauthner axons in the spinal cord of the tench, Tinca tinca L. *Journal of Neurocytology* **19**: 111-126.

Zottoli, S.J. (1977) Correlation of the startle reflex and Mauthner cell auditory responses in unrestrained goldfish. *Journal of Experimental Biology* **66**: 243-254.

SUBJECT INDEX

207, 209, 210
Enkephalinergic, 103, 107
Enkephalins, 103, 104, 105, 106, 107
Ensemble discharge, 43, 45, 46, 47
Equilibrium, 31, 90, 131, 151, 191, 193
Equilibrium point hypothesis, 151
Evolution, 41, 79, 80, 291
Excitatory Amino Acids (EAA's), 299
Excitatory Post-Synaptic Potential (EPSP), 137,
    138, 145, 147, 149, 156, 205, 206, 277
Extensor muscles, 22, 56, 92, 93, 98, 135, 136,
    137, 138, 139, 140, 143, 144, 145, 155, 182
Extracellular potassium, 32
Eye
    Movements, 127, 241, 243
    Muscles, 113, 241
Eyeblink response, 117

—F—

Facial nucleus, 119, 121
Fastigial nucleus, 228, 229, 242
Feedback
    Control, 131, 215, 217, 261, 287
    Loop, 12, 215, 216, 219, 220, 259, 267
    Velocity, 154
Femorotibialis, 197
Ferrets, 117, 118, 119
Fictive
    Escape, 296
    Locomotion, 46, 130, 143, 144, 145, 147, 148,
        190
    Swimming, 207, 270, 271, 273, 299, 302
    Walking, 291
Finger, 54, 56, 57, 61, 62, 63, 70, 73, 74, 76, 77,
    79, 80, 81, 82, 83, 84, 169
Fishes, 273, 295, 296, 297, 299, 301, 302
Flexion reflex, 64, 97, 99, 100, 101
Flexor Reflex Afferent pathways, 103, 104, 105,
    106, 107
Flexor Reflex Afferents (FRA), 100, 103, 147
Flight oscillator, 286, 291
Flocculus, 241, 242, 245, 246, 247
Fluorenscence, 195, 196, 198
Forebrain, 118
Frog, 31, 161, 204, 297, 299, 302
Fusimotor
    Axons
        Dynamic, 12, 28
        Static, 11
    Control, 19
    Drive, 20, 22, 23, 36, 89, 92, 93
    Fibres, 11, 12, 13, 17, 152
    Innervation, 11, 27, 89
    Neurones, 19, 22, 89, 90, 91, 92, 93, 94, 152,
        153
    Stimulation, 12, 13, 15, 16, 28, 29, 38, 39
    System, 12, 19, 22, 23, 73, 91, 152

—G—

Gain control, 153, 154, 156, 253, 259
Gallamine, 39
Gamma Amino Butyric Acid (GABA), 125, 131,
    154, 187, 191, 193, 253, 257, 269, 271, 273,
    277
Gastric mill, 279, 282
Glabrous skin, 73, 74, 77
Globus pallidus, 173
Glutamatergic transmission, 129
Glycine, 129, 269, 271
Golgi cells, 119
Golgi Tendon Organs (GTOs), 39, 43, 135, 136,

139, 145, 182
Granule cells, 243

—H—

Hairy skin, 73, 74, 90
Halothane, 61, 64, 118
Hand, 14, 23, 31, 37, 39, 54, 56, 61, 73, 74, 75,
    77, 79, 80, 82, 83, 84, 160, 168, 169, 171,
    215, 218
Heaviness, 80, 82, 83
Hindbrain, 271, 273, 296
Hip, 97, 98, 99, 100, 140, 143, 145, 160, 161,
    162, 181
Human brain, 169
Human subjects, 19, 20, 21, 22, 54, 56, 57, 73, 79,
    80, 83, 100, 153, 155, 170
Hyperextension, 63, 66
Hyperflexion, 66
Hyperkalaemia, 31, 32
Hypermobility syndrome, 61, 63, 64
Hypoxia, 111, 112

—I—

Impulse activity, 16, 17
Inferior dental arch, 111
Inferior olive, 118, 220, 221, 225, 226, 228, 229,
    231, 241, 242, 243, 245, 246, 247
Inflammatory process, 62, 64
Information theory, 67
Inhibitory Post-Synaptic Potential (IPSP), 145,
    147, 198, 205, 228, 231, 271, 273, 277
Insects, 135, 138, 140, 290, 292
Interneurones, 146, 147, 148, 195, 197, 198, 199,
    208, 257, 259, 288, 289, 290, 291, 299
    Excitatory, 129, 130
    Ib, 47, 48
    Inhibitory, 129, 204, 271, 288, 297
    Lateral (L), 129
    Network, 129
    Premotor rhythm generating, 205
    Sensory, 273
    Spinal, 297
    Spinal reciprocal inhibitory, 271
Interosseus membrane, 89, 90, 91
Interphalangeal joint, 56, 61, 73, 74, 75
Interpositus nucleus, 117, 118, 225
Intrafusal muscle fibres, 27, 40
    Bag1, 3, 4, 11, 12, 13, 14, 15, 17, 28, 29, 30,
        31, 32, 35, 36, 37, 38
    Bag2, 3, 4, 5, 6, 7, 11, 12, 13, 14, 16, 17, 27,
        28, 29, 30, 31, 32, 35, 36, 37, 41
    Chain fibres, 3, 4, 5, 6, 7, 8, 11, 12, 14, 16, 17,
        27, 28, 31, 37, 39
Intrinsic hand muscles, 61
Ischaemia, 57

—J—

Jaw Jerk Reflex (JJR), 110, 111
Joint
    Ankle, 99, 155
    Capsule, 54, 57, 64, 92
    Deformity, 61, 64, 66
    Disease, 61, 62, 64
    Elbow, 70, 217
    Hip, 181
    Knee, 56, 61, 62, 64, 89, 92, 98
    Metacarpophalangeal, 73, 74, 75, 76
    Rotation, 64, 74, 75, 76, 77
    Sense, 53
    Temporomandibular, 111

Joystick, 168, 216